BOTANY ILLUSTRATED

SECOND EDITION

Introduction to Plants, Major Groups, Flowering Plant Families

by Janice Glimn-Lacy
and Peter B. Kaufman

 Springer

Janice Glimn-Lacy
6810 Shadow Brook Court
Indianapolis, IN 46214-1901
USA
janglimn@umich.edu

Peter B. Kaufman
Department of Molecular, Cellular, and
    Developmental Biology
University of Michigan
Ann Arbor, MI 48109-1048
USA
pbk@umich.edu

Library of Congress Control Number: 2005935289

ISBN-10: 0-387-28870-8   eISBN: 0-387-28875-9
ISBN-13: 978-0387-28870-3

Printed on acid-free paper.

Printed in the United States of America.   (TB/MVY)

9 8 7 6 5 4 3 2 1

springer.com

Preface

This is a discovery book about plants. It is for everyone interested in plants including high school and college/university students, artists and scientific illustrators, senior citizens, wildlife biologists, ecologists, professional biologists, horticulturists and landscape designers/architects, engineers and medical practitioners, and physical therapists and their patients. Here is an opportunity to browse and choose subjects of personal interest, to see and learn about plants as they are described. By adding color to the drawings, plant structures become more apparent and show how they function in life. The color code clues tell how to color for definition and an illusion of depth. For more information, the text explains the illustrations. The size of the drawings in relation to the true size of the structures is indicated by × 1 (the same size) to × 3000 (enlargement from true size) and × n/n (reduction from true size).

The contents reflect a balanced selection of botanical subject matter with emphasis on flowering plants, the dominant plants of the earth. After a page about plant names and terms, the book is divided into three sections. The first is an *introduction to plants*, showing structure and function; then, *major groups*, providing an overview of the diverse forms; and lastly, one-seventh of the *flowering plant families*, with the accent on those of economic importance. The sequence in the sections is simple to complex (cell to seed), primitive to advanced (blue-greens to flowering plants), and unspecialized to specialized (magnolias to asters and water-plantains to orchids). Where appropriate, an "of interest" paragraph lists ways these genera are relevant in our lives (categories include use as food, ornamentals, lumber, medicines, herbs, dyes, fertilizers; notice of wild or poisonous; and importance in the ecosystem). "Of interest" sections in *Botany Illustrated*, second edition, have been expanded to include many more topics of interest.

Evolutionary relationships and the classification of plants have been undergoing many changes in the past two decades since the first edition. In this edition controversial categories have been eliminated allowing individuals to be exposed to current thinking on plant classification. Classification from this second edition may be found in the Index under "Fungi Kingdom" and "Plant Kingdom." Pages on bacteria have been eliminated and two new pages on plant fossils, with accompanying illustrations, have been added. Every text page has undergone extensive revision.

For those interested in the methods used and the sources of plant materials in the illustrations, an explanation follows. For a developmental series of drawings, there are several methods. One is collecting several specimens at one time in different stages of development; for example, several buds and flowers of a plant (see 29) and button to mature forms of mushrooms (see 50, 51). Then, some are cut open to observe parts and decide how to present them, while others are to use for final drawings. Another method is waiting for the plant to change, which involves "forcing" stems (see 14), germinating seeds (see 40), watching one leaf expand (see 69), and drawing a flower in one season and its mature fruit in another (see 104, 109, 110, 111). An alternative to waiting for fruit is to use a collection of dry or frozen specimens, so that as spring flowers appear, the later maturing fruits can be seen at the same time (see 102, 105, 106).

In the first section, *introduction to plants*, there are several sources for various types of drawings. Hypothetical diagrams show cells, organelles, chromosomes, the plant body indicating tissue systems and experiments with plants, and flower placentation and reproductive structures. For example, there is no average or standard-looking flower; so, to clearly show the parts of a flower (see 27), a diagram shows a stretched out and exaggerated version of a pink *(Dianthus)* flower (see 87). A basswood *(Tilia)* flower is the basis for diagrams of flower types and ovary positions (see 28). Another source for drawings is the use of prepared microscope slides of actual plant tissues. Some are traced from microscope slide photographs such as cross-sections, vascular bundles, and transections. Scanning and transmission electron micrographs are traced for chloroplasts, amyloplasts, trichomes, internodes, and pollen grains. Preserved museum specimens provide the source for animals in the pollination series. The remainder of the drawings are from actual plants found in nature, the grocery store, plant nurseries, farm fields, botanical gardens, florist shops, and suburban yards.

In the *major groups* section, three pages have hypothetical diagrams, indicated in the captions. Other microscopic forms are from observations of living material or prepared microscope slides. For plants not locally available, dry-pressed herbarium specimens are measured for drawings (*Stylites, Helminthostachys, Gnetum,* and *Ephedra*) or machine-copied and traced (habit drawings of filamentous algae) or chemically revived to three dimensions (bryophytes) for drawing with the use of a dissecting microscope. Drawings are also made from liquid-preserved specimens (*Tmesipteris* habit, *Welwitschia* and *Ginkgo* reproductive structures). For the majority of this section's drawings (including *Welwitschia* habit), living organisms are used.

For the *flowering plant families* section, except for two indicated diagrams in the grass family, all the drawings are from actual plants gleaned from fields, forests, roadside ditches, bogs, neighbors' yards, botanical conservatories, florist shops, grocery stores, and our gardens. The bumblebee arrived of its own accord.

Janice Glimn-Lacy
Peter B. Kaufman

Color Code Clues

The illustrations may be colored by using the easy-to-follow *Color Code*. Each drawing has lines from structures to letters, duplicated in the *Color Code*. All structures similar to the one with a line and letter are the same color; for example, only one of five petals may be identified. Colored pencils are recommended for pleasing results. My personal preference is Berol Prismacolor pencils. The colors needed are 2 shades of red, green, and blue, and 1 each of orange, yellow, pink, purple, brown, and black. Sometimes the *Color Code* lists double colors such as yellow-green, red-brown, or purple-green. Color the same area with a light touch of both colors for the closest resemblance to the plant's natural color. A black pencil used lightly provides a gray color. White is to remain blank. Using pink over light purple results in a lavender color. So as not to color areas that are to remain white or have small areas of color, follow the order listed in the *Color Code* for an individual drawing. Letters, missing from the *Color Code*, are in the text. They usually indicate structures shown as black outlines or colorless areas.

When drawing a 3-dimensional object, scientific illustrators traditionally use an upper-left light source on the object, which casts shadows on the lower right. Converting those shadows into ink dots (stipple) and lines (hatch) within the drawing outline creates an illusion of depth. Another aid to realism in black-and-white drawings is to show foreground structures as darker and larger than background ones. Obvious examples using this technique are the pitcher-plant leaves (see 91) and the iris flower and leaves (see 129). We see colored objects as brighter and more distinct when near and as muted and less distinct when distant. Accordingly, by the use of bright shades of colors for foreground structures and muted shades for background structures, your colored pictures will appear more 3-dimensional. Darker shades of a color on the dot-and-line shadow areas will add to the effect.

To avoid having a green book, the illustration of foliage has been kept to a minimum with an abundant variety of colors indicated throughout. In the *introduction to plants* section, there is uniformity of a designated color. For example, water-conducting tissue and cells (xylem) are always color-coded blue, so that as you look at your colored pages, you see how water moves within a plant. For enlarged flower drawings in the *flowering plant families* section, true colors of flower parts are indicated in the *Color Code*. Dotted lines indicate where structures have been cut, show "see-through" areas, or separate color areas. Some lines are close together, others are single lines to color, but a color overlap of the lines will not detract from the over-all result.

Begin with the first page, which explains the names used, then choose any page that strikes your interest. If a flowering plant structure is new to you, it would be most helpful to look over the structure of the stem (14), leaf (21, 23), flower (27, 28), and fruit types (38, 39). Then, you will be ready to explore. Some discoveries are seeing how an apple flower matures into an apple (see 99), finding that grape flowers have flip-top petal caps (see 104), and learning that inside a common garden sage flower is a mechanical "arm" for sticking pollen on visiting bees (see 113).

Acknowledgements

We wish to thank the following colleagues of The University of Michigan for their contribution and review of specific topics: Dr. Joseph Beitel, lycopods; Dr. Michael Carleton, bats; Dr. Allen Fife, lichens; Dr. Michael Huft, euphorbs; Dr. Thomas Rosatti, heaths; Dr. Ann Sakai, maples; Dr. Alan Simon, birds; Dr. Wm. Wayt Thomas, sedges; Dr. Kerry S. Walter, orchids. Also, we would like to thank the following University of Michigan Botany Faculty in the Department of Botany (now, Department of Ecology and Evolutionary Biology and Department of Molecular, Cellular, and Developmental Biology); Dr. Howard Crum, bryophytes; Dr. Harry Douthit, bacteria; Dr. Robert Shaffer, fungi; Dr. William R. Taylor, algae; Dr. Edward G. Voss, grasses; and Dr. Warren H. Wagner, Jr., ferns, plus access to his extensive plant collections; the staffs of the University Herbarium, Museum of Natural History and University Science Library for advice, loans of resource materials, use of microscopes, books and other printed materials. Dr. P. Dayananadan was especially generous with his excellent scanning electron micrographs. For assistance in obtaining rare and out-of-season living plants, we thank William Collins, Jane LaRue, and Patricia Pachuta of the University of Michigan Matthaei Botanical Gardens. We also thank Dr. David Darby of the University of Minnesota for his review of geological information. For palms, we thank Dr. Ackerman of the Garfield Conservatory, Chicago. For locating regional plants, we appreciate the assistance of Robert Anderson, Patricia and Terrance Glimn, Marie Mack, Lorraine Peppin, Ellen Weatherbee, and Catherine Webley. Also, we wish to thank Dr. Peter Carson, Dr. Tomas Carlson, Dr. Michael Christianson, Dr. Michael Evans, Dr. Jack Fisher, and Dr. James Wells.

For reviewing the text format and page design, reading and coloring sample pages, and suggesting many helpful changes, we thank the 80 beginning biology students of James Lenic and James Potoczak of the West High School, Garden City, Michigan, 1980, and the 93 practical botany students of Dr. Eric Steiner, Dr. Kit Streusand, Dr. Lucinda Thompson, Dr. Gay Troth, and Dr. James Winsor of the University of Michigan. From the School of Art and Design at The University of Michigan, we are indebted to Douglas Hesseltine for book design suggestions, to Duane Overmyer for typography lessons, and William L. Brudon as both instructor of biological illustration and as an advisor and friend.

Belatedly, we wish to gratefully thank our editor at Van Nostrand & Reinhold, Susan Munger, for the first edition. Now with the publication of the second edition, we wish to wholeheartedly thank our editors Jacco Flipsen, Keri Witman, and Shoshana Sternlicht at Springer. Also, many thanks to Robert Maged and his Springer production staff.

Lastly, we express our special appreciation to our families—Jack, John, and Jim Lacy and Hazel, Linda, and Laura Kaufman.

Contents

About the Authors

Janice Glimn-Lacy has a **Bachelor of Science in Botany** from The University of Michigan, 1973, and plans to have a **Bachelor of Fine Arts** from Herron School of Art and Design, Indiana University-Purdue University at Indianapolis (IUPUI), 2008. She is a member of the University of Michigan Alumni Association, the Indiana Native Plant and Wildflower Society, the Indianapolis Museum of Art, the Indianapolis Museum of Art Horticultural Society, and the Watercolor Society of Indiana. While a Landscape Designer, she wrote and illustrated *What Flowers When with Hints on Home Landscaping* (the Flower and the Leaf, Indianapolis, 1995). She was a co-author and illustrator of *Practical Botany* (Reston), illustrator of *Michigan Trees* (University of Michigan Press), *Plants Their Biology and Importance* (Harper & Row), several Ph.D. theses, and many botanical journal articles. For several years she taught botanical illustration in the University of Michigan Adult Education Program, at the Indianapolis Art Center, and at the Indianapolis Museum of Art.

Peter B. Kaufman, Ph.D. Botany, obtained degrees at Cornell University and the University of California, Davis. A plant physiologist, now Professor Emeritus of Biology, and currently, Senior Research Scientist in the University of Michigan Integrative Medicine Program (MIM), he has been at the University of Michigan since 1956. He is a Fellow of the American Association for the Advancement of Science (AAAS). He is past President of the Michigan Botanical Club, past Chairman of the Michigan Natural Areas Council, a co-author of *Practical Botany* (Reston), *Plants, People and Environment* (Macmillan), *Plants Their Biology and Importance* (Harper & Row), *Creating a Sustainable Future, Living in Harmony With the Earth* (Researchco Book Center), *Natural Products from Plants* (CRC Press)/Taylor and France's Group, LLC, *Handbook of Molecular and Cellular Methods in Biology and Medicine* (CRC Press), *Methods in Gene Biotechnology* (CRC Press), and the author of numerous articles in botanical journals. He hosted a 20-part educational TV series '*The House Botanist*' (University of Michigan) and twice was a guest on ABC's program 'Good Morning America' to talk on gardening topics.

BOTANY ILLUSTRATED

SECOND EDITION

Introduction to Plants,
Major Groups,
Flowering Plant Families

Names and Terms

Division. The major category is called a division and is equivalent to phylum in the animal kingdom. Division name endings indicate plants (-phyta) and fungi (-mycota). In the top illustration are some examples of Magnoliophyta, the flowering plant division. The endings of lesser categories indicate class (-opsida), subclass (-idea), order (-ales), and family (-aceae). Classification of one specific plant is shown.

Class. Flowering plants are divided into 2 classes: Magnoliopsida, the dicots (2 seed leaves), and Liliopsida, the monocots (1 seed leaf). A seed leaf is the first leaf to emerge from an embyro in the seed (see 40). In the second illustration are examples of Liliopsida, the monocot class.

Family. The third group of plants shows some monocots in the lily family, Liliaceae (see 128).

Genus Species. Next are 3 lilies in the **genus** *Lilium*. A plant type is given a 2-part scientific name – a generic name, the genus, which is a noun and is always capitalized, and a specific epithet, the **species**, which is usually an adjective. For example *officinalis* means medicinal and is commonly used for herbs. Sometimes a species is named for a person. A species that is not named is written as sp. (plural: spp.).

In a written list of species in the same genus, it is only necessary to fully write the genus once at the beginning. The genus of species that follow is indicated with the first letter of the genus followed by a period. For example: *Mammillaria albescens*, *M. bocasana*, *M. columbiana*.

The 2 parts of the name, called a **binomial**, are italicized in print and underlined when handwritten. The binomial is sometimes followed by a name abbreviation of the person who first described the plant in botanical literature. For example, the most common is Linnaeus, the Swedish botanist, indicated by an L.

One of the reasons the scientific name of a plant may change is because it is discovered that someone else named the plant earlier. The newly found name is published in Latin. Another reason a name may change is that the plant is reclassified into a different group. There may also be many forms of a plant with separate species names, when later, it is decided they are all the same species.

The specific lily here is *Lilium michiganense*, with the English common name Michigan Lily. A spontaneously occurring variation of a species in nature also may have a **variety** (var.) name (see 101). An X between the genus and new specific epithet (see 77) indicates a hybrid between 2 species. Hybrids may occur in nature or be derived by plant breeding. Intergeneric hybrids have a combined name of the genera with a capital X at the beginning. For example, X *Heucherella* is a hybrid between *Herchera* (pink to red flowers) and *Tiarella* (white flowers) and has pale pink flowers.

Common Names. As names differ from region to region, a plant may have many common names. For example, *Amelanchier* may be commonly named serviceberry, shadbush, or June-berry. Sometimes a common name is used for unrelated plants. So to be universally understood, botanists use the binomial system of naming plants in Latin. A plant variety that exists only through cultivation has a common **cultivar** name (cv., **culti**vated **vari**ety) (see 93).

Habit vs. Habitat. The gross **habit** refers to the appearance of an individual plant in its entirety. Where a plant grows in its environment is called the **habitat**, such as a beech-maple forest, a pond, a prairie, or a desert.

Herbaceous vs. Woody. Herbs have little or no secondary vascular tissues and their above-ground shoots usually die back at the end of the growing season. Herbs are considered to be more advanced in evolution than woody plants (shrubs and trees) with secondary vascular tissues (secondary xylem and phloem) present, which persists in growth year after year in stems and roots.

Macroscopic vs. Microscopic. If a plant structure can be seen with the unaided eye, it is designated macroscopic; if a microscope is used to observe it, it is microscopic in size.

COLOR CODE

orange:	tepals (a) (see 27)
brown:	stamens (b)
light green:	stigma (c)
green:	peduncle (d), leaves (e), stem (f)

Classification of a Species

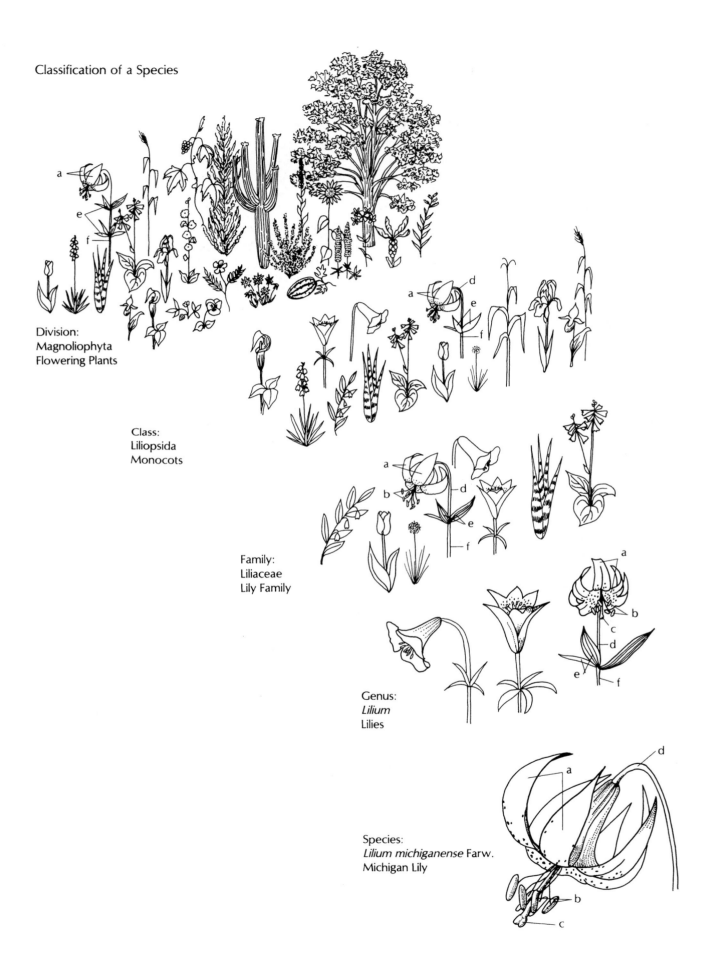

Division:
Magnoliophyta
Flowering Plants

Class:
Liliopsida
Monocots

Family:
Liliaceae
Lily Family

Genus:
Lilium
Lilies

Species:
Lilium michiganense Farw.
Michigan Lily

Cell Structure

Parts of a Cell

Cell Wall. The cell wall (a) encloses and protects the cell contents and plays a vital role in cell division and cell expansion. Composed of overlapping cellulose microfibrils, other polysaccharides and varying amounts of lignin, the cell wall is a relatively rigid structure in mature cells. It may vary in thickness and has pits that that function in communication between cells. The region between the primary walls of adjacent cells, the middle lamella (b), is composed of a cementing substance called pectin.

Other substances that may be present in the cell wall are gums, resins, silica, calcium carbonate, waxes and cutin, and both structural protein and enzymes (which are also proteins). There may be intercellular spaces (c) between walls of bordering cells.

Pits. Primary pit fields (d) are thin areas in the cell wall with tiny strands of cytoplasm, called plasmodesmata (s), connecting one cell with another. Pits are important in facilitating the flow of water and mineral nutrients between conducting cells in the xylem vascular tissue (see 11).

Plasma Membrane. A semipermeable membrane (e) encloses the cytoplasm within a cell. It is composed of variable amounts of fat type molecules (lipids) and proteins, and has within it channels for the movement of ions such as potassium (K^+), calcium (Ca^{2+}), and hydrogen (H^+).

Cytoplasm. The cytoplasm (cytosol, f) is a liquid, gel-like substance and contains several types of organelles; smooth (g) or rough endoplasmic reticulum (h), rough referring to attached ribosomes (i) on the ER (endoplasmic reticulum) and free ribosomes.

Vacuole. In a mature plant cell, one large vacuole usually occupies most of the space within the cell. It is surrounded by a single-layered membrane, the tonoplast (j), and contains cell sap composed of water, sugars, and various organic and inorganic solutes. It may, in some cells such as in beet roots and flower petals, contain water-soluble pigments. The vacuole functions in regulation of osmotic balance and turgidity of the cell, and it stores secondary metabolites.

Organelles. Within the cytoplasm are mitochrondria (k), dictyosomes (Golgi bodies), microbodies, and microtubules (m). Microtubules are represented by an array of parallel tubular tracks and facilitate movement of proteins and organelles within the cell. The ER is a system of tubes and sacs, that work with dictyosomes to produce and secrete compounds and deliver specific proteins and membrane lipids to their proper locations within a cell.

Also, there may be plastids such as chloroplasts (n), leucoplasts, and chromoplasts; and non-living substances of water-soluble products or reserve substances such as oil droplets, protein bodies, and crystals (see 3).

Nucleus. The nucleus is enclosed by a double membrane (o) that has pores (p) in it to allow communication with the cytoplasm (f). Within the nucleus are chromosomes, which contain DNA needed to create proteins within the cell. Chromosomes are only visible during cell division (see 6 and 7). Also present in the nucleus are one or more nucleoli (q) containing RNA. The rest of the nucleus is filled with nucleoplasm (r). The information needed to create the entire plant is within the nucleus, mitochondria, and chloroplasts of each cell.

COLOR CODE

tan:	cell wall (a), middle lamella (b), dictyosome (l)
colorless:	intercellular space (c), pit field (d), plasma membrane (e), vacuolar membrane (j)
purple:	smooth ER (g), rough ER (h)
black:	ribosome (i)
green:	chloroplast (n)
blue:	mitochondrion (k)
orange:	mitrotubules (m)
red:	nuclear membrane (o), pore (p)
gray:	nucleolus (q)
pink:	nucleoplasm (r)
yellow:	cytoplasm (f), plasmodesmata (s)

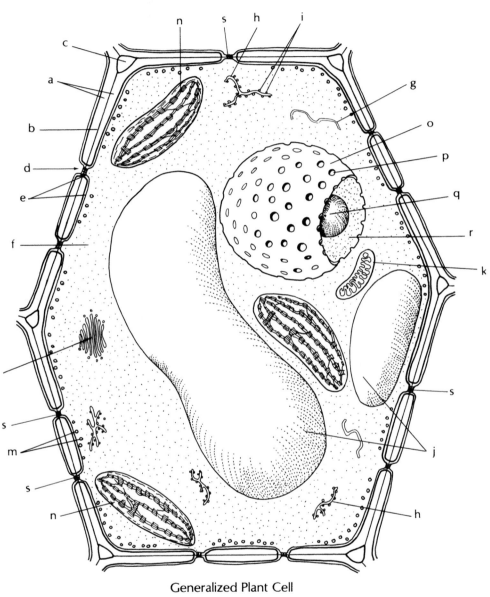

Generalized Plant Cell
diagram

Cell Organelles

The cytoplasm (cytosol) of a plant cell contains various organelles with specific functions. With the use of a transmission electron microscope (TEM), which shows structures magnified thousands of times, these organelles can be viewed individually. The organelles represented here are not shown in proportion to one another.

Plastids. Plastids are classified by the primary pigment they contain. Young cells have undifferentiated plastids (proplastids), which can multiply by simple division. They develop into the various kinds of plastids characteristic of mature cells.

Chloroplast. This plastid (a) is where photosynthesis takes place. In higher plants it is usually oval-shaped and is surrounded by a double membrane (b). Within it are sac-like structures (shown cut in half) called thylakoids (c). A stack of thylakoids is called a granum (d), and this is where green chlorophyll pigments are located. Chlorophyll and proteins bound to the thylakoids use light energy to make simple sugars from carbon dioxide and water (photosynthesis, see 24).

Extensions from some thylakoid membranes form interconnections between grana. The thylakoids also contain accessory pigments, carotenoids, and xanthophylls. In the stroma (e) of the chloroplast are DNA, RNA, oil droplets (f), ribosomes (g), and other materials such as starch grains, found in chloroplasts of green plant tissues that have been actively photosynthesizing.

Leucoplast. Such colorless plastids contain storage products, which include oils, protein bodies, or starch grains. In plant parts with a high starch content, such as potato tubers or rice grains, a leucoplast that contains starch grains (h) is called an amyloplast (i).

Chromoplast. These plastids are colored red (k), orange (l), or yellow (m), depending on the pigments they contain. In the changing colors of ripening fruit such as a tomato or a red pepper, the chloroplasts (green) differentiate into chromoplasts (orange to red). As fruit color changes, so do plastid structure, pigment types, and content.

Mitochondria. These organelles are surrounded by a double membrane (n). The inner of these two membranes (o) has infoldings, called cristae (p), that protrude into the cavity within. Mitochondria are the primary sites of enzymes controlling cell respiration (a chemical release of energy from sugar or other metabolites). They can replicate by simple division and, like chloroplasts, they contain DNA.

Ribosomes. Ribosomes (q) contain ribosomal RNA and function in protein synthesis. They are also found in cytoplasm, and associated with the endoplasmic reticulum (r) in the cytoplasm.

Dictyosomes (Golgi Bodies). These organelles appear as a stack of flattened sacs (s) and associated vesicles (t). Dictyosomes produce and secrete cell wall polysaccharide precursors and complex carbohydrate substances that are secreted out of root cap cells. This results in less injury to the growing root as it penetrates the soil.

COLOR CODE

green:	chloroplast (a), thylakoid (c), granum (d)
white:	double membrane (b, n), stroma (e), starch grain (h), amyloplast (i)
yellow:	lipid (f), chromoplast (m)
black:	ribosome (g, q)
blue:	inner membrane (o), crista (p)
red:	chromoplast (k)
orange:	chromoplast (l)
purple:	endoplasmic reticulum (r)
tan:	dictyosome sacs (s), vesicles (t)

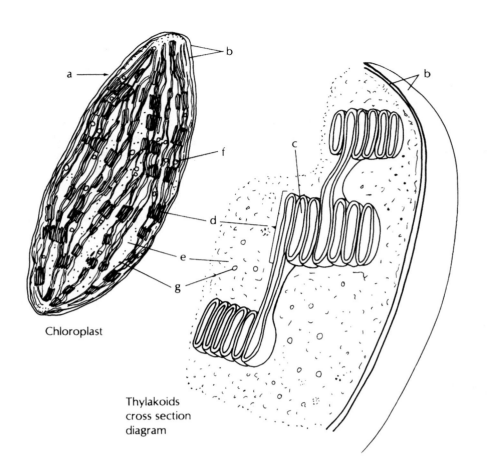

Chloroplast

Thylakoids
cross section
diagram

Cell Organelles

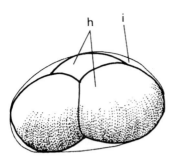

Amyloplast
with compound starch grain
× 3000

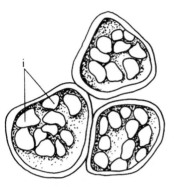

Amyloplasts
in parenchyma cells

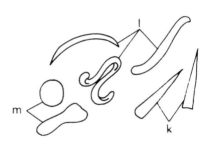

Chromoplasts

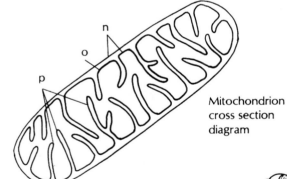

Mitochondrion
cross section
diagram

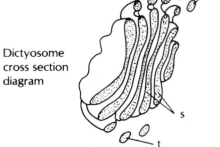

Rough ER
ribosomes on endoplasmic reticulum
diagram

Dictyosome
cross section
diagram

Cell Pigments

Chlorophylls. The chlorophylls are oil-soluble. **Chlorophyll a** (greenish-yellow in solution) is the primary photosynthetic pigment in green plants for the transfer of light energy to a chemical acceptor. Light that is absorbed provides the energy for photosynthesis (see 24). A green leaf absorbs blue light (mostly at 430 nm) and red light (mostly at 660 nm). It reflects the green wavelengths so as to appear green to us. Chlorophyll a, alone, is found in blue-greens and in some red algae. Accessory pigments in photosynthesis transfer light energy to chorophyll a. One of these is **chlorophyll b** (blue-green solution), found in higher plants and green algae with chlorophyll a. **Chlorophyll c** is also an accessory pigment found with chlorophyll a in brown algae and diatoms. **Chlorophyll d,** together with chlorphyll a, is found in some red algae.

Carotenoids. These pigments are considered as accessory pigments in photosynthesis, when found in chloroplasts associated with chlorophyll, and as color pigments when found in chromoplasts. Carotenoids, like the chlorophylls, are not water-soluble. They absorb mainly violet and blue light between 400 and 500 nm, and they reflect red, orange, yellow, brown, and the green color of avocado fruit. They give color to carrot roots, tomato fruit, and many yellow flowers. Carotenoids of red and yellow are revealed in autumn leaves (h, l) after the green chlorophyll pigments begin to break down. This occurs when daylengths start to shorten and cooler temperatures prevail in temperate regions of the world. **Carotenes** (yellow or orange non-oxygen containing pigments) seem to function in the prevention of chlorophyll destruction in the presence of light and oxygen. β-carotene is the most important of the carotines. **Xanthophyll** pigments are oxygen-containing carotenoids. They transfer energy to chlorophyll from light. Some of the xanthophylls include lutein, zeaxanthin, violxanthin, and fucoxanthin, which is found in brown algae.

Phycobilins. These pigments are water-soluble. As accessory pigments they absorb light and transfer excitation energy to chlorophyll a. The red pigment, **phycoerythrin,** is found in red algae. The blue pigment, **phycocyanin,** is present in blue-greens and some red algae.

Phytochrome. This pigment plays an important role in regulating many processes of plant growth and development. Phytochrome is found in two reversible forms. One form, P_r, absorbs red light (mostly at 660 nm) and reflects a blue-green color. The other form, P_{fr}, absorbs far-red light (mostly at 730 nm) and reflects a light-green color. The highest amounts of phytochrome are found in meristermatic tissues (see 9).

Flavinoids. These water-soluble pigments accumulate in cell vacuoles. They absorb ultraviolet wavelengths of light. **Anthocyanins** are phenolic pigments that are found in most fruits and in many flowers. These red-, purple-, and blue-reflecting pigments are seen in the red color of apple (*Malus*) fruit and geranium (*Pelargonium*) flowers; the blue of cornflowers (*Centaurea*) and larkspur (*Delphinium*); and the red and purple of *Fuchsia* (l, m) flowers. The function of these pigments can only be hypothesized, but study of plants having these pigments is useful in determining evolutionary relationships. **Flavins** often appear to us as yellow or ivory-colored flowers. Some flavins, such a riboflavin (vitamin B_2), act as co-factors in enzyme reactions and some are thought to be receptor pigments in the bending of plants toward light (phototropic responses; see 20).

Betalains. These are water-soluble, nitrogen-containing reddish pigments that are found in only 9 families of the flowering plant subclass, Carophyllidae (see 75). These pigments are water-soluble and found in the vacuoles of cells. **Betacyanin** appears as blue-violet to red and is found in beet (*Beta*) roots (n) and red cactus flowers. **Betaxanthin** pigments reflect yellow, orange, and orange-red.

COLOR CODE

purple:	column a, petal anthocyanin (l)
blue:	column b
green:	column c, leaf chlorophylls (g), ovary (j), peduncle (k)
yellow:	column d, leaf carotenoids (i)
orange:	column e
red:	column f, leaf carotenoids (h), sepal anthocyanin (m)
blue-purple-red:	root betacyanin (n)

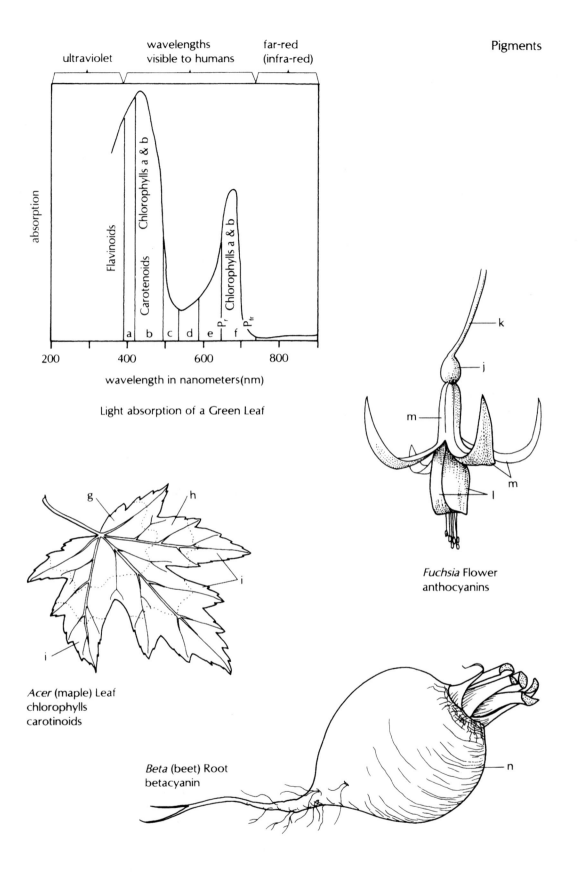

ultraviolet

wavelengths
visible to humans

far-red
(infra-red)

absorption

Flavinoids

Carotenoids

Chlorophylls a & b

Chlorophylls a & b

P_r P_{fr}

a   b   c   d   e   f

200   400   600   800

wavelength in nanometers(nm)

Light absorption of a Green Leaf

Fuchsia Flower
anthocyanins

Acer (maple) Leaf
chlorophylls
carotinoids

Beta (beet) Root
betacyanin

Cell—Water Movement

More than 90% of most plant tissues is composed of water. Water is needed for most life processes to take place. The movement of water (a) from cell vacuole (b) to another cell's vacuole (c) within the plant is by diffusion through the cell's semipermeable plasma membrane (d), which encloses the cytoplasm (e). The flow is from a region of high concentration of water to a region of low concentration of water, a process that is called **osmosis.** The more substances (solutes) dissolved in the water, the lower the concentration (water potential) of water. For example, in a leaf cell that is producing sugar (a solute, f) from photosynthesis, water passes into the cell. An osmotic equilibrium is reached when water no longer enters the cell.

Turgor Pressure

As water (a) passes into a cell, it expands the plasma membrane (d), exerting pressure against the cell wall (g). This is called **turgor pressure**. In fully inflated cells, turgor pressure is one of the key "driving forces" in plant growth. In cells of developing fruits, turgor pressure is very important for fruit enlargement, and indicates the necessity of keeping plants well watered during this period. Turgor pressure maintains the size and shape of plant parts such as leaves and flowers.

Plasmolysis

Water (a) is lost from a cell when there is a lower water concentration in adjacent cells that have more solutes present. Turgor pressure within the cell then decreases. When there is no turgor pressure, the plasma membrane (d) shrinks from the cell wall (g), leaving through the cell wall. This process is termed **plasmolysis** and is reversible when water is returned to the cytoplasm and vacuole through osmosis. Plasmolysis of cells results in wilting of plant shoots and is the opposite of fully turgid cells, where plants are fully "inflated" or crisp, like fresh celery.

"Wilting"

When excessive cell water (a) is lost through evaporation to surrounding dry air, "wilting" occurs. In this condition cell turgor pressure is lost, the plasma membrane (d) shrinks from the cell wall, and cell wall (g) contracts because there is no external solution to fill in between. This condition can result in plants that are "permanently wilted" if the wilting persists over a period of 10 to 12 days. If wilted plants are watered before they reach the permanent wilting condition, they regain turgor in several hours time.

COLOR CODE

dark blue:	water (a)
colorless:	vacuole membrane (b, c), plasma membrane (d)
yellow:	cytoplasm (e)
light blue:	sugar solution (f)
tan:	cell wall (g)
red:	nucleus (h)
gray:	external solution (i)

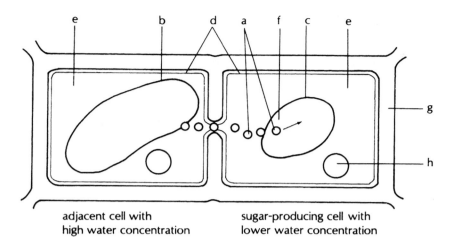

Water Movement
Between Cells

adjacent cell with
high water concentration

sugar-producing cell with
lower water concentration

Osmosis

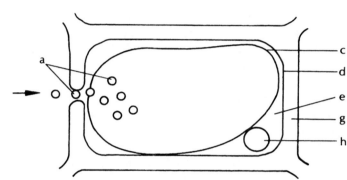

Cell inflated or Turgid
due to turgor pressure

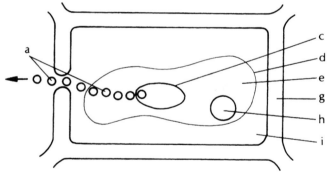

Cell flaccid
due to plasmolysis

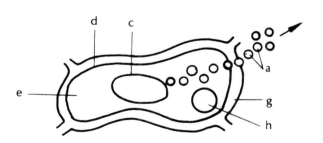

Cell from permanently
wilted plant

Cell Chromosomes

Chromosomes (a) are the darkly stained microscopic bodies in the nucleus which contain genetic information (genes) that determine inheritance characteristics of an organism. A chromosome basically consists of a core of DNA (deoxyribonucleic acid) surrounded by a jacket of basic protein called histone. Bacteria and blue-greens have no nuclei or chromosomes, but instead, have a DNA-bearing structure called a nucleoid (see 43). Some of their DNA may be in circular forms called **plasmids** (b).

Chromosome Structure. Chromosomes are composed of chromatin, which can be seen microscopically with the use of stains such as the Feulgen reagent or acetocarmine. Chromosomes are best observed during mitotic or meiotic division, when the chromatin is condensed and chromosomes visible (see 7, 30). Chromosomes have a simple external structure. A chromosome (a) has a small body called a **centromere** (c) with "arms" (d) on either side. The part of the centrome where spindle fibers are attached is called a **kinetochore.** Small segments, almost circular, that occur near the ends of a chromosome are called satellites (e).

During mitosis and meiosis, the chromosome replicates into two strands called **chromatids** (f) attached together at the centromere. **Chiasma (crossing-over)** is when chromatids (f) overlap, break, and exchange segments with other chromatids (g) during meiosis. It may involve 2, 3 or all 4 chromatids. This can give rise to genetic variability or **mutations.**

Chromosome Number. Chromosome number is characteristic for the majority of vegetative (somatic) cells of an organism. This is called the 2n or **diploid number.** Organisms whose somatic cells contain multiple sets of chromosomes are called **polyploids.** These occur as a result of hybridization between sexually compatible parents with different chromosome numbers. Polyploidy is widely evident in plants, but is rare in animals.

COLOR CODE

purple: chromosomes (a), centromere (c), "arms" (d), satellite (e), chromatids (f)
blue: plasmid outline (b)
red: chromatids (g)
green: chromosomes (h)
black: chromosomes (i)

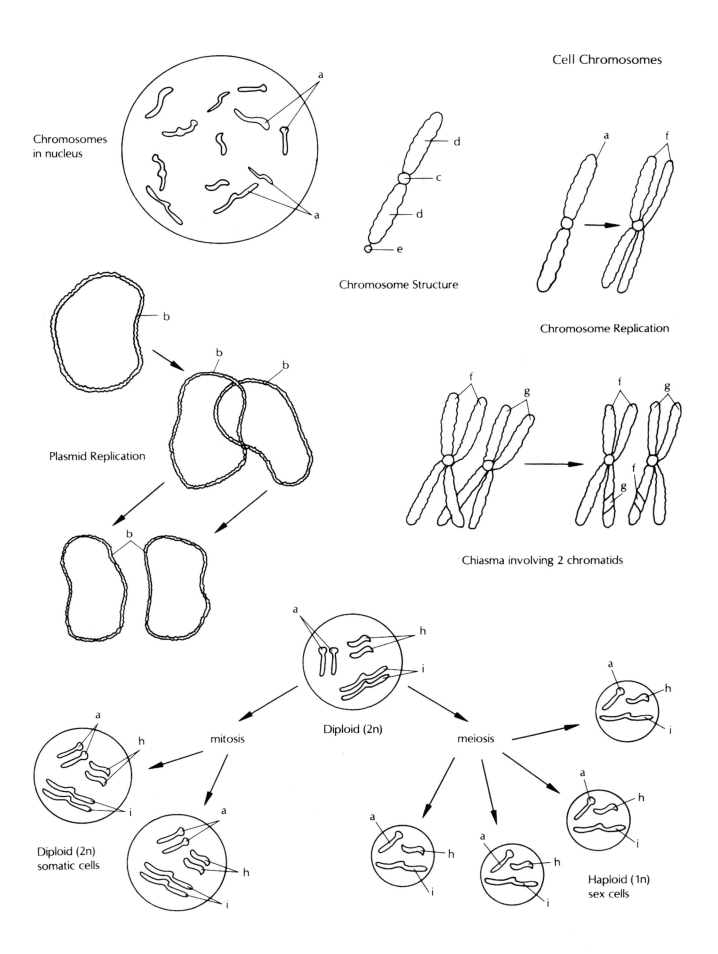

Cell Chromosomes

Chromosomes in nucleus

Chromosome Structure

Chromosome Replication

Plasmid Replication

Chiasma involving 2 chromatids

Diploid (2n)

mitosis

meiosis

Diploid (2n) somatic cells

Haploid (1n) sex cells

Cell—Mitosis

Mitosis is a process of nuclear division in which chromosomes divide lengthwise, separate, and form 2 identical nuclei. Mitosis is usually accompanied by cell division, which involves the formation of a new cell wall between the 2 identical nuclei. In plants, good sites to observe mitosis are in root tips, shoot apices, and newly developing leaves. This is done by making thin sections of the tissues, mounting them on slides, staining them with acetocarmine or Feulgen stains, then observing the sections with the high power or oil immersion lens of an optical (light) microscope.

The Cell Cycle

The period when the nucleus is between divisions is called **interphase**. During interphase is a period of replication of cytoplasmic organelles, followed by a period when chromosomes (a) are duplicated, then a period when spindle fibers (bundles of microtubules) are made. Phases of mitosis are **prophase, metaphase, anaphase, and telophase.** Interphase and the 4 phases of mitosis constitute the cell cycle.

Phases of Mitosis

Prophase. At the first stage of mitosis, the chromosomes (a) become visible as long threads. For clarification only 3 chromosomes are shown. On each strand is a small body called a **centromere** (f). Then the chromosomes start to shorten and thicken and divide into 2 helical coiled strands called **chromatids** (g). DNA is replicated half to each chromatid.

Metaphase. In stained cells on microscope slides, the nuclear membrane and nucleolus (b) can no longer be seen during metaphase in vascular plants. Metaphase is marked by the appearance of the spindle (h). It is at this time that the chromosomes migrate to the spindle and the centromeres (f) align in a flat equatorial plane. The pairs of chromatids (g) are held together at the centromeres (f).

Anaphase. Anaphase is initiated with division and separation of the centromere, providing each chromatid with centromere (f). This phase ends with the chromatids, which are now called chromosomes, moving to opposite poles (i) of the spindle (h).

Telophase. This phase begins with the chromosomes (a) completing their movement to the poles. It ends with the chromosomes once more becoming diffuse, as in interphase. A new nuclear membrane (c) forms around each group of chromosomes (a) and nucleoli (b) reappear. During interphase, a new cell wall (e) forms between the nuclei.

COLOR CODE

purple:	chromosomes (a), centromere (f), chromatid (g)
gray:	nucleolus (b)
red:	nuclear membrane (c)
yellow:	cytoplasm (d)
tan:	cell wall (e)
orange:	spindle (h)

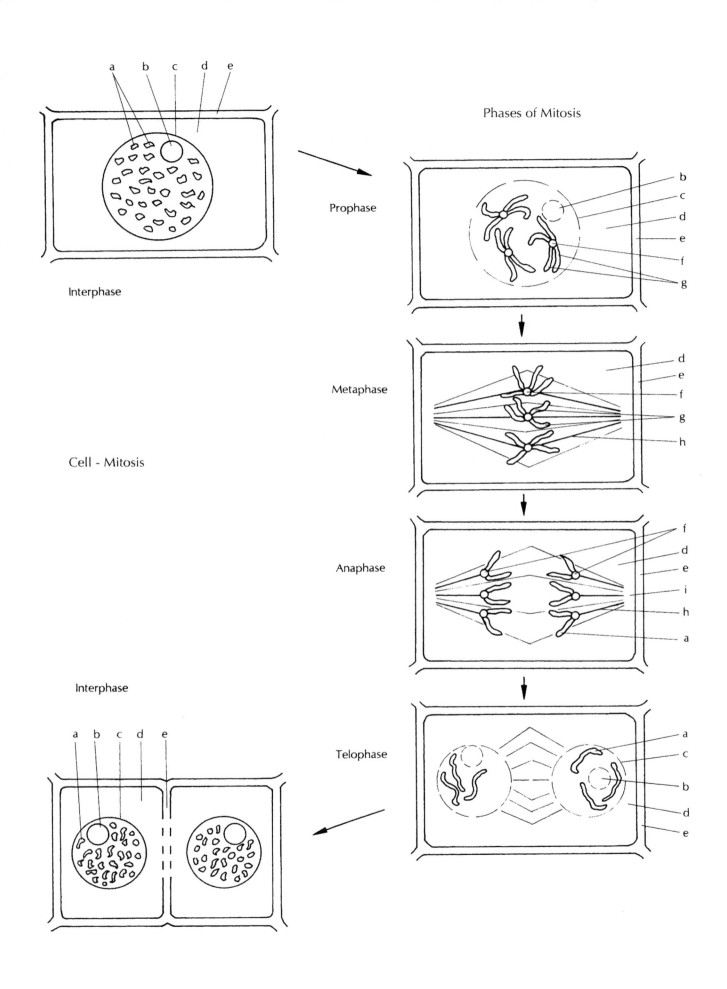

Interphase

Cell - Mitosis

Interphase

Phases of Mitosis

Prophase

Metaphase

Anaphase

Telophase

Cell Types

Parenchyma Cells

Parenchyma cells (a) make up the major portion of the primary plant body. They are usually thin-walled and vary in shape from spherical with many flat surfaces, to elongated, lobed, or folded. As living cells that are unspecialized initially, they later differentiate to more specialized cells. Parenchyma cells are found in photosynthetic tissue of green leaves and green stems, in epidermis, below the epidermis in cortex (b), in pith, and in the vascular system.

As food storage cells, they occur in specialized organs such as bulbs and tubers, in seeds (as endosperm), and in seed leaves (cotyledons). Specialized parenchyma tissue (also called aerenchyma) with intercellular air spaces aids water plants in floating. Parenchyma cells may appear as secretory forms such as glandular and stinging hairs, nectaries, and salt glands.

Collenchyma Cells

Collenchyma cells (c) provide elastic support to stems and leaves due to variously thickened primary walls (d) containing cellulose, hemicellulose, pectin, and water. These closely arranged, living cells are short or elongated in shape. They are usually found near the surface in the cortex around vascular bundles (e) of leaf petioles and stems.

Sclerenchyma Cells

Sclerenchyma tissue cells function in mechanical support due to thick lignified secondary walls (f), which contain large amounts of cellulose and lignin. At maturity, some sclerenchyma cells no longer have living protoplasts. Fibers and sclerids are types of sclerenchyma cells.

Fibers (g) are elongated cells with pitted cell walls (h). They are found in water-conducting tissue, xylem (i) and food-conducting tissue, phloem (j), along leaf veins and margins, and surrounding vascular bundles in stems. Examples of commercial fibers are hemp, flax, jute, rattan, and cotton, used in making rope, mats and baskets.

Sclereids are dense (lignified), short cells which may look like stones, rods, bones, stars, or branched structures. Familiar forms, with dense cell layer of sclereids occur in nut shells, fruit pits, and the seed coats in legume tree pods (Fabaceae, see 100). Pear (k) and quince fruits contain "stone cells" (l), that is, nests of sclereids in the fleshy mesocarp (m) tissue (see 39).

Vascular Cells

Cells of the xylem tissue, **tracheids** (n), are elongated, have bordered wall pits (o) for water conduction, and are aligned side by side. Also, water- and mineral-conducting cells of the xylem, namely, the **vessel elements** (p), have bordered pits (q) in their cell walls. The vessel elements are aligned end-to-end to form long tubes. The xylem sap passes vertically through the vessel elements via end perforations that may be parallel slits (r) or a single large opening (s). Conifers and primitive woody flowering have only tracheids for water conduction.

Sieve cells are enucleate (minus a nucleus), found in the phloem of conifers and primitive vascular plants such as ferns. The sieve cells are elongated and thin-walled. **Sieve-tube elements** (t) are enucleate and found in more advanced flowering plants. Both sieve cells and sieve-tube elements form long end-to-end columns called sieve tubes. Sieve plates (u), consisting of primary pit fields, occur in the end walls of sieve tube elements. In sieve cells, the walls, over their surfaces, contain localized sieve areas containing many pores that allow for cell-to-cell solute transport. Cell-to-cell connecting strands of cytoplasm pass through the sieve plates.

Companion cells (v), a specialized type of parenchyma, may be present in varying numbers in association with sieve tube elements.

COLOR CODE

green:	parenchyma cells (a), collenchyma cells (c)
light green:	parenchyma cortex (b)
white:	vascular bundles (e), xylem (i), phloem (j), mesocarp (m)
tan:	fibers (f, g, h)
yellow:	pear epidermis (k), sclereids (l)
blue:	tracheid (n, o), vessel elements (p, q, r, s)
orange:	sieve-tube elements (t), sieve plate (u), companion cells (v)

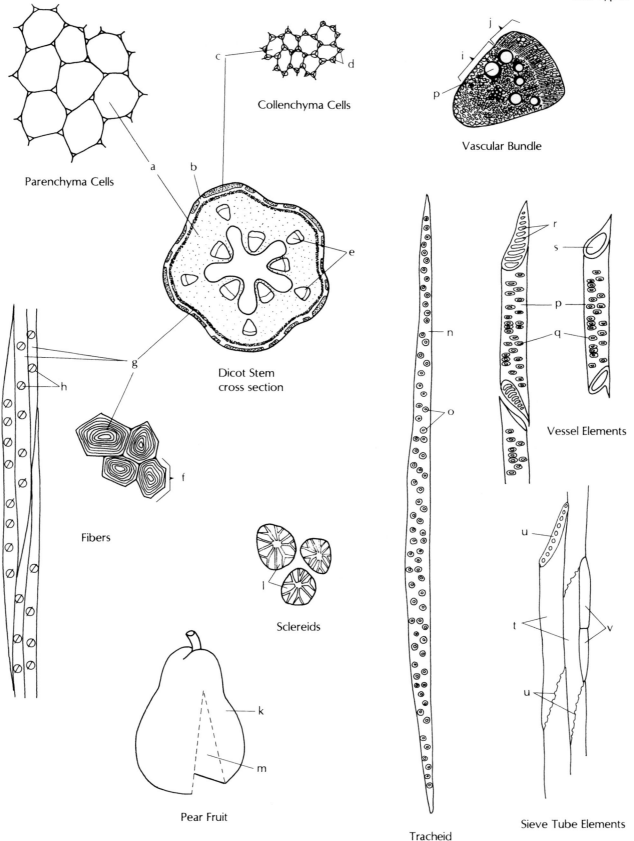

Parenchyma Cells

Collenchyma Cells

Cell Types

Vascular Bundle

Dicot Stem
cross section

Fibers

Sclereids

Pear Fruit

Tracheid

Vessel Elements

Sieve Tube Elements

Tissue Systems of the Plant Body

Cells in plants are arranged in tissues such as epidermis, cortex, and pith. Several tissues also make up a tissue system such as the vascular system.

Meristems

Throughout the life of a plant, new cells are continuously being formed at sites called **meristems**. Meristems consist of undifferentiated cells that are found at shoot tips, at root tips, in the vascular cambium, and in the cork cambium. Meristems produce cells that differentiate into specialized tissues of three systems: dermal, ground, and vascular. Only primary growth (left column), resulting from the activity of the shoot and root apical meristems, is found in some herbaceous dicots, most monocots, and lower vascular plants. Conifers and woody dicot shrubs and trees exhibit secondary growth (right column).

Shoot Tissue Systems

The shoot is the portion of the plant above the roots and is composed of stems, leaves, flowers and fruits. The stem apical meristem (a) forms leaf primordia (b) on its flanks and young stem tissues below during the vegetative phase, and in flowering plants, flower primordia during the reproductive phase.

Dermal Tissue System. The outer layer of the apical meristem gives rise to the epidermis (c) of the primary plant body. In the stems of plants with secondary growth, the epidermis is replaced by periderm (d), commonly called the "outer bark." Periderm consists of the meristematic cork cambium producing cork (phellem) outward and the pelloderm inward (see 15).

Ground Tissue System. The cells and tissues of the ground tissue system are derived from the apical meristem. In the diagram, ground tissue is manifest as cortex (e), located between epidermis and vascular bundles, and pith (f) in the center of the stem.

Vascular Tissue System. Also derived from the apical meristem, the procambium (g) initiates the vascular system with cells that differentiate inside into primary xylem (h), the water- and nutrient-conducting tissue, and outside into primary phloem (i), the food-conducting tissue. In plants with secondary growth, a persistent cambium, the vascular cambium (j), makes possible added layers of secondary xylem (k) and phloem (l), resulting in an increase in stem diameter.

Root Tissue Systems

The root apical meristem (m) produces new root cap (n) cells ahead (below) of the root apex, as well as cells of the protoderm (young epidermis), ground meristem, and procambium back of the root apex.

Dermal Tissue System. With primary growth, roots are covered with epidermis (o). With secondary growth, the epidermis and root hairs (derived from epidermal cells) are sloughed off because of formation of periderm (p) on the outside.

Ground Tissue System. Cortex (q) lies between the epidermis and vascular system in portions of roots undergoing primary growth. The innermost layer of cortex, the endodermis (r), bounds the pericycle tissue (s), which surrounds the primary vascular system. In portions of roots undergoing secondary growth, the cortex, like the epidermis, is shed as the periderm develops cork cells.

Vascular Tissue System. In the root, unlike the stem, the primary vascular system is in a central cylinder (t) with xylem and phloem arranged in an alternate radial manner. Primary xylem "arms" (u) radiate out from the center with alternate poles of primary phloem (v) in the outer portion of the cylinder. Between them is procambium (w). In older portions of roots, where secondary growth is evident, the vascular cambium (x) produces successive layers of secondary xylem (y) and phloem (z).

COLOR CODE

red dot:	apical meristem (a, m)
red line:	procambium (g, w), vascular cambium (j, x)
colorless:	epidermis (c, o), pith (f), root cap (n), endodermis (r), central cylinder (t)
green:	young leaf (b), shoot cortex (e)
blue:	primary xylem (h, u)
light blue:	secondary xylem (k, y)
orange:	primary phloem (i, v)
light orange:	secondary phloem (l, z)
tan:	root cortex (q), periderm (d, p)
yellow:	pericycle (s)

Primary Growth   Secondary Growth

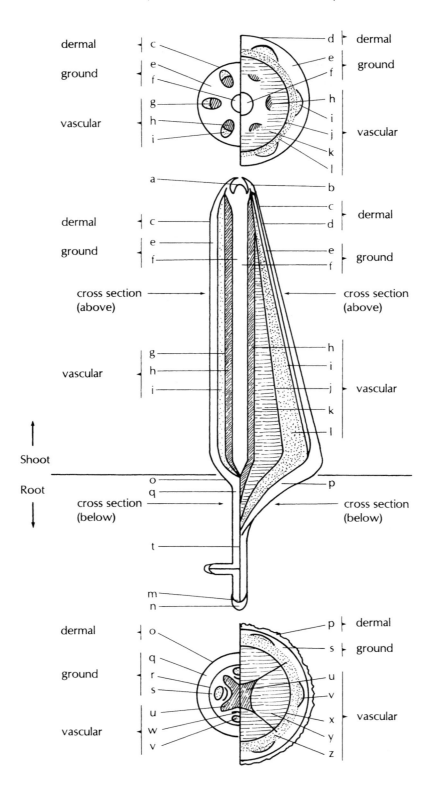

Tissue Systems of the Plant Body

Tissue—Epidermis

Usually a single layer of cells (a) makes up the epidermis that covers roots, stems, leaves, and fruits. At the outer surface of the epidermal cells is a continuous layer (cuticle, b) made up of fatty material (cutin). It is sometimes overlaid with a protective, waterproof coating of wax. Oil, resin, and salt crystals may also be deposited on the surface. Functions of the epidermis include mechanical support, protection from desiccation (drought) and against attack by virulent pathogenic organisms and insects, gas exchange, restriction of water loss by evaporation (transpiration) through stomates and water and mineral storage.

In a surface view, the epidermal cells appear hexagonal, elongated, or wavy-margined (c) like jigsaw puzzle pieces. Shape varies with the plant part being examined, such as root, stem, leaf (d), flower parts (e), or fruit. Some plants such as those in the grass family (Poaceae), have parallel rows of elongated cells (f) alternating with short specialized silica (g) and cork (h) cells, stomata (i) and trichomes. Pits (j) are present in the elongated cells.

Guard Cells. In stomata, pairs of guard cells (k), specialized epidermal cells, control air exchange (CO_2 and O_2) and water loss from plants by changing the size of the pore (stoma, l) that separates their inner walls. A pair of guard cells can expand or contract as a result of changes in turgor pressure. Usually, as guard cells develop high turgor pressure, stomata open, and with reduced turgor pressure, they close. The stomatal pore (l) allows gas exchange to occur between outside air (m) and inner plant (n) tissues. Other cells, called subsidiary cells (o), may be associated with guard cells.

Trichomes (Hairs). The epidermis may have various extensions of one- or many-celled hairs (trichomes, r) some of which may be glandular. Glandular excretions include terpenes (essential oils, carotenoids, saponins, or rubber), tannins, or crystals (such as salt).

Fragrance in flowers is provided when essential oils in the petals vaporize. Non-glandular hairs may be found on any shoot surface as well as on seeds such as cotton (see 90) and willow (see 95).

Menthol in mints is produced in hairs on the leaves. It is the primary flavoring constituent of peppermint too. Many members of the mint family (Lamiaceae, see 113) and aster family (Asteraceae, see 119) produce terpenes in hairs in leaves and stems that repel herbivores like deer. The glandular hairs on bracts of the female flowers of hops produce humulone that is responsible for the bitter principle in beer.

The leaves and stems of hemp or marijuana (*Cannabis sativa*) are covered with hairs that contain over 20 different cannabinoids responsible for hallucinogenic properties of this plant. These constituents also help in treating patients with glaucoma and allaying the adverse side effects of chemotherapy in treating cancer.

Nectar Glands. Nectar glands, found in various flower structures and on stems and leaves, produce a sugar solution. Some flowers have specialized scent glands (osmophors) in the form of flaps, hairs, or brushes. Glands on the surface of carnivorous plants produce nectar (s) or enzymatic digestive fluids (t).

Root Hairs. Root epidermal cells (u) function in the uptake of water and dissolved minerals. Most plant roots produce extensions from the epidermal cells of trichomes called root hairs (v). They occur above the elongation zone of the root. At maturity, root hairs collapse as the epidermal cells are sloughed off. Alternatively, they may remain intact due to lignification of their cell walls. But even these root hair cells may disappear as bark forms on the root in more mature regions.

COLOR CODE

colorless:	cells (a, c, d, e, f, k, o, t, u), pits (j)
yellow:	cuticle (b)
gray:	silica cells (g)
tan:	cork cell (h)
green:	chloroplast (p)
red:	nucleus (q)
optional:	trichomes (r, s, v)

Epidermis

Trichomes

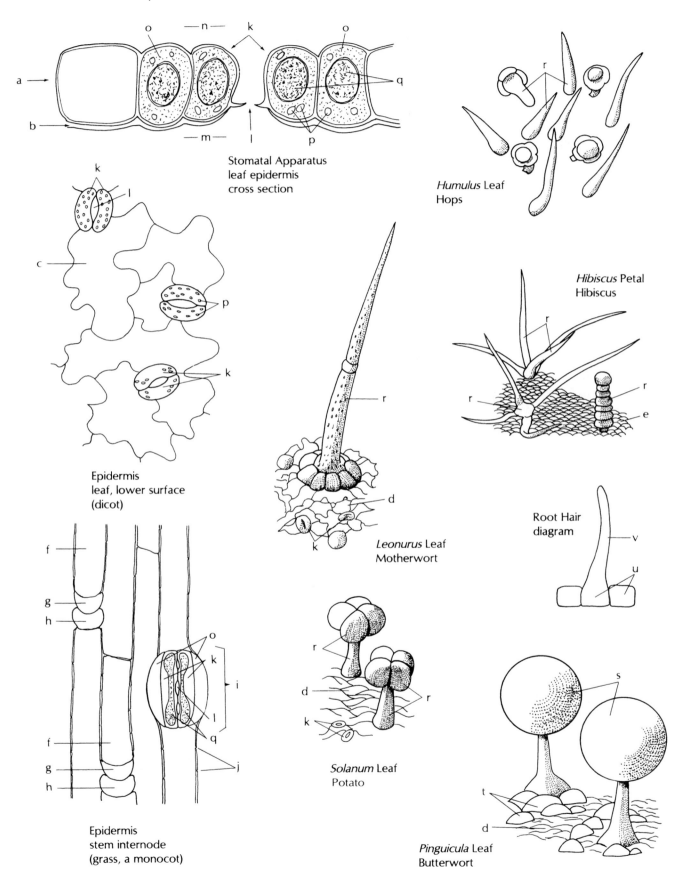

Stomatal Apparatus
leaf epidermis
cross section

Humulus Leaf
Hops

Hibiscus Petal
Hibiscus

Epidermis
leaf, lower surface
(dicot)

Leonurus Leaf
Motherwort

Root Hair
diagram

Epidermis
stem internode
(grass, a monocot)

Solanum Leaf
Potato

Pinguicula Leaf
Butterwort

Tissue—Primary Vascular System

In the primary plant body, the vascular system is made up of vascular bundles composed of conducting elements, interspersed with fibers for support and parenchyma cells for food storage. Procambial cells (a) give rise to primary xylem toward the inside and primary phloem toward the outside of the plant. In most monocots, after differentiation of vascular elements, no procambium remains.

Xylem: Water-conducting Tissue

Several types of cells (living and non-living) make up xylem tissue: tracheary elements (b), composed of tracheids (c) and vessel elements (d), which conduct water and nutrients; fibers (e), which provide support; and living parenchyma cells (f), which store food. Sclereids may also be present. In monocots, vessel elements, whose walls have been stretched and broken by elongation, result in spaces (lacunae, g) in the xylem.

Bordered Pits. The primary walls of tracheids and vessel elements have depressions called primary pit fields. When secondary walls (h) are formed, the bordered holes (pit apertures, i) consist of a pit chamber (j) and a pit membrane (k). Of two adjacent cells, the pits of each are at the same level forming pit-pairs. Composed of the primary wall of each of the two adjacent cells, the pit membrane is permeable and allows passage of water and mineral nutrients from cell to cell.

Phloem: Food-conducting Tissue

Phloem tissue is composed of sieve elements of sieve cells or sieve tube elements (o) for food conduction, fibers (m), and parenchyma cells (n). Companion cells (s) are associated with sieve tube elements. Latex-producing cells (laticifers) and sclereids may be present.

(Dicot vascular bundle adapted with permission: Esau, K., *Plant Anatomy,* 1967, John Wiley & Sons.)

COLOR CODE

red:	procambial cells (a)
colorless:	parenchyma cells (f, n), lacuna (g), pit membrane (k), secondary wall (h)
blue:	tracheary elements (b, c, d)
tan:	fibers (e, m)
orange:	sieve tube elements (l), companion cells (o)

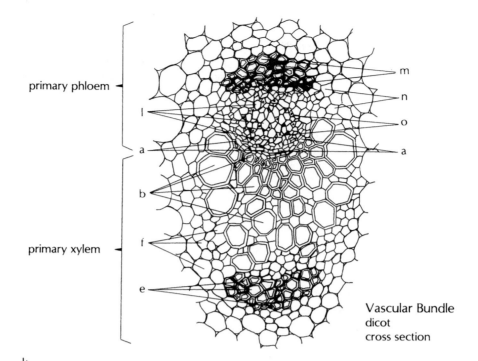

primary phloem

m

n

l

o

a · a

b

primary xylem

f

e

Vascular Bundle
dicot
cross section

Bordered Pit

h   k

i →   → j

pit-pair

h

k

2 adjacent cells

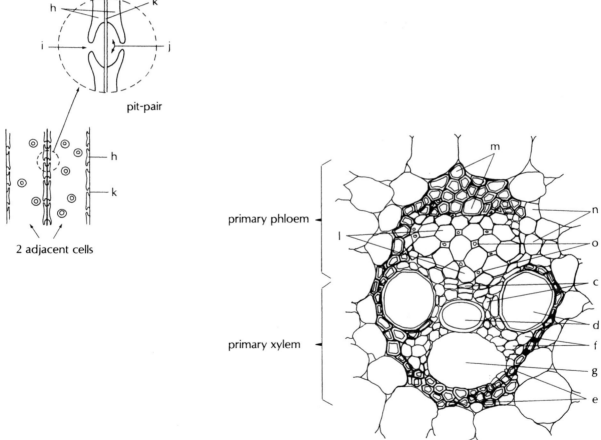

m

primary phloem

l   n

o

c

primary xylem   d

f

g

e

Vascular Bundle
monocot
cross section

Root Types and Modifications

The **root** is the underground organ of the plant. Its primary functions include uptake of water and minerals and anchorage of the above-ground (aerial) portions of the plant.

Root Types

In conifers and dicots, a primary root (a), called a radicle in the seed embryo, develops lateral branching roots (b). As this system develops further, the individual roots may show secondary growth due to cambial activity. The **cambium** is a meristem that produces secondary xylem, and secondary phloem (see 9).

In many monocots, the first-formed roots of the seedling, called **seminal roots**, eventually die. They are replaced by **adventitious roots** (d), which emerge from the stem. Dicots may also have accessory adventitious roots, for example, the aerial roots on vines such as ivy (see 16).

Root Modifications

Storage Roots. Plants may develop specialized, thickened roots, as those found on carrot (*Daucus,* f), beet (*Beta*), sweet potato (*Ipomea batatas*, see 112), manroot (*Ipomea pandurata*), tuberous begonia (*Begonia*), and dahlia (*Dahlia*). These roots are an adaptation for food storage and are seen in many biennials and perennials. A **biennial plant** is one that produces leaves the first year and flowers and fruit the second year, then dies. A **perennial plant** lives for more than two years.

Prop Roots. Some plants develop supporting roots as found in corn (*Zea,* h) and mangrove (*Rhizophora*). Such roots are especially effective in anchoring the shoot system in the soil, preventing them from capsizing in strong winds, heavy rains, and impact from predators and human activities.

Contractile Roots. Contraction, or shortening of some roots, helps pull down and anchor the plant more firmly in the soil. Contractile roots (i) are common in herbaceous dicots and monocots, and occur in taproots, adventitious roots, and lateral roots, and on roots of underground storage stems such as bulbs and corms. (An herbaceous plant or herb is nonwoody and can die to the ground in freezing climates.)

Symbiotic Relationships in Roots

Nodules Involved in Nitrogen Fixation. Some plants are stimulated by certain genera of bacteria to develop extra tissues in the roots in the form of nodules (l), where nitrogen (N_2) in the air is "fixed," that is, converted to other forms, mainly ammonia (NH_3), which can then be converted to organic forms such as amino acids. The bacteria and plant exchange nutrients and metabolites in a mutually beneficial relationship. Examples are found in alfalfa (*Medicago*), clover (*Trifolium*), garden pea (*Pisum sativum*), and garden bean (*Phaseolus vulgaris*).

Mycorrhizae. These are mutually beneficial associations between fungi (n) and plant roots (o). For the plant, the fungi increase the uptake of minerals, particularly phosphorus, and also water. For the mycorrhizal fungus, the plant root cells provide sugars, amino acids, vitamins, and water. Mycorrhizae are classified according to whether the fungal strands (hyphae) invade (endomycorrhizae) or surround (ectomycorrhizae) the root cortical cells. A third class is where fungi both invade and surround (ectendomycorrhizae) root cortical cells. Most plants have mycorrhizal associations, which help them compete with other plants. Mycorrhizae develop best in nutrient-poor soils.

COLOR CODE

tan:	dicot root system (a, b), scale leaves (j), stem axis (k), nodules (l)
green:	leaves (c, e, g), prop roots (h)
white:	monocot root system (d), roots (i, m, o), roots swollen with mycorrhizal fungal strands (n)
orange:	storage tap root (f)

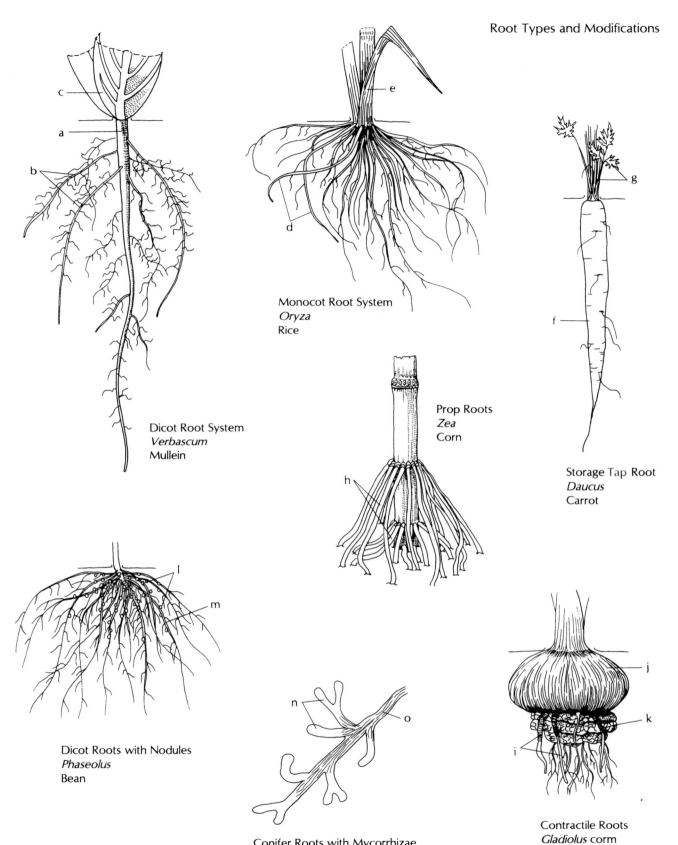

Root Types and Modifications

Dicot Root System
Verbascum
Mullein

Monocot Root System
Oryza
Rice

Prop Roots
Zea
Corn

Storage Tap Root
Daucus
Carrot

Dicot Roots with Nodules
Phaseolus
Bean

Conifer Roots with Mycorrhizae
Picea
Spruce

Contractile Roots
Gladiolus corm

Root Tissues

Basic Differentiation Regions of the Root

Apical Meristem. The root apical meristem (a) is located at root tips. It produces new root cap (b) cells below, and above, it produces cells that contribute to protoderm (c), ground meristem of the cortex (d), and procambium (e), the three primary meristems that initiate tissues.

Root Cap. The function of the root cap (b) is to protect the apical meristem and to aid the developing root as it penetrates the soil as the root elongates. It is composed of parenchyma cells whose walls, along the surfaces of the root cap, are mucilaginous to provide easier passage through soil particles. As cells (f) are sloughed off, more cells are produced by the apical meristem.

Elongation Region. Back of the apical meristem is the region (g) where most elongation of the root takes place.

Mature Root. In primary growth, the epidermal cell walls of the mature root (j) become cutinized or suberized with wax-like substances. With secondary growth, the cork cambium replaces epidermis (k) and cortex with root bark tissues, cork and phelloderm.

Non-vascular Tissue Systems

Epidermis. The epidermis is the outermost tissue of the root. It is usually a single layer. In the root elongation zone are root hairs (h), lateral protrusions of epidermal cells (i), which are active in water and mineral nutrient uptake.

Cortex. Basically, cortex tissue is made up of parenchyma cells. In conifers and dicots with secondary growth, the cortex (d) is shed. Monocots, without secondary growth, retain their cortex and many develop sclerenchyma tissue. Food, mostly as starch, is stored in the cortical cells in amyloplasts (see 3). Proplastids may develop into chloroplasts in cortical cells if the root is exposed to light.

Endodermis. The innermost layer of cortex (d) that encircles the vascular tissue is called the endodermis (l).

Along the radial and transverse walls of endodermal cells is a suberin-rich fatty layer called the **casparian strip** (m). It blocks absorbed soil water from passing between the cells to the xylem; thus, water and mineral nutrients must pass through the selectively permeable membranes of endodermis cells instead, an effective filtering system (see 17).

Vascular Tissue System

Pericycle. The central cylinder (n) of vascular tissue is surrounded by thin-walled parenchyma tissue called pericycle (o). The pericycle is the site where vascular cambium and lateral roots originate. As a lateral root (p) penetrates through the cortex, its apical meristem and its derivative tissue systems become organized. Adventitious roots, which develop from stems, are initiated in parenchyma near differentiating vascular tissue.

Xylem. If no xylem is formed in the center of the root, parenchyma tissue develops there. Primary xylem arms (q) radiate from the center and vary in number. The dicot types include diarch (2 arms), triarch (3 arms), or tetrarch (4 arms, q), and the monocot type is polyarch (many arms).

Phloem. Procambium cells (e), between the radiating arms of primary xylem, produce primary phloem (r) outward and xylem tissue inward. In secondary growth, the vascular cambium (s), which originates from the pericycle, extends completely around the tips of the xylem arms and inside the primary phloem poles. It produces secondary xylem (t) inward and secondary phloem (u) outward. (For a further stage in secondary growth, see 9, root cross-section.) (Lateral root adapted with permission: Esau, K., *Plant Anatomy, 1967,* John Wiley & Sons.)

COLOR CODE

red:	apical meristem (a), procambium (e), vascular cambium (s)
colorless:	root cap (b), protoderm (c), root regions (g, h, i, j), cortex (d), cells (f), epidermis (k), lateral root (p)
tan:	endodermis (l), casparian strip (m)
yellow:	pericycle (o)
blue:	central cylinder (n), primary xylem (q)
light blue:	secondary xylem (t)
orange:	phloem (r)
light orange:	secondary phloem (u)

Root Tissues

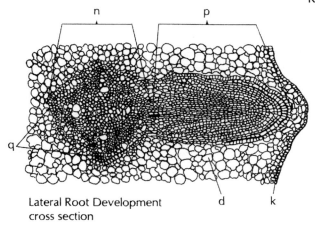

Lateral Root Development
cross section

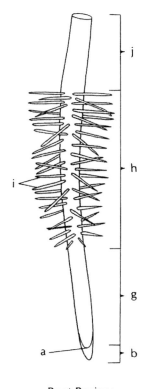

Root Regions

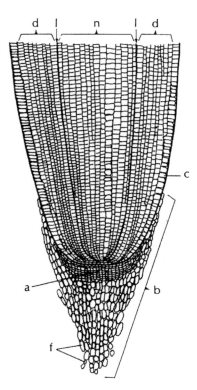

Root Tip
vertical section

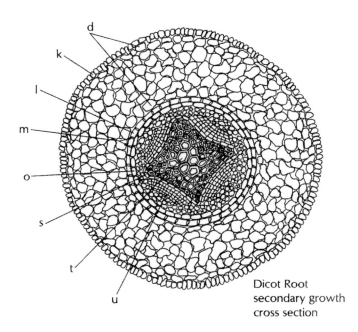

Dicot Root (tetrach)
primary growth
cross section

Dicot Root
secondary growth
cross section

Stem Structure

Functions of the stem include: photosynthesis; long-distance transport of hormones and organic metabolites in the phloem; transport of water and mineral nutrients in the xylem; support of the shoot system; sites for responses of shoot to gravity (**gravitropism**) and unilateral light (**phototropism**); food storage; and sites of initiation of adventitious roots, new leaves and inflorescences (flowers).

Parts of a Stem

Shoot Tip. The stem shoot apex produces leaf primordia (a) that develop into leaves (b).

Node. The site where a leaf or leaves, as well as axillary buds, arise on the stem is termed a **node** (c).

Internode. The elongated portion of a stem between two nodes is called an **internode** (d).

Bud. A **terminal bud** (e), which develops at the apical end of the stem, encloses a meristematic stem axis (f) with its apical meristem (g). **Axillary buds** (h) are lateral buds that form at the leaf-stem axils (nodes) and may develop into lateral vegetative shoots or flowers.

Bud Scales. Small, modified leaves or stipules that cover and protect the bud are called **bud scales** (i). Some buds lack scale coverings and are called naked buds.

Bud Scale Scar. Perennial woody stems exhibit scars (j) at the sites where terminal bud scales were shed during previous years (k, l).

Leaf Scar. A scar remains on the stem where a leaf stalk (petiole) was attached. Leaf scars (m) and buds are species-specific, and therefore, are useful for winter plant identification.

Bundle Scar (Leaf Trace). The vascular bundles (j) from a shed leaf appear as dots within a leaf scar.

Lenticel. Openings for gas and water exchange in the stem surface, called **lenticels** (o), are found on some plants with secondary growth. Lenticels are very prominent, for example on cherry (*Prunus*) tree bark.

COLOR CODE

red:	apical meristem (g), maple bud scales (i), procambium (p)
green:	leaf primordia (a), leaf (b), node (c), stem internode (d), bean terminal bud (e), bean axillary bud (h)
white:	stem axis (f)
tan:	lilac bud scales (i), leaf scar (m)
dark brown:	bud scale scars (j), bundle scars (n), lenticels (o)
gray-pink:	stem growth, last year (k)
gray:	stem growth, 2 years ago (l)

Stem Structures

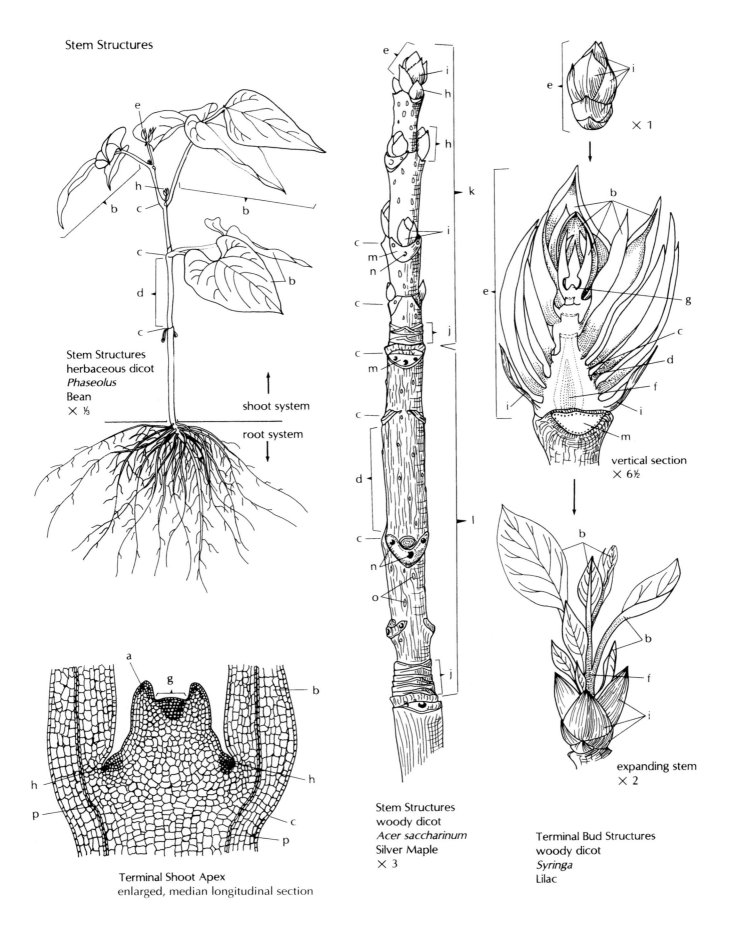

Stem Structures
herbaceous dicot
Phaseolus
Bean
× ⅓

shoot system

root system

Terminal Shoot Apex
enlarged, median longitudinal section

Stem Structures
woody dicot
Acer saccharinum
Silver Maple
× 3

× 1

vertical section
× 6½

expanding stem
× 2

Terminal Bud Structures
woody dicot
Syringa
Lilac

Stem Tissues

Meristems

In the primary state of stem growth, the **apical meristem** at the shoot apex produces cells that differentiate into the three primary meristems. They are the **ground meristem**, which gives rise to pith, cortex of the stem and the mesophyll of leaves, the **procambium**, which produces primary xylem and phloem of the vascular system, and **protoderm,** which produces the epidermal system. In plants with secondary growth, the protoderm is replaced by the **vascular cambium.** An **intercalary meristem** (a) is temporary, located between tissues that already have differentiated. Examples include internodes of grasses and *Equisetum*; leaves of conifers and *Welwitschia* (see 74); fruiting stalks of peanut; and stipes of kelps (see 60).

Ground Tissue System

Dicot and Conifer Stems The ground tissue between vascular tissue (d) and epidermis (e) is **cortex** (f). It consists of parenchyma tissue with intercellular spaces, but may include sclerenchyma and collenchyma tissues. If the vascular tissue forms a ring between the cortex and the center of the stem, in the center is pith (g). In conifers, the cortex (f) may have resin ducts (h).

Monocot Stem. With usually dispersed vascular bundles (i) in the ground tissue, the parenchyma (j) is not differentiated into pith or cortex.

Vascular Tissue System

Dicot and Conifer Stems. Vascular bundles (d) are arranged in a ring around the stem. In primary growth, all cells of the procambium differentiate into primary vascular tissue (k, l). In plants with secondary growth, some of the meristematic cells remain and usually form a contiguous ring of vascular cambium (m), which begins to produce secondary phloem (n) outside, and secondary xylem (o) inside. The yearly additions of secondary xylem are commonly called "growth rings." The ages of trees are determined by the number of growth rings in the trunk near the base of the tree trunk. Rings can be counted visually or by use of an increment borer.

The cambium forms two systems of vascular tissue, one vertically up the axis, and the other horizontally across the axis (called a ray system). Inward from the cambium, xylem rays (p) are produced, and outwardly phloem rays (q) are produced. Associated with the rays are gum ducts in dicots, which contain resins, oil, gums, and mucilages. Comparable to **gum ducts**, some conifers have **resin ducts**. Secondary xylem tissue, "wood," may be composed of sapwood (r) and heartwood (s). Sapwood consists of live xylem components, active in the transport of water and mineral nutrients. The inner heartwood is composed of inactive xylem that stores secondary metabolites. Organic compounds such as oils, gums, resins, and tannins fill the cells that prevent wood rot caused by fungi and other organisms.

Monocot Stem. Monocots have no vascular or cork cambia and, therefore, no secondary growth. They exist in a primary state of growth (herbaceous plants). In large plant bodies such as palms, a thickening of the trunk occurs by multiple divisions of parenchyma cells of the procambium. In some members of the lily family (Liliaceae) such as *Yucca* and *Dracaena,* divisions outside the vascular bundles account for thickening of the stem axis.

Secondary Dermal Tissue

In woody dicots and conifers, the epidermis is replaced by periderm (t), "outer bark." The **periderm** is composed of **cork cambium** (phellogen, u), **cork** (phellum, v) and **phelloderm** (w). The cork cambium (u) produces phellem (v) toward the outside and phelloderm (w) toward the inside of the periderm. Cork cells at maturity are dead, and most are filled with suberin. The cork covering of stem surfaces functions for protection. It is compressible and water- and oil-resistant, and protects against insect and fungal attacks, cold, and, in some trees (e.g. redwood and sequoia), fire. Openings in the cork tissue, **lenticels** (x), serve, the same function as stomata in gas exchange. Secondary phloem (n) and periderm (t) together are commonly called "bark."

COLOR CODE

red:	intercalary meristem (a), vascular cambium (m), cork cambium (u)
green:	ridge (b), grooves (c), cortex (f), parenchyma (j)
colorless:	epidermis (e)
yellow:	resin ducts (h)
orange:	primary phloem (k)
blue:	primary xylem (l)
light orange:	secondary phloem(n), phloem rays (q)
light blue:	secondary xylem (o,r,s) xylem rays (p)
tan:	pith (g), periderm (t), cork (v, x), phelloderm (w)

Stem Tissues

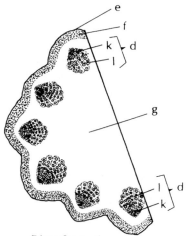

Dicot Stem, ½
primary growth
cross section

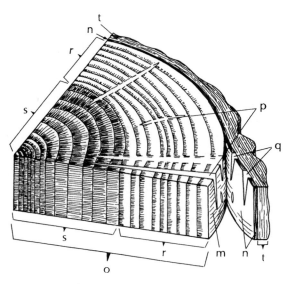

Dicot Stem, ¼
secondary growth
cross section

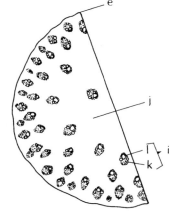

Monocot Stem, ½
primary growth
cross section

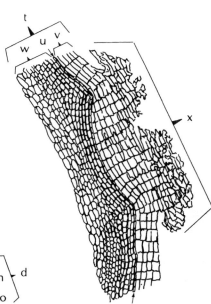

Woody Dicot Outer Bark
lenticel
periderm

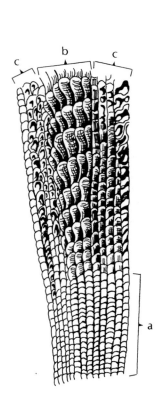

Equisetum Stem Internode
greatly enlarged
lower portion is intercalary meristem

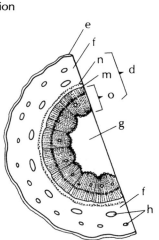

Conifer Stem, ½
secondary growth
cross section

Stem Modifications

Stems may be of various forms to serve different functions, such as for food or water storage, for subterranean or aerial anchoring devices, as a means of asexual reproduction, or for climbing.

Bulb. A bulb is a short, underground, food-storage stem axis with extremely reduced internodes and surrounding fleshy scale leaves (a). An onion is the example here.

Corm. A solid, bulb-like, underground stem without fleshy scales forms a corm. It has greatly shortened internodes. Examples are the food-storage, reproductive corms of *Gladiolus* and *Crocus* (shown).

Pseudobulb. Many orchids grow on branches or trunks of other plants (**epiphytes**). These plants develop a fleshy stem internode with water storage parenchyma. Other orchids, in contrast, are **terrestrial** (grow in the soil.)

Rhizome. Found near or below the soil surface, a **rhizome** is an underground stem that produces scale-like leaves and adventitious roots at the nodes. Rhizomes are in *Iris* (see 129), *Equisetum* (see 20, 66), and Irish potato. A potato rhizome with a developing, edible tuber is shown.

Spur Shoot. Some trees have short, woody stems with shortened internodes, such as apple (see 99) and *Ginkgo* (see 72). In apple, the spur shoots produce flowers and fruits. In *Ginkgo*, they produce pollen-producing male strobili on male trees or fruits from female strobili on female trees (dioecious condition, see 72).

Stolon. A lateral stem, "runner" (k), from the base of a plant develops long internodes, and where the apex touches the soil, a new plant with shoots and adventitious roots forms at a node. Strawberry (see 20) and strawberry begonia (shown) form stolons.

Succulent Stem. Fleshy water storage stems of parenchyma are found in the spurge (see 103) and cactus (see 85) families. A fleshy cactus stem is shown. Other families containing representative genera with fleshy stems are found in the aster family (see 119) and geranium family (see 108).

Tuber. Tubers are swollen, underground food storage stems arising at the tips of rhizomes. They bear buds, "eyes," at the nodes on the potato tuber. These "eyes" develop into potato sprouts (shoots) when the potato starts to grow.

Vine. Vines are stems with long internodes and may have one of various types of climbing devices such as tendrils opposite leaves at a node as in grape (see 104), modified stipule tendrils (w) as in green briar, disc-tipped tendrils (t) as in Virginia creeper, adventitious roots as in philodendron and ivy (q), twining leaf tips as in gloriosa lily, and twining leaf petioles as in clematis, or the entire vine may twine as in wood rose (*Ipomoea tuberosa*).

Of interest... A bizarre wide flattening of the stem in some plants is called **fasciation** (not shown), caused by viruses, some bacterial pathogens, or herbicidal (2, 4-D) drift. The most familiar example is in crested cockscomb (*Celosia cristata*). The condition can occur in herbaceous and in woody plants. Flowers may crowd together in a mass at the end of the flat stem.

Witches'-broom is the term used to describe an abnormal bushy growth of stems that seem to originate from one point. A dense cluster of branches with no apical dominance forms. There are various causes depending on the species of herb or woody plant afflicted: insects, bacteria, viruses, or fungi.

COLOR CODE

white:	fleshy scale leaves (a), leaf buds (c), roots (j, q), potato rhizome and tuber
yellow-tan:	outer scale leaves (b)
tan:	onion stem axis, bud scale (d), scale covering (e, f), stem (o, r), tendril (t)
light yellow:	corm internodes, flower petals (i)
green:	leaf (g, n, s,), stem (m)
purple-green:	orchid pseudobulb (internode), stipe (h), leaf (l)
purple-red:	stolon (k)
dark green:	leaf (p)
light green:	stem (u), leaf (v), tendril (w)

Stem Modifications

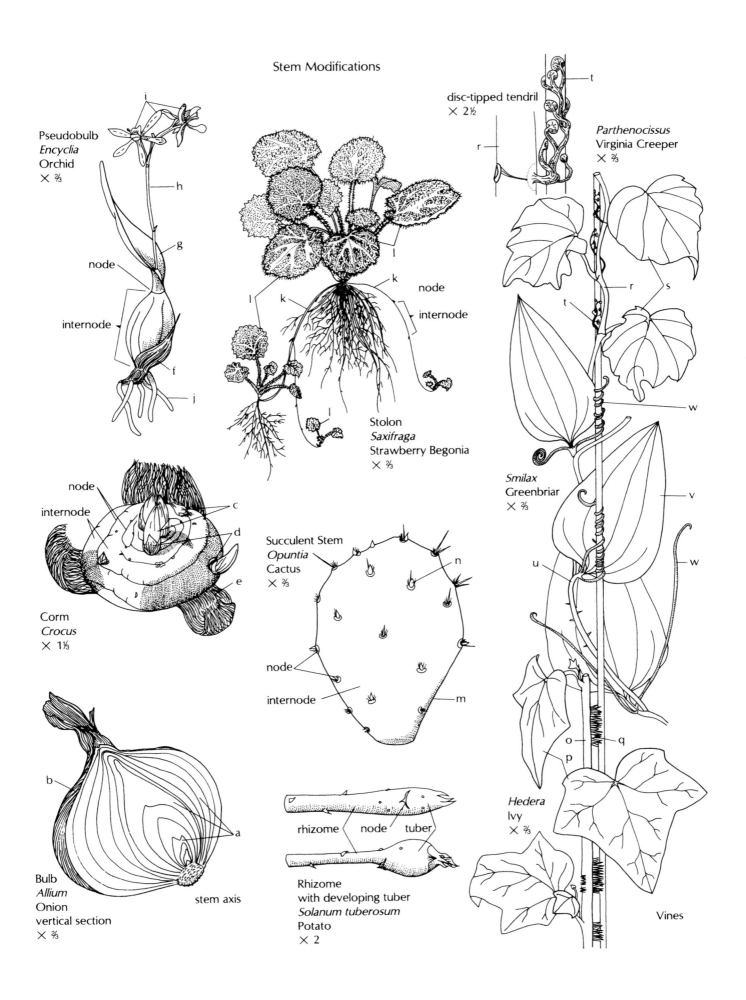

Pseudobulb
Encyclia
Orchid
× ⅔

node

internode

disc-tipped tendril
× 2½

Parthenocissus
Virginia Creeper
× ⅔

node

internode

Stolon
Saxifraga
Strawberry Begonia
× ⅔

node

internode

Smilax
Greenbriar
× ⅔

Corm
Crocus
× 1⅓

Succulent Stem
Opuntia
Cactus
× ⅔

node

internode

Bulb
Allium
Onion
vertical section
× ⅔

stem axis

rhizome   node   tuber

Rhizome
with developing tuber
Solanum tuberosum
Potato
× 2

Hedera
Ivy
× ⅔

Vines

Stem—Water Transport

Transpirational "Pull"

Water has a natural cohesiveness, surface tension, so will rise in a tube (capillary action), but not very high. How then does water rise from the soil to the leaves of tall trees? The air surrounding mesophyll cells in leaf tissue (see 22) has a humidity of 100% and, thus, the tissue air is saturated with water. Air outside the plant usually has a lower humidity unless it is raining. The evaporation of moist air as it escapes the leaf tissue, via the pores (stomata), provides a "pull" for water movement within the plant.

Loss of water by evaporation from leaf surfaces is called **transpiration**. Water (a) moves upward in the plant from a region of higher water concentration in the soil to a region of lower water concentration, the dry air outside the leaves.

Pathway of Water Transport

Water (a) in the soil is absorbed by the epidermal root hairs (b). It then moves across the root through living cortical cells (c), is "filtered" through the semipermeable membranes of endodermal cells (d) passes through the pericycle (e), and enters the water-conducing cells (tracheids and vessel elements) of the xylem (f). Once in the xylem of the vascular system, water rises up the root to the stem and into the leaves.

The maximum rate of water rise recorded in the roots of several oak species is 60 meters per hour. From the xylem, water enters the mesophyll tissue of the leaves and evaporates into the atmosphere via epidermal stomata when they are open. Much smaller amounts of water are lost from leaf cuticle and bark lenticels in the stems.

Environmental Influences

The rate of water transport in most plants is at a maximum during the daytime when air temperatures are high, relative humidity is low, and stomata are fully open. When the rate of water loss by transpiration exceeds the rate of water uptake from soil, the stomata usually close. Many succulent plants that live in the desert open their stomata only at night in order to conserve water. Thus, water transport and transpiration in these plants occur primarily at night.

COLOR CODE

blue:	arrows, water (a), xylem (f)
colorless:	root hair (b), cortex (c)
tan:	endodermis (d)
yellow:	pericycle (e)
orange:	phloem (g)

Stem – Water Transport

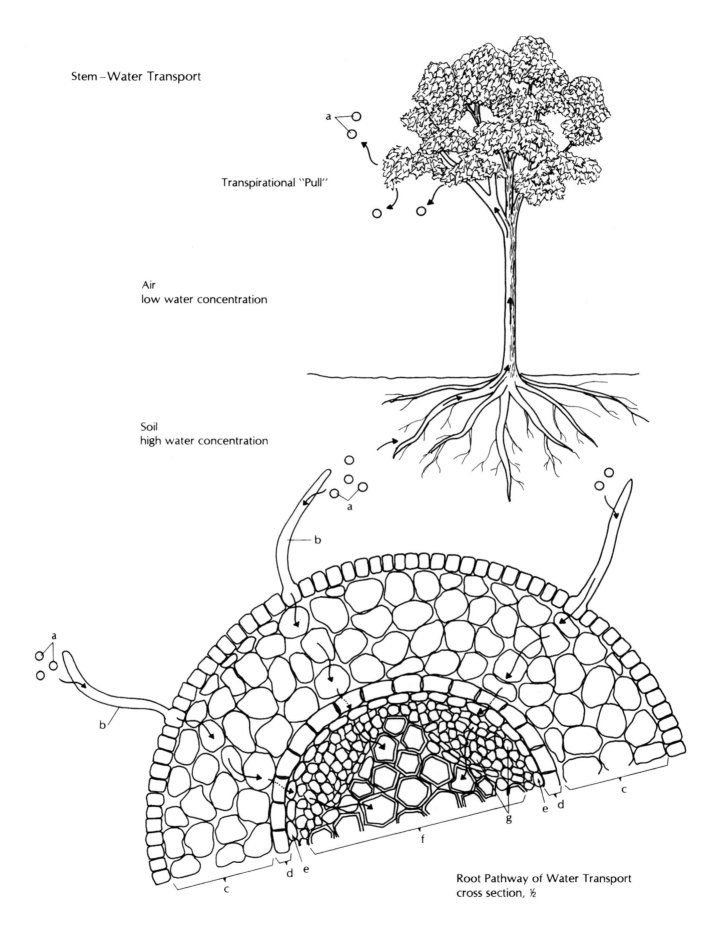

Transpirational "Pull"

Air
low water concentration

Soil
high water concentration

Root Pathway of Water Transport
cross section, ½

Stem—Food Transport

Translocation is the term used to designate transport of soluble organic materials (food) from one part of a plant to another. These organic materials are mainly sugars (from photosynthesis) and amino acids and amides (products of nitrogen metabolism). Primarily, they come from leaves, which we call sources, and are transported to areas called sinks, where they are either metabolized or stored. Metabolic sinks include young, growing shoot tips (terminal and lateral), developing flower buds, growing stems, and elongating roots. Storage sinks include tubers, tuberous roots, bulbs, corms, fruits, or stem parenchyma tissue. The transport of organic solutes occurs in the phloem of the vascular system (see 11).

At maturity, the sieve tube elements (see 8) of the phloem contain cytoplasm and cytoplasmic organelles but usually lack nuclei. Connecting strands, containing cytoplasm, pass through sieve areas on the end walls of adjacent sieve-tube elements, which provide channels for flow of organic solutes from one sieve tube element to the next.

Solute Transport in the Phloem

Münch Pressure-Flow Theory

One of the most satisfactory explanations that has been proposed on long-distance transport of sugars in the phloem is the **Münch Pressure-Flow Hypothesis.** The diagram shows how it operates. Cells in the leaf (a) produce high amounts of sugar (b), which are actively (using ATP as an energy source) loaded into sieve tube elements (termed **"phloem loading,"** c) in the veins.

Because of the high concentration of sugar and low concentration of water, water (d) enters the sieve tube elements (e) from the xylem by osmosis (see 5). This causes a high water pressure, which drives the sugar "sap" (b–d) through the sieve tube elements to a sink such as a fruit (f) or a storage root (g). At the sink, the sugars are actively "unloaded" (h) into cells where they are used or stored as starch (a glucose polymer) or fructan (a fructose polymer). The high concentration of water (d) that remains diffuses back into the xylem (i).

COLOR CODE

orange: circles of sugar (b, c, h)
blue: arrows of water (d)

Stem - Food Transport

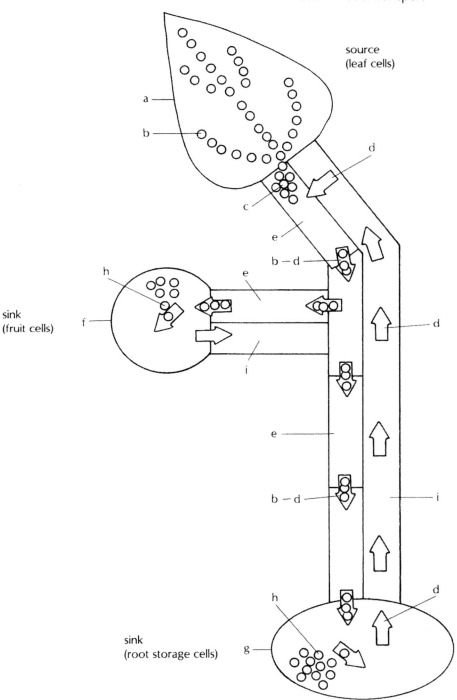

source
(leaf cells)

sink
(fruit cells)

sink
(root storage cells)

Münch Pressure-Flow System

Stem—Apical Dominance

Apical dominance is the inhibition of the growth of axillary buds by the terminal bud of a shoot. When apical dominance is lost, axillary buds start to grow. This often occurs during flowering. In plants such as oats (*Avena sativa*) and sunflower (*Helianthus annuus*, a, b) with a very strong apical dominance, axillary buds may never develop during the vegetative phase of shoot development in both species, or in sunflower even when the plant is flowering (c).

In plants such as wheat (*Triticum*) or *Coleus*, where apical dominance is weak, axillary buds may start to grow. In *Coleus*, the lowest lateral buds on the primary shoot (d) develop into lateral shoots (e) first, and those near the top (f) last.

In horticulture, **"pinching off"** the shoot tip of a dicot plant removes the terminal bud, breaking apical dominance, and results in a "bushy" plant as the axillary buds form shoots. Removing terminal buds produces thickly branched hedge plants, more flower shoots in plant such as chrysanthemums and asters and encourages more foliage growth in dicot houseplants.

In grasses, the shoots that develop from lateral buds are called **tillers** (h). Cereal grasses, such as wheat and rice (*Oryza*), that produce tiller shoots during the vegetative phase of shoot development, produce more "heads" of flowers (inflorescences) during the reproductive phase, and hence, yield more grain (i) per plant than cereals that do not tiller until the reproductive phase is underway or after harvesting, such as oats and annual rye. Even corn produces **"suckers"** (tiller shoots), which are often removed to enhance yield in "ears" of the main shoot.

Apical dominance is regulated by two growth hormones: auxin (IAA, indole-3-acetic acid) from the terminal bud (k) and **cytokinin** (e.g., zeatin) from the roots. When the ratio of auxin to cytokinin decreases in favor of more cytokinins, apical dominance is lost and axillary buds are released to expand.

To demonstrate how auxin is involved, three alike plants with strong apical dominance can be used. One plant is used as a standard of comparison, the control. In another plant, the terminal bud is removed and lanolin paste (l) is applied to the cut surface. For the third plant, the apical bud is also removed and lanolin paste containing 0.1% auxin (m) is applied. After a period of time, the results show that in the control plant and the plant with the auxin application, the axillary buds (n) have remained suppressed. But in the plant without terminal bud or auxin application (lanolin only), apical dominance has been lost; the axillary buds are released and develop into elongated shoots (o).

COLOR CODE

dark green:	shoot (a), leaves (b)
yellow:	flower (c), lanolin (l)
purple:	shoots (d, e, f), leaves (g)
green:	tiller (h), leaves (j), terminal bud (k)
	lateral buds (n), lateral shoots (o)
tan:	grain (i)
orange:	lanolin with auxin (m)

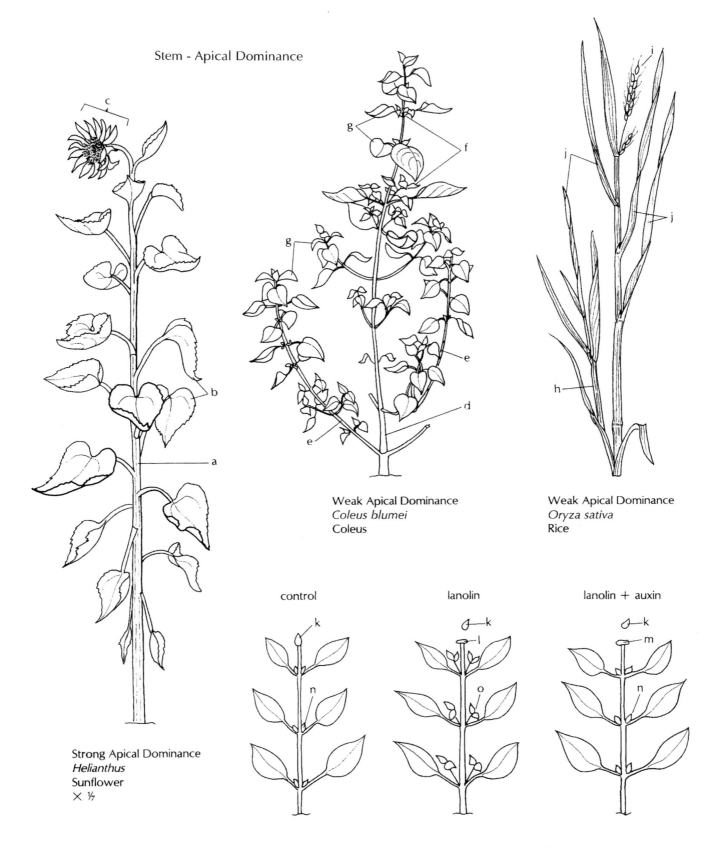

Stem - Apical Dominance

Weak Apical Dominance
Coleus blumei
Coleus

Weak Apical Dominance
Oryza sativa
Rice

Strong Apical Dominance
Helianthus
Sunflower
× ½

control

lanolin

lanolin + auxin

Apical Dominance Experiment Results

Stem—Growth Movements

A **tropism** is growth in response to an environmental stimulus. Growth in response to unequal light intensities is called **phototropism**, while growth stimulation by gravity is called **gravitropism**. When movement is toward the source of stimulation, it is positive, and when away from the stimulation source, it is a negative response.

Phototropic Responses

Positive Phototropism. Shoots exhibit positive phototropism by growing towards light. When a dark-grown oat seedling (a) is exposed to a light source from the right, the shoot curves to the right (b) by cell elongation on the dark side. It was discovered that a transmissible substance must be produced by the tip of the outermost seedling leaf, the coleoptile (c).

If a piece of mica (d) is inserted in the tip, no curvature occurs when a seedling is placed in a light source from the right. With the coleoptile tip cut off (e), the shoot (f) does not curve toward the light. It is known that the hormone, **auxin**, is produced in the coleoptile tip. When a tip (g) is placed on an agar block (h) with a glass cover slip (i) dividing it, a greater concentration of auxin (j) arrives on the side of the block away from the light source.

Negative Phototropism. Roots (l) exhibit negative phototropism by growing toward the center of gravity. **Skototropism** is a form of negative phototropism exhibited by tropical vine shoots in which initial growth is toward dark shadows cast by trees blocking light. After the vine has climbed the tree and reaches a light-exposed area, it becomes positively phototropic.

Gravitropic Responses

Orthotropism. (Ortho = straight, correct, vertical.) Shoots (m) not only curve toward light (positively phototropic) but straight away from the earth's center of gravity (negatively gravitropic). Roots (n) and some fruit stalks show positive gravitropism by growing vertically toward the center of gravity. The root cap perceives the stimulus for the positive response, and auxin and other hormones regulate this growth process in the root.

Plagiotropism. Plant parts that grow at an oblique angle from the vertical exhibit plagiotropism. Examples are lateral roots (p) and stems (q) that diverge from the main vertical axis.

Diagravitropism. When plant parts grow at right angles to the line of gravity, they exhibit diagravitropism. Rooted rhizomes (s), stolons (v), horizontal branches, and horizontal leaf orientation are some examples.

Agravitropism. Some plant parts are neutral to gravity and grow in various directions, such as the aerial roots of epiphytes (plants that grow upon other plants) and the flowering shoots of some plants that orient themselves in response to stimuli other than gravity.

COLOR CODE

white:	shoots (a, b, e, f, g, m), roots (l, n, p), flower (y)
tan:	agar (h), seed coat (k), rhizome (s), shoot (t)
green:	leaf cutting (o), stem (q), leaves (r, w), shoots (u), stolon (v)
red:	strawberry fruit (x)

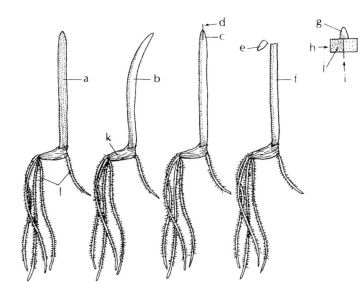

light source

Phototropic Responses of Seedlings
Avena
Oat

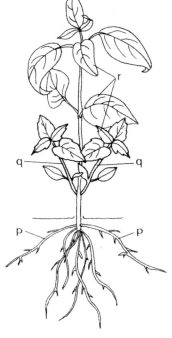

Plagiotropism of Lateral Shoots
Fuchsia

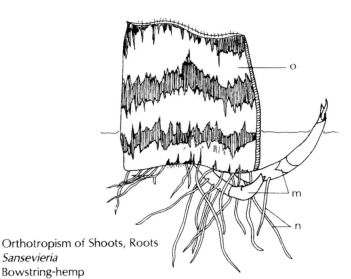

Orthotropism of Shoots, Roots
Sansevieria
Bowstring-hemp

Gravitropic Responses

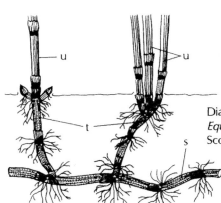

Diagravitropism of Rhizome
Equisetum
Scouring Rush

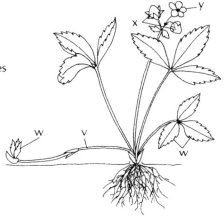

Diagravitropism of Stolon
Fragaria
Strawberry

Leaf Types and Arrangement

Persistence

Deciduous. A perennial plant is one that lives more than two years. Leaves that fall off at the end of a growing period are called deciduous. Examples are found in maples (*Acer*), birches (*Betula*), and dogwoods (*Cornus*).

Evergreen. Leaves that remain on a plant for more than a year are called evergreen. Pines (*Pinus*), spruces (*Picea*), rhododendrons (*Rhododendron*) are examples.

Leaf Types

Conifer. The leaves of conifers are needle-like (a) as on pines or scale-like as on junipers (*Juniperus*). They are usually evergreen, although some conifers have deciduous leaves such as larches (*Larix*). Each conifer leaf has a single vein.

Ginkgo. This unique plant (see 72) has fan-shaped leaf blades (c) with dichotomous venation in which the veins branch into two equal or unequal lengths. The leaves are deciduous and shed within a 24-hour period.

Dicot. Most dicot blades have pinnate (feather-like) venation (e) in which major veins diverge from one large mid-vein, with smaller network connections between. Some dicot blades have palmate (hand-like) venation (i) where several large veins diverge from the petiole to the margins. Some dicots have parallel venation. An example is plantain (*Plantago lanceolata*).

Monocot. Most Temperate Zone monocots have narrow strap-shaped blades (f) with sheathing bases (g) surrounding the stem (h). With parallel venation, major veins arise at the base, remain more or less parallel, and converge at the tip with small vein interconnections.

Some monocot leaf veins converge at the leaf margins instead of the tip, such as philodendron (*Philodendron*) and some monocots have pinnate venation with Swiss cheese plant (*Monstera*) as an example.

Leaf Attachment

Petiolate. The leaf blade is attached to the stem by a stalk called a petiole (d).

Sessile. When the leaf blade is attached directly to the stem without a petiole, it is sessile (a, f, n).

Leaf Form

Simple. A simple leaf has one blade, which may be broad (i) as in the maple leaf shown, narrow, or needle-like (a).

Compound. A compound leaf is one that has two or more blade-like leaflets such as sumac (*Rhus*, j, pinnately compound) or clover (*Trifolium*, k, palmately compound) attached to a petiole (d).

Leaf Arrangement

Opposite. Two leaves (l) emerge opposite each other on a stem. The example shown is milkweed (*Asclepias syriaca*). Generally maples, ashes, dogwoods, and members of the honeysuckle Family (Caprifoliaceae) have opposite leaves. MAD Cap is an easy way to remember.

Alternate. In an alternate arrangement, single leaves (m) are attached spirally along the stem. Lamb's quarters (*Chenopodium album*) is shown (see 88). Most genera have alternate leaves.

Whorled. In a whorled arrangement, several leaves (n) emerge together around a stem node. Michigan lily (*Lilium michiganense*) is the example shown (see 128).

COLOR CODE

blue-green:	blade (a)
green:	shoot peg (b), petiole (d), blade (l, n)
dark green:	blade (e)
yellow:	blade (c)
light green:	blade (f, m), sheath (g), stem (h), leaflets(k)
orange:	blade (i)
red:	leaflets (j)

Leaf Types and Arrangement

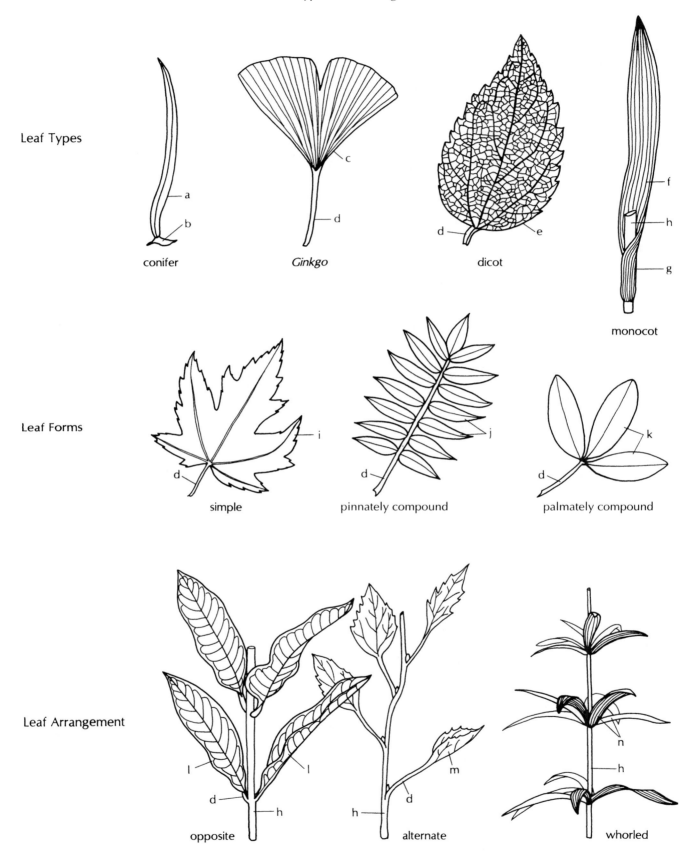

Leaf Types

conifer

Ginkgo

dicot

monocot

Leaf Forms

simple

pinnately compound

palmately compound

Leaf Arrangement

opposite

alternate

whorled

Leaf Tissues

The leaf blade's outer surface is epidermal tissue and the ground tissue within is photosynthetic mesophyll, interspersed with "veins" (vascular bundles) of conducting (vascular) tissue. Vascular connections between blade and stem pass through the petiole.

Dicot Leaf

Epidermis. There may be one or more layers of epidermal cells (a) with cutinized outer walls that form a waterproof outer surface. Stomata of guard cells (b) are present on the upper and lower surface or only on the lower surface as shown here. Various hairs and glands may also be present (see 10).

Ground Tissue. The parenchyma cells of the ground tissue (c) are divided into palisade (d) and spongy (e) mesophyll. Usually the columnar cells (f) of the palisade layer lie under the upper leaf epidermis and contain most of the chloroplasts of the ground tissue. The spongy parenchyma (e) has irregular-shaped cells (g) with air spaces (h) between and lies above the lower leaf epidermis.

Vascular Tissue. Four small vein transections are shown which are part of the vast, interconnecting network between major veins. A leaf vein (vascular bundle) usually has xylem (i) on the upper side and phloem (j) on the lower side. The reverse may be true in major veins. A layer of parenchyma cells called a bundle sheath (k), which functions in moving materials between the vein and the mesophyll tissue, usually encloses small veins.

Monocot Leaf (Grass)

Epidermis. Along with the epidermal cells (a), guard cells (b), hairs, silica, and cork cells, grass leaves have **bulliform cells** (l). The bubble-shaped, water-filled bulliform cells control leaf rolling (when the leaf is dry) and unrolling (when the leaf is turgid or filled with water). Under drought conditions, bulliform cells collapse and the leaf rolls up.

Ground Tissue. In the temperate region, grass mesophyll (c) consists of only spongy parenchyma tissue whose cells contain chloroplasts—sites of photosynthesis.

Vascular Tissue. Grass leaves have bundle sheaths (k) surrounding each vein. Sclerenchyma fiber strands (m) may be associated with the veins. (Leaf transection adapted with permission: Esau, K., *Plant Anatomy*, 1967, John Wiley & Sons.)

Conifer Leaf

Epidermis. The usually small, triangular or flat "needle" leaves of conifers are adapted to dry conditions. Epidermal cells (a) have thick walls of cuticle, and sunken stomatal guard cells (b) overlapped on the surface by subsidiary cells (n). Stomata may be on one or all sides of the leaf surface. Under the epidermis may be a layer of thick-walled sclerenchyma cells called **hypodermis** (o).

Ground Tissue. Mesophyll tissue (c) is not usually differentiated into palisade and spongy layers. Resin ducts (p) are present.

Vascular Tissue. There are one or two veins in the center surrounded by transfusion tissue (q) and endodermis (r) containing tannins and resins.

COLOR CODE

colorless:	epidermal cells (a), air space (h), bulliform cells (l), subsidiary cells (n), hypodermis (o), transfusion tissue (q)
green:	guard cells (b), monocot and conifer mesophyll (c) palisade parenchyma cells (f)
light green:	spongy parenchyma cells (g), bundle sheath (k)
blue:	xylem (i)
orange:	phloem (j)
tan:	sclerenchyma fibers (m), endodermis (r)
yellow:	resin duct (p)

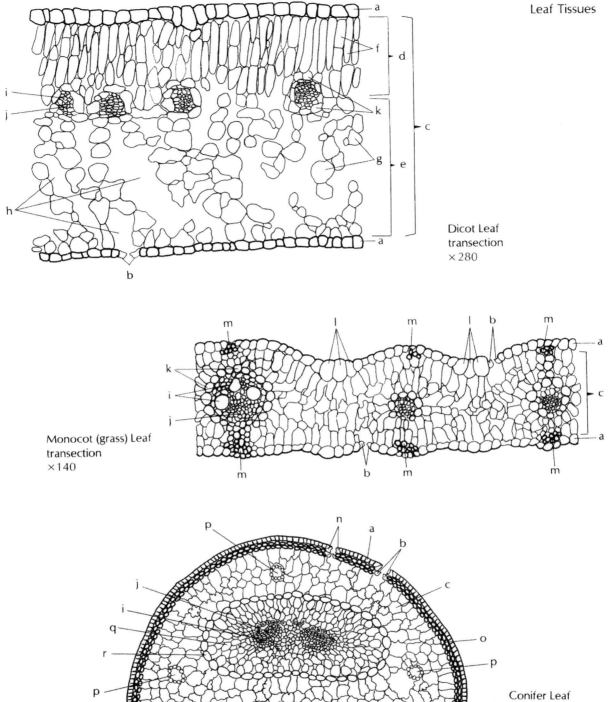

Leaf Tissues

Dicot Leaf
transection
×280

Monocot (grass) Leaf
transection
×140

Conifer Leaf
transection
×78

Leaf Modifications

Leaves may become highly specialized in function, exhibiting bizarre forms. Only a few of the many leaf modifications are shown here.

Bract. Associated with a flower, **bracts** are reduced or modified leaves. Very small bracts are found in sedge and grass flowers (see 122, 123), while large colorful bracts (a) subtend flowers (b) such as *Bougainvillea* and poinsettia (*Euphorbia*). The **spathe** is an elaborate bract found in the arum family (see 127).

Carnivorous Leaves. Carnivorous plants live in low-nitrogen environments and enzymatically digest animals (mainly insects) in their leaves to supplement nutritional needs. The Venus'-flytrap (*Dionaea*) with a two-lobed leaf blade (c, d) has trigger hairs (e) that set off a response that brings the spine-covered leaf margins (f) together in a closing trap. The leaf has a wide petiole (g). On butterwort (*Pinguicula*, see 10) and sundew (*Drosera*) leaves (h), long, sticky-bulbed, glandular hairs (i) entrap and digest insects.

In pitcher-plants (*Sarracenia*), the hairy (j) lower surface of the leaf blade (k) is curved inward, forming a well of water and digestive enzymes. Nectar glands attract insects, while downward pointed hairs keep them in the leaf enclosure (see 91). The carnivorous leaf colors of the above examples attract flies (see 34).

A water plant, bladderwort (*Utricularia*) has bait and trigger hairs (o) surrounding the mouth (p) of a bladder (q), a modified leaf. Movement of the hairs entices small crustaceans to investigate. By negative pressure within the bladder, victims are sucked in and digested, sealed in by a trap door at the mouth of the bladder.

Phyllode. This is a widened petiole that appears blade-like. In the acacia (*Acacia melanoxylon*) plant, some leaves consist of only petioles (**phyllodes**, r), while others develop compound leaflets (s).

Spine. Spines are highly reduced leaves or stipules found on many herbs, shrubs, and trees. Often a plant exhibits a series of forms (**homology**) from leaf (t) to spine (u).

Stipule. Usually in pairs, **stipules** (v) are basal appendages to the petiole (w). They may appear as hairs, leaves, tendrils, or spines.

Succulent Leaves. An adaptation to dry environments, some plants store water in leaf mesophyll tissue. The epidermis may have thick cutinized and lignified walls to reduce water loss. In some succulents, the stomata reverse the usual condition and open only at night. The leaf epidermis (y) of living stones (*Lithops*), an example of mimicry, appears sand-colored. In the cut-away drawing, green mesophyll tissue (z) is exposed, surrounding the newly formed leaf in the center.

COLOR CODE

red:	bracts (a), inner blade surface (c), hairs (i, j), veins (l)
white:	flower (b), hairs (e)
yellow-green:	blade margins (f), petiole (g), blade (d)
green:	leaf (h, n), leaflets (s)
light green:	petiole (m), blade (k), leaf (t), stipule (v), mesophyll (z)
tan:	hairs (o), bladder (q), spine (u), epidermis (y)
dark green:	petiole (r, w), blade (x)

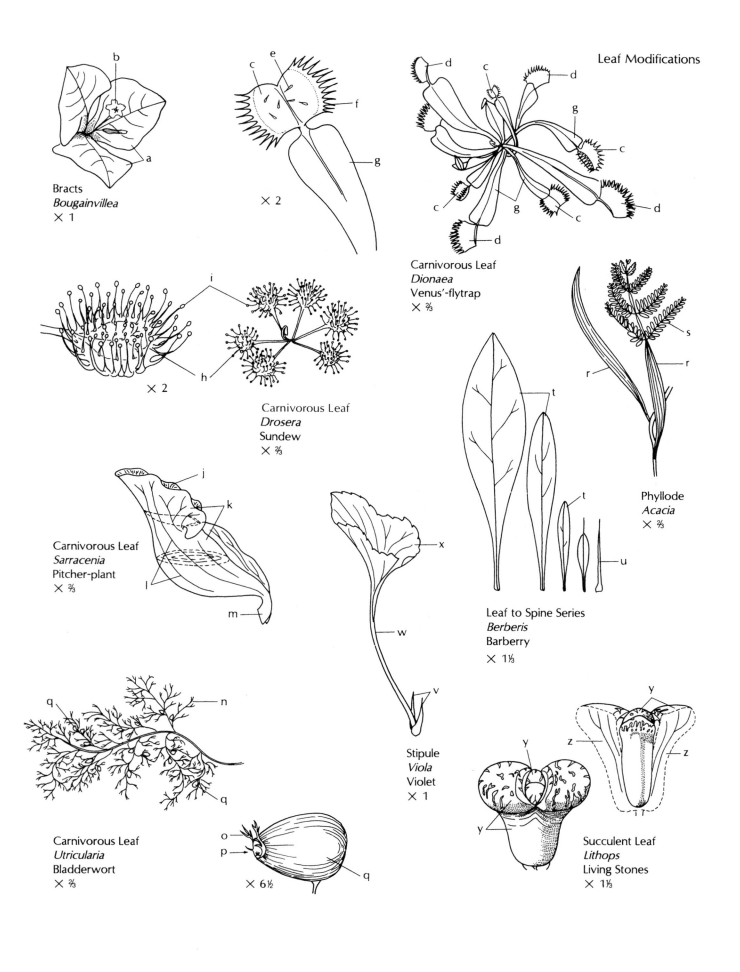

Leaf Modifications

Bracts
Bougainvillea
× 1

× 2

Carnivorous Leaf
Dionaea
Venus'-flytrap
× ⅔

× 2

Carnivorous Leaf
Drosera
Sundew
× ⅔

Carnivorous Leaf
Sarracenia
Pitcher-plant
× ⅔

Phyllode
Acacia
× ⅔

Leaf to Spine Series
Berberis
Barberry
× 1⅓

Carnivorous Leaf
Utricularia
Bladderwort
× ⅔

× 6½

Stipule
Viola
Violet
× 1

Succulent Leaf
Lithops
Living Stones
× 1⅓

Leaf—Photosynthesis

Plant Energy Source

Photosynthesis is the process in plants in which light energy is converted to useful chemical energy. In the process, carbon dioxide and water, in the presence of chlorophyll and light energy, form sugars, oxygen, chemical energy (ATP—adenosine triphosphate), and water. A general equation is shown.

With the exception of fungi, photosynthesis occurs in all major groups of organisms discussed in this book, including certain bacteria. Most plants that photosynthesize have chloroplasts, in which photosynthesis occurs. Other cells, as in photosynthetic bacteria and blue-greens, have no chloroplasts. Instead photosynthesis occurs in pigments located in the cytoplasm.

Significance of Photosynthesis

Photosynthesis is responsible for the conversion of carbon from carbon dioxide into organic compounds in plants. It allows the plant to make organic building blocks for sugars, amino acids, nucleic acids, fatty acids, new cells, starch, protein, DNA and RNA, hormones and vitamins, and many secondary compounds. Without this photosynthetic process, life would not exist on earth. Plants provide, directly or indirectly, food for all animals and all of our atmospheric oxygen. They also "fix" CO_2 in photosynthesis, which helps to reduce global warming.

Photosynthesis consists of two basic parts: a series of light reactions and a series of dark reactions. In the light reaction (a), chlorophyll a (and accessory pigments) absorb light, which "excites" the electrons in the chlorophyll molecule. The electrons are passed along through a series of carriers (acceptors), and energy (ATP) is produced, which catalyzes the dark reaction. During the light reactions, water is split, giving off oxygen.

In the dark reaction (b), carbon dioxide, products of the light reaction (reducing power), and energy are used. Carbon dioxide is "fixed" or converted into a cyclic production of a series of carbon sugars, the most important of which ultimately is sucrose, made up of 6-carbon sugars, glucose, and fructose.

C$_3$ and C$_4$ Plants

In regard to photosynthesis, plants are divided into two types: C_3 and C_4 plants. In the leaves of C_3 plants, which include most plants, light-driven carbon fixation reactions involve the formation of 3-carbon organic acids as the first stable products. In many tropical grasses C_4 photosynthesis, such as sugar cane, corn, and sorghum, the photosynthetic process is more efficient. The grass bundle sheath (see 22) produces 3-carbon acids and sugars, whereas the mesophyll tissue produces 4-carbon acids. These plants are more efficient in producing sugars because they have little or no sugar loss during **respiration** (a chemical release of energy from sugars) in the light (**photorespiration**).

The illustration shows where the photosynthetic process takes place in the leaf (c) of a C_3 plant. In the chloroplast (i), chlorophylls and accessory pigments are located in the thylakoids (j), where the light reaction takes place. Dark reactions take place in the stroma (k) of the chloroplast (see 3).

Environmental Regulation of Photosynthesis

Photosynthetic rate is regulated by the intensity of light, daylength, amounts of carbon dioxide and oxygen in the air (or water for water plants), temperature, and the level of air pollution.

COLOR CODE

green:	light reaction (a), leaf (c), guard cells (e), palisade cells (h) with chloroplasts (i), thylakoid (j)
gray:	dark reaction (b), stroma (k)
blue:	water (xylem) in vein (d)
orange:	sugars (phloem) in vein (d)
colorless:	epidermis (f), spongy cells (g)

Leaf—Photosynthesis

$$\text{carbon dioxide} + \text{water} \quad \xrightarrow[\text{chlorophyll}]{\text{light}} \quad \text{sugar} + \text{oxygen} + \text{water}$$

carbon dioxide + water $\quad$ sugar + oxygen + water

$(CO_2) \qquad (H_2O) \qquad\qquad\qquad (O_2) \qquad (H_2O)$

General Equation of Photosynthesis

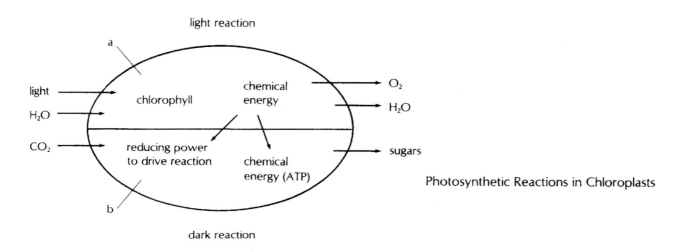

light reaction

a

light $\rightarrow$

H_2O $\rightarrow$

chlorophyll $\qquad$ chemical energy $\rightarrow O_2$

$\rightarrow H_2O$

CO_2 $\rightarrow$

reducing power to drive reaction $\qquad$ chemical energy (ATP) $\rightarrow$ sugars

b

dark reaction

Photosynthetic Reactions in Chloroplasts

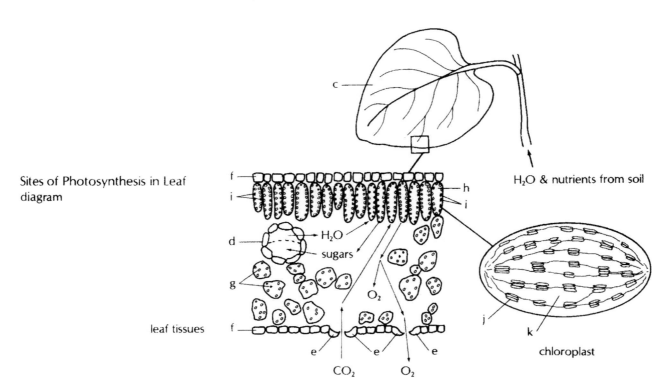

Sites of Photosynthesis in Leaf diagram

c

H_2O & nutrients from soil

f

i $\qquad$ h

$\qquad$ i

d $\rightarrow H_2O$

sugars

g

O_2

leaf tissues $\qquad$ f

j

k

e $\qquad$ e $\qquad$ e

$CO_2 \qquad O_2$

chloroplast

Leaf—Nutrient Deficiency Symptoms

The essential elements that are required for the healthy growth of higher plants are carbon, hydrogen, oxygen, and those listed below. Plants take in carbon in the form of atmospheric carbon dioxide, obtain hydrogen and oxygen from water, and take up the remaining elements from the soil. By growing plants in an aqueous nutrient solution (**hydroponics**) and omitting an essential nutrient, specific deficiency symptoms can be observed.

Nitrogen. Decaying organic matter provides soil nitrogen. Also, nitrogen-fixing bacteria on some plant roots convert atmospheric nitrogen into organic forms used by plants. Nitrogen deficiency results in yellowing (**chlorosis)** of older leaves.

Phosphorus. Plants deficient in this element exhibit dark green older leaves and stunted growth. Red anthocyanin pigments sometimes accumulate.

Potassium. This element is found in wood ashes (potash). Deficiency symptoms are exhibited in older leaves that yellow and have dead (necrotic) tissue in spots, and roots are easily infected with pathogenic organisms.

Magnesium. A lack of magnesium results in a yellowing of the leaves between the veins with the tips and margins turning upward.

Sulfur. Usually, a sulfur deficiency affects the younger leaves first with the veins becoming light green and a yellowing of the tissue between them. As sulfur compounds are found in most soils, this deficiency is rare.

Calcium. Lack of calcium affects meristematic regions, resulting in the death of terminal buds and root tips.

Iron. A deficiency of iron results in a chlorosis of young leaves with the yellowing occurring between veins.

Chlorine. Without chlorine, leaves wilt, turn yellow, show dead spots, and then become a bronze color. Roots become stunted and club-shaped near the tips.

Manganese. Although a rarely found deficiency, without manganese, younger leaves form yellow speckles between the veins.

Boron. Lack of boron produces death of shoot and root apical meristems, beginning with young leaves becoming light green at the base, and root tips becoming swollen.

Zinc. A deficiency of zinc results in formation of small leaves and shortened stem internodes. The leaf margins may become distorted.

Copper. A rare deficiency, lack of copper produces dark green, twisted young leaves, which often have dead spots.

Molybdenum. Lack of this element produces chlorosis of old and midstem leaves, and eventually affects the young leaves.

COLOR CODE

green:	terminal bud (a), new leaf (b), middle leaf (c), old leaf (d), veins (e)
tan:	roots (f)
light green:	chlorosis (g)
yellow:	late-stage chlorosis (h)
brown:	dead tissue (i)
dark green:	pigment accumulation (j)
red-purple:	anthocyanin pigmentation (k)
red-brown:	damaged tissue (l)

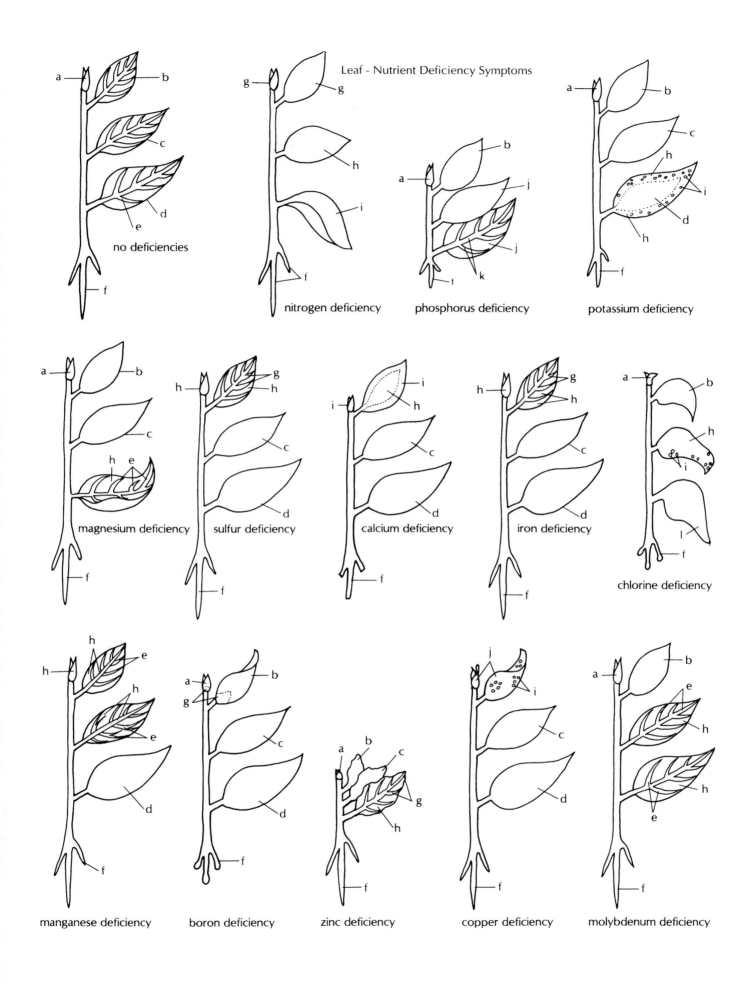

Leaf - Nutrient Deficiency Symptoms

no deficiencies

nitrogen deficiency

phosphorus deficiency

potassium deficiency

magnesium deficiency

sulfur deficiency

calcium deficiency

iron deficiency

chlorine deficiency

manganese deficiency

boron deficiency

zinc deficiency

copper deficiency

molybdenum deficiency

Flower Initiation in Response to Daylength

Plants are classified as short-day, long-day, or day-neutral in terms of their flowering response to daylength (photoperiod). Most plants are **day-neutral.** They are not induced to flower by either long or short photoperiods. In the tropics, where photoperiods do not change nearly as much as in temperate regions, most plants are day-neutral or long-day in their response to photoperiod.

Short-day Plants

It has been found that rather than the length of daylight time, it is actually the length of the dark period (night length time) that induces flowering in short-day plants. If poinsettias are briefly illuminated during their long-night period, the flowering response is not induced. **Short-day** plants grow vegetatively (a) during long-day periods and flower (b) during short-day periods. Some examples of short-day plants are poinsettia (*Euphorbia pulcherrima*), thanksgiving cactus (*Schlumbergera truncata*), chrysanthemum (*Chrysanthemum*), lamb's quarters (*Chenopodium album*), cocklebur (*Xantium strumarium*), strawberry (*Fragaria*), *and many varieties of* rice (*Oryza*).

The nature of the flowering stimulus in short-day plants is not yet known. However, the stimulus most likely involves an interaction between flower-inhibitor and flower-promoter type hormones. The evidence that initiation of flowers involves hormones is that the signal can be perceived by one mature leaf.

Using cocklebur plants, if one mature leaf (c) is exposed to short days while the rest of the plant is exposed to long days, flowering will result. Furthermore, by grafting a branch from a cocklebur plant given short days onto a cocklebur plant given long days, flowering is induced in the stems given long days.

Long-day Plants

Plants that require **long days** (short nights) for flowering include spinach (*Spinacia oleracea*), winter barley (*Hordeum vulgare*), winter wheat (*Triticum aestivum*), oats (*Avena sativa*), perennial ryegrass (*Lolium perenne*), clover (*Trifolium pratense*), coneflower (*Rudbeckia*), dill (*Anethum graveolens*), rose-of-sharon (*Hibiscus syriacus*), petunia (*Petunia*), and all **biennials.** Biennials are plants with a 2-year growth cycle. They require cool winter temperatures the first year, and long days the second year for flower induction.

Some plants cannot be induced to flower by exposure to critical daylength periods until a certain age is reached. Depending on the species, some trees are not "flower-mature" until 5 to 40 years of age. For example, a sugar maple tree (*Acer saccharum*) may not flower until it is 30 years old, but it can live 200 to 300 years.

COLOR CODE

red: tepals (d)
green: stems (e, f, l), leaves (g, j), peduncle (k)
tan: fruit(h)
yellow: rays (l)
brown: flowers (m)

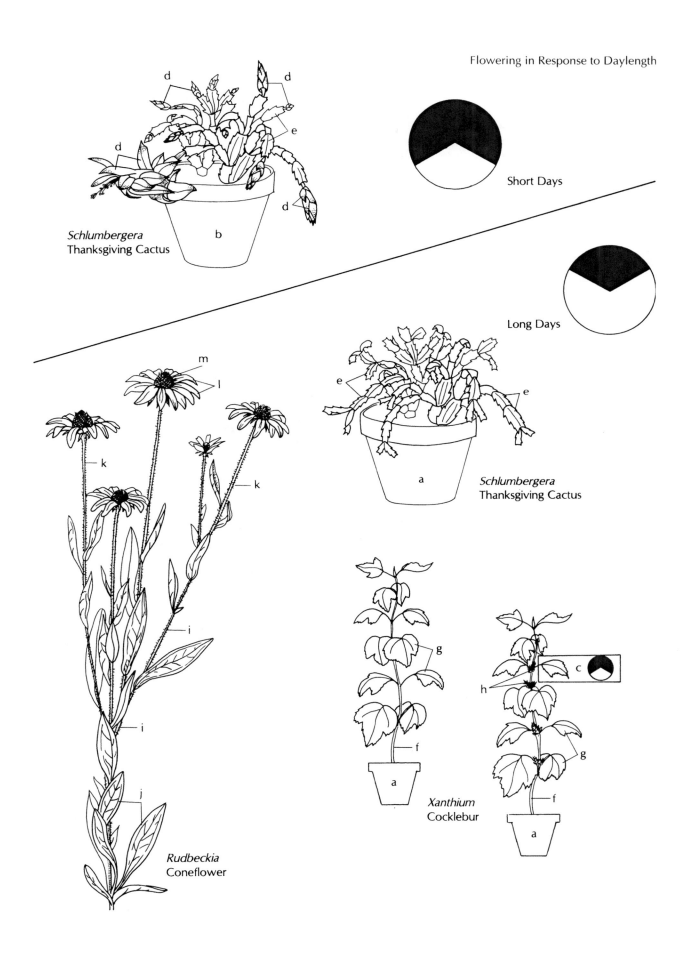

Flowering in Response to Daylength

Schlumbergera
Thanksgiving Cactus

Short Days

Long Days

Schlumbergera
Thanksgiving Cactus

Rudbeckia
Coneflower

Xanthium
Cocklebur

Flower Structure

Parts of a Flower

Peduncle. The stalk that attaches the flower to the plant axis is called a **peduncle** (a). A **pedicel** is the stalk of a single flower in a cluster (**inflorescence**) with the pedicels attached to a peduncle (see 98, 100, 109).

Receptacle. The enlarged part of the flower axis where floral parts are borne is the **receptacle** (b).

Sepals. The outermost whorl of floral parts are leaf-like structures called **sepals** (c). While usually green, they may be colored in some species (see 129, 130). Sepals protect the developing flower in the bud stage. All of the sepals, collectively, make up the **calyx.**

Petals. The second whorl of floral appendages are **petals** (d), which may be separate or partially fused together. Usually, they are showy and may be colored. All of the petals, collectively, make up the **corolla.**

Tepals. When the calyx and corolla look alike and are petal-like, they are called **tepals** (see 126, 128).

Perianth. This term is used to refer to the calyx and corolla together.

Stamen. A stamen consists of a stalk-like **filament** (e) with an **anther** (f), made up of, usually, 2 pollen sacs, at the upper end. The **connective** (g) joins the **pollen sacs** containing **pollen grains** (h) (see 31). All of the stamens, collectively, are termed the **androecium** (= male household).

Pistil. A pistil is made up of stigma, style, and ovary. At the apex of the pistil is the **stigma** (i), where pollen grains adhere to the sugary, liquid excretions on the surface. Between the stigma and ovary is an elongated part of the pistil called the **style** (j). The **ovary** (k) consists of one or more **carpels,** which contain the ovules (see 28). Within the **ovule** (l), megaspores are formed, one of which develops into the embryo sac containing an egg (see 31).

After fertilization, the ovules develop into seeds (see 37). That part of the carpel where the ovules are attached is the **placenta** (see 28). A **locule** (m) is the chamber or space within a carpel. All of the pistils within a flower are called, collectively, the **gynoecium** (= female household).

Number of Parts

In a broad generalization, dicot flowers tend to have floral parts in whorls of 4 to 5 and monocots in series of 3. For example, a dicot flower might have 4 sepals, 4 petals, and 8 stamens or 5 sepals, 5 petals, and 15 stamens. A monocot flower might have 3 sepals, 3 petals, and 6 stamens. When observing flowers, it is interesting to determine number patterns. Numbers of floral parts are also used in "keying out" plants (plant identification).

COLOR CODE

light green:	peduncle (a), stigma (i), style (j)
white:	receptacle (b), filament (e), ovary (k), ovules (l)
green:	sepals (c)
optional:	petals (d)
yellow:	anther (f), connective (g), pollen (h)

Flower Structure

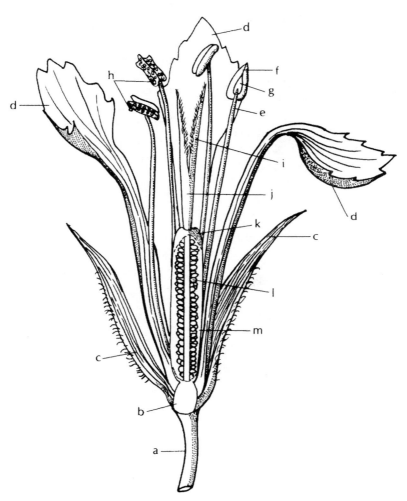

Parts of a Flower
vertical section
diagram

Flower Structure Variations

Types of Flowers

Unisexual (Imperfect). A unisexual flower has either staminate (male) (a) or pistillate (female) (b) structures.

Bisexual (Perfect). A bisexual flower has staminate (a) and pistillate (b) structures.

Monoecious (= of one house). An individual plant with separate (unisexual) staminate and pistillate flowers is termed **monoecious** (see 94).

Dioecious (= of two houses). The term, **dioecious,** is used when a species has separate staminate-flowered plants and pistillate-flowered plants (see 105).

Polygamomonoecious. An individual plant with bisexual flowers plus staminate or pistillate unisexual flowers is **polygamomonoecious** (poly = many; gamete = sex cell).

Complete. A **complete flower** has staminate (a) and pistillate (b) structures, plus sepals (c) and petals (d).

Incomplete. An **incomplete flower** lacks one or more floral parts. The incomplete flower shown lacks petals.

Ovary Positions

Superior. The **superior ovary** (f) of the pistil is above the site of attachment of other floral parts (stamens, sepals, petals).

Inferior. The **inferior ovary** (f) is below the site of attachment of other floral parts.

Half-inferior or **Half-superior.** These are flowers that show varying degrees of ovary (f) position from partially superior to partially inferior.

Placentation

Axile. By fusion of 2 or more carpels (g), the ovules (h) are attached to the placenta (i) at the central axis **(axile placentation).**

Free Central. Fused carpels (g) have lost their inner walls and the placenta (i) is located in the center **(free central placentation).** Examples are members of the pink family (Caryophyllaceae, see 87) and nightshade family (Solanaceae, see 111, pepper).

Basal. The placenta (i) is located at the base of the ovary **(basal placentation).**

Parietal. The placenta (i) is located on the inner walls of carpels, which do not extend into the center of the ovary **(parietal placentation).**

COLOR CODE

yellow:	stamen (a)
light green:	pistil (b), ovary (f), carpel wall (g), ovules (h)
green:	sepals (c), peduncle (e)
blue:	petals (d)
white:	ovules (h), placenta (i)

Flower Structure Variations

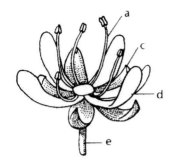

Unisexual
(imperfect)
staminate

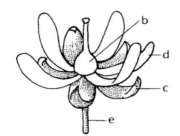

Unisexual
(imperfect)
pistillate

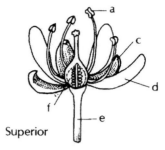

Superior

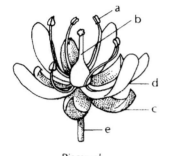

Bisexual
(perfect)
complete

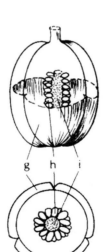

Bisexual
(perfect)
incomplete

Types of Flowers

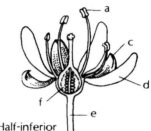

Half-inferior

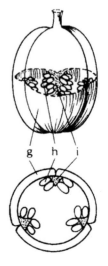

Inferior

Ovary Positions

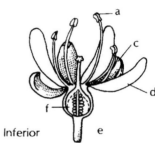

Axile

Free Central

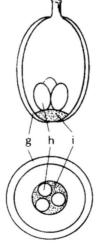

Basal

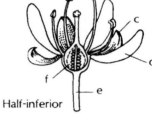

Parietal

Placentation

Flower Development

Flower Bud

Early Stage. Above the peduncle (a), upright bracts (b) cover the sepals (c). The sepals enclose the developing petals (d), anthers (e), and the column of fused staminal filaments (f).

Intermediate Stage. At this stage, the receptacle (g) expands and a young pistil (h) within the column (i) can be seen. At the top of the pistil, stigmas (j) form, the style (k) begins to elongate, and the ovary (l) enlarges.

Late Stage. The folded petals (m) emerge from the sepals (n). Within, the filament column (p) has elongated and the stigmas (q) have emerged as the style (r) elongates. The superior ovary (l) has developing ovules (s) within the carpels.

Mature Flower

Bracts (b) are separate from the basally fused sepals (n) and the five petals (m) are expanded. The fully-formed fused filament column (t) protrudes above the petals. Anthers (o) are fully mature on short filaments (u), which separate from the column. Five stigmas (q) on a divided style (v) rise above the column.

Anthesis. is the term used for the time when a flower comes into full bloom. It is a time when the flower is fully expanded.

COLOR CODE

dark green:	peduncle (s), bracts (b), sepals (c)
light green:	petals (d), filament column (i)
white:	anthers (e), filament column (f, p), receptacle (g), pistil (h), stigma (j), style (k, r), ovary (l), ovules (s)
green:	sepals (n)
red:	petals (m)
yellow:	anthers (o)
dark red:	stigmas (q), filament column (t), filaments (u), style (v)

Flower Development

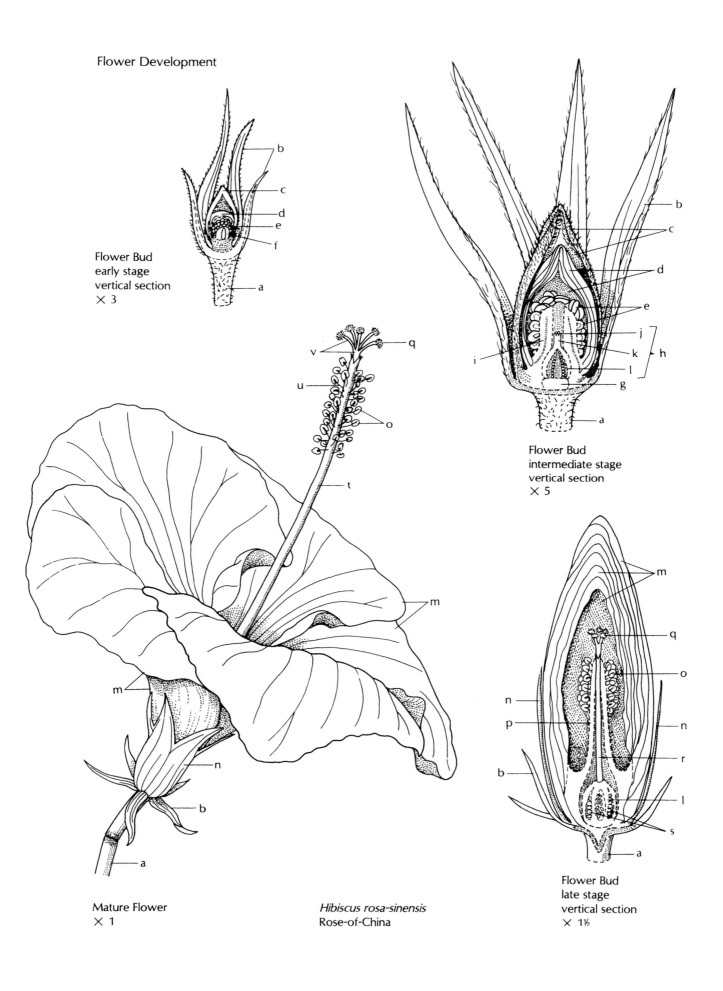

Flower Bud
early stage
vertical section
× 3

Flower Bud
intermediate stage
vertical section
× 5

Flower Bud
late stage
vertical section
× 1⅓

Mature Flower
× 1

Hibiscus rosa-sinensis
Rose-of-China

Flower—Meiosis

Each species has a specific number (2n) of chromosomes in the nucleus of each of its body (somatic) cells. Half of the chromosomes (1n) come from one parent and half (1n) from the other parent. For new body cells to form, 2n (diploid) cells divide by mitosis into more 2n cells.

For an individual to become a parent, it must produce 1n (haploid) cells that can fuse with 1n (haploid) cells from another individual. The process of producing haploid cells from diploid cells is called **meiosis**.

In higher plants, the end products of meiosis are haploid pollen grains in the anthers of the stamen and a haploid egg in each ovule of the pistil. In lower plants, the end products of meiosis are 1n (haploid) gametes (= sex cells).

Stages of Meiosis

Meiosis has 2 basic stages: (I), the reduction stage, and (II), the division stage. Unlike mitosis, during meiosis, chromosomes come together in pairs, called homologous pairs. Each chromosome of the pair has the same gene alignment and each represents a different parent. This pairing is significant because, during meiosis, "crossing over" (chiasmata) with an exchange of genes is possible (see 6).

It is now obvious that the 2n (diploid) number is an even number for normal pairing to occur. The cells that undergo meiosis are called "mother cells." Within each of the 2 stages of meiosis are 4 phases: prophase, metaphase, anaphase, and telophase.

Prophase I. During early prophase, the 2n chromosomes (a, b, c) become visible as threads and then come together in homologous pairs. (For simplicity, only 3 pairs of chromosomes are shown within a nucleus.) Then each chromosome separates to form

2 chromatids held together at the centromere (d). "Crossing over" (e) of chromatid strands results in chiasmata (sing.: chiasma), which are visible sites of exchanged genetic material (f).

Metaphase I. At this phase, the chromatids are positioned so that the centromeres are aligned on either side of the equatorial plane of the spindle apparatus (g).

Anaphase I. Next, the paired chromatids separate and move to opposite poles (h).

Telophase I. Lasting only a short time, telophase completes the reduction stage with the original number (2n) of chromosomes reduced to half (1n).

Prophase II. The second division stage of meiosis is essentially like mitosis (see 7).

Metaphase II. During this phase, the chromatids align in the center of the plane of the spindle.

Anaphase II. The centromeres (d) divide and chromosomes move to opposite poles (h) of the spindle.

Telophase II. By this phase, the chromosomes have completed their migration to the spindle poles. As a result, 4 1n products are formed. The chromosomes (i, j, k) now have a different genetic makeup than the chromosomes of the original 2n cell.

This meiotic description is characteristic of higher plants when 4 haploid products are formed. In flowering plants, in the ovule of the pistil, 3 of these nuclei usually disintegrate and the 4th divides (mitotically) repeatedly to form the nuclei of the embryo sac, one of which is the 1n egg nucleus. In the anther sac of the stamen, all 4 products of meiosis remain alive and develop into tetrads (4's) of 1n (haploid) pollen grains.

COLOR CODE

orange:	chromosomes (a)
yellow:	chromosomes (b)
green:	chromosomes (c)
purple:	centromere (d)
colorless:	spindle (g)
blue:	chromosomes (i, j, k)

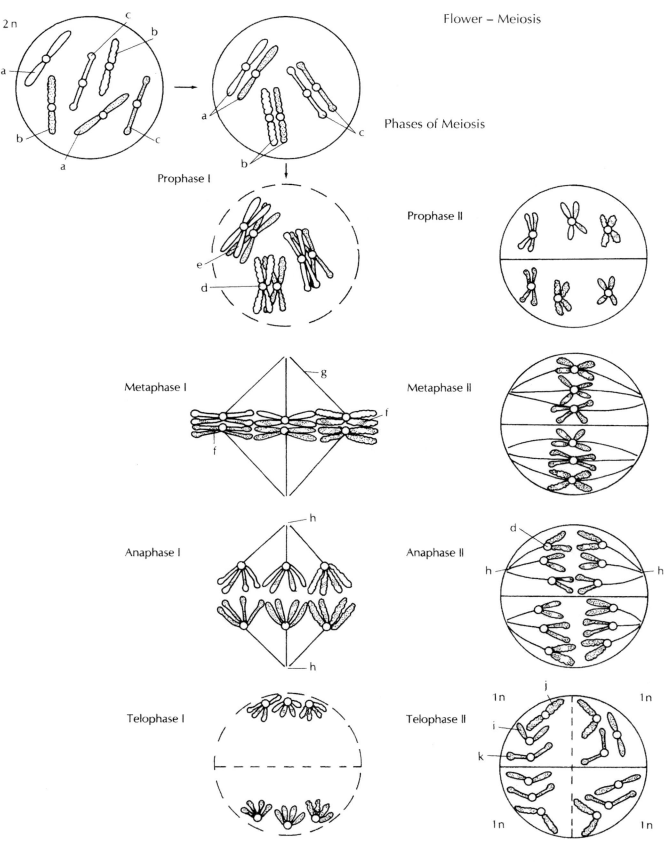

Flower – Meiosis

2 n

Phases of Meiosis

Prophase I

Prophase II

Metaphase I

Metaphase II

Anaphase I

Anaphase II

Telophase I

Telophase II

1n   1n

1n   1n

Reduction Stage

Division Stage

Flower—Pollen Development

Young Stamen

Once flower primordia are initiated, stamens begin to develop. The stamen in a flower consists of a filament (a) and an anther (b) with, usually, 2 pollen sacs (c). Pollen formation takes place within the chambers (locules) of the pollen sac (microsporangium). Each chamber is surrounded by a layer of cells called the **tapetum** (d), which functions as a source of nutrition for the 2n microspore mother cells (e). During a double division (meiosis), each mother cell forms a tetrad of 4 1n (haploid) microspores (f), which develop into 4 pollen grains (male gametophytes).

Pollen Grain

In each pollen grain (g), nuclear division (mitosis) produces a large 1n vegetative nucleus (h) and a smaller 1n generative nucleus (i), which divides to form two 1n sperm nuclei, usually after the pollen grain germinates to form a pollen tube (see 37).

Mature Stamen

Depending on the stamen type, the anther wall (j) spontaneously opens (dehisces) at a slit (k) to liberate pollen (l), or the anther wall breaks down, releasing pollen, or a portion of the anther lobe separates, leaving a pore (poricidal dehiscence).

Pollen Grain Types

Pollen grains have one or more furrows (m) or thin strips where the pollen tube may emerge. In some pollen grains, the pollen tube may emerge from a pore in the furrow or a pore not associated with a furrow.

In flowering plants, the main types of pollen grains (n) are uniaperaturate with one furrow and triaperaturate with 3 furrows. Monocots and some primitive dicots have uniaperaturate pollen grains. Most dicots have triaperaturate pollen grains. Pollen grains vary from species to species in size, shape, and decoration of the outer wall (exine). Some pollen grains lack an exine. The study of pollen grains of past and present-day plants is called **palynology.**

COLOR CODE

white:	filament (a), microspore mother cell (e), microspore (f), anther wall (j)
light green:	anther pollen sacs (c)
orange:	tapetum (d), vegetative nucleus (h), generative nucleus (i)
yellow:	pollen grains (g, l, m,n)

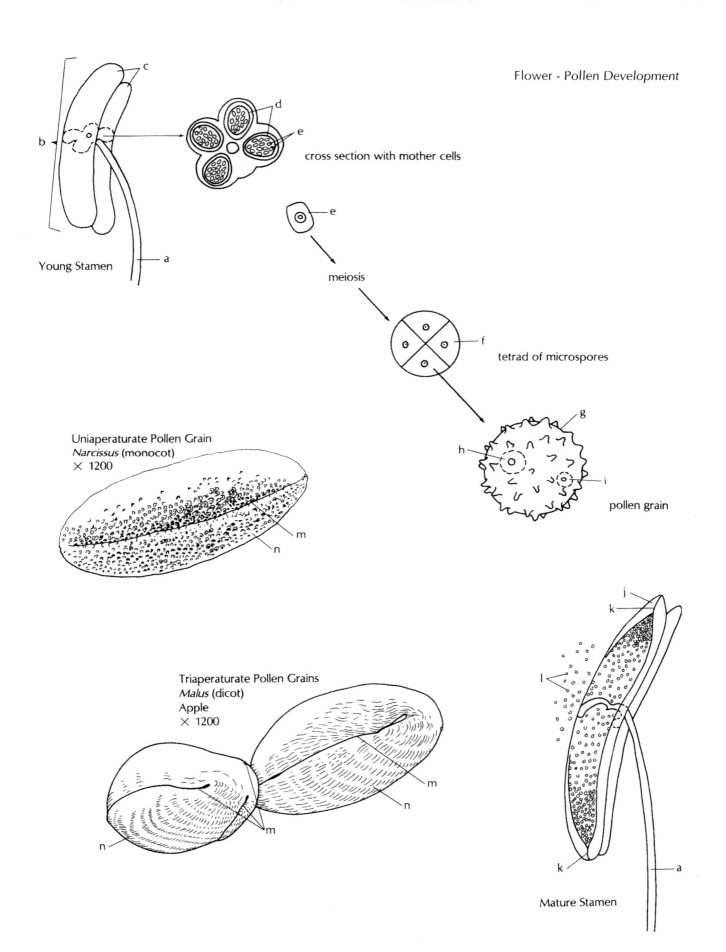

Young Stamen

cross section with mother cells

meiosis

tetrad of microspores

pollen grain

Uniaperaturate Pollen Grain
Narcissus (monocot)
× 1200

Triaperaturate Pollen Grains
Malus (dicot)
Apple
× 1200

Mature Stamen

Flower—Ovule Development

Young Pistil

When flowering is initiated, the pistil begins to develop in the center of the flower. It consists of stigma (at top), style and ovary (at base). The ovary (a) of the pistil has one or more separate or fused carpels (b). Within the carpel are locules (c) where ovules (d) develop. Ovule development begins with meiosis in a 2n megaspore mother cell (e) within the center portion of a young ovule (megasporangium). The megaspore mother cell undergoes a double reduction/division (meiosis) to form a linear tetrad of 1n megaspores. Of these, usually 3 abort (f) with one megaspore (g) remaining.

Embryo Sac

The nucleus of the remaining megaspore divides repeatedly (mitosis) to form the 8 haploid (1n) nuclei of the embryo sac (female gametophyte, h). The central part of the ovule is vegetative tissue, nucellus (i). Surrounding the embryo sac are one or two layers of cells, the integuments (j), with a small opening at the base called the micropyle (k).

Mature Ovule

The 3 nuclei of most embryo sacs align into positions so that 1 near the micropylar end is the egg cell (l), 2 are nuclei of synergid cells (m) with associated filiform apparatus (n), 2 are polar nuclei (o), and 3 are nuclei of antipodal cells (p). The mature ovule within the pistil consists of the embryo sac (h) surrounded by nucellus (i) and integuments (j). It is attached to the placenta (q) by a stalk called the funiculus (r).

COLOR CODE

light green:	ovary (a), integuments (j) placenta (q), funiculus (r)
white:	carpel (b), locule (c), ovule (d), aborted megaspores (f), synergids (m) filiform apparatus (n) polar nuclei (o), antipodals (p)
gray:	mother cell (e) megaspore (g), embryo sac (h)
tan:	egg (l)
orange:	nucellus (i)

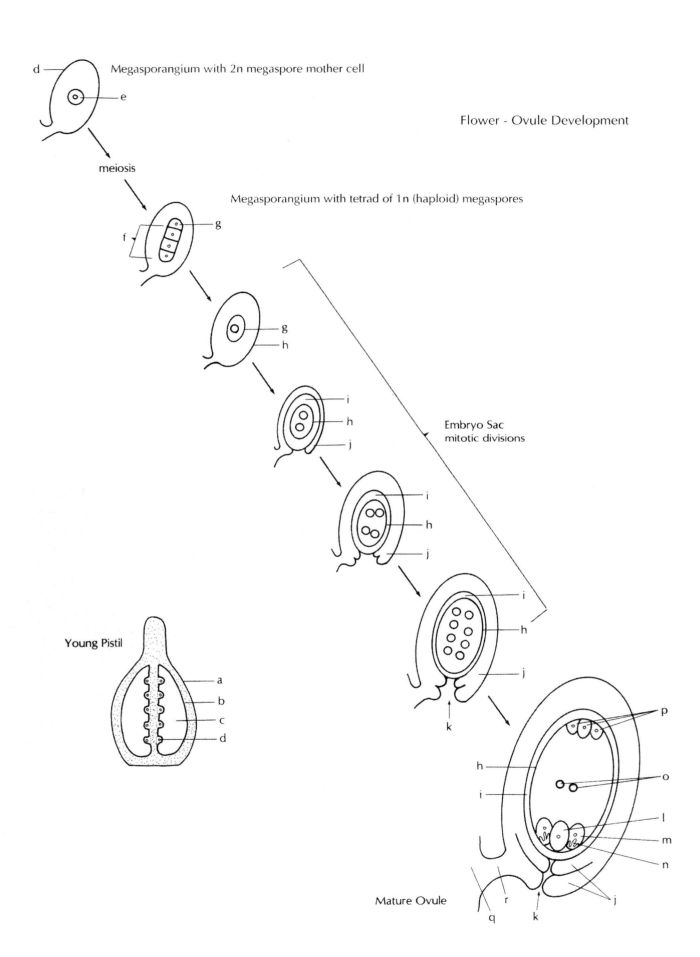

Megasporangium with 2n megaspore mother cell

meiosis

Megasporangium with tetrad of 1n (haploid) megaspores

Flower - Ovule Development

Embryo Sac
mitotic divisions

Young Pistil

Mature Ovule

Flower Pollination by Insects

The transfer of pollen from an anther to the stigma of a pistil is called **pollination.** Some flowers are self-fertile and self-pollinated. **Cross-pollination** takes place when pollen is transferred to another plant. This provides genetic diversity. Within flowers of an individual plant, the time of pollen and stigma maturity may not coincide. Thus, all the pollen may be shed before the stigma matures and pollen must be transferred from another plant for fertilization to take place.

Insect Pollination

Although primitive seed plants, such as conifers, are wind-pollinated, fossil evidence shows that when flowering plants appeared, primitive insects such as beetles pollinated them. Beetles have biting mouthparts, and pollination of primitive flowers involved devouring pollen and other floral parts. Flowering plants developed structures that attracted other types of insects. As flowering plants diversified, there was a parallel diversity of insect specializations. Insects developed various mouthpieces, some in the form of hollow tubes (probosci).

Nectar. In some plants, nectar (a sugar-water solution) was secreted. Then, concealed nectar at the bases of petal, stamens, or ovaries appeared in some plants. Some plants developed unisexual flowers (see 28) secreting nectar, although most flowers pollinated by insects are bisexual (see 28). Some flowers developed long nectar tubes or spurs (see 115, j), that attract insects with long probosci such as moths and butterflies.

Petals. Many flowers secrete a strong scent. Scent produced in the petals differs among types of plants. In contrast to a green background, other colors appeared in flower petals. Some flowers, such as *Iris*, developed lines or hairs on petals (see 129, g) directed toward the nectar glands (see 101, k). These are commonly called **nectar guides (**see 130, n).

Specific insects were attracted to plants that developed corolla structures in the form of "landing platforms" (see 113, d), "overnight traps" (as in some arum family, Araceae, plants), and those that mimic the shape, scent, and color of female insects, a common strategy in the orchid family (Orchidaceae).

Sepals. Some plant flowers developed stout sepals with overlapping bracts around the ovary. This is apparent in the arum family (see 127).

Stamen. Plants produced pollen grains with a sticky coating and also the capacity to produce excessive pollen as food to lure pollinating insects. Some flowers developed fertile and sterile pollen on separate anthers to prevent wasteful expenditure of resources.

Pistil. As plants diversified, there was a change in ovary position from an exposed superior position to one lower or below other floral parts (inferior ovary) (see 28). Pistils with inferior ovaries may have long styles, which elevate the stigma (see 113, sage flower).

Butterfly-pollinated Flowers

Butterflies are attracted to flowers aggregated in a head of many small flowers. The butterfly tongue is a long coiled tube adapted to sucking nectar from long-tubed flowers. Pollen is dusted on the tongue and head of the butterfly. The flickering appearance of flowers shaken by the wind is associated with the presence of nectar to butterflies. Flowers attractive to butterflies have sweet scents and are usually blue, dark pink, yellow-red, and purple. The illustrated example is red clover (*Trifolium pratense*) with a female black swallowtail butterfly (*Papilio polyxenes asterius*).

COLOR CODE

white:	sepal ribs (a), petal tube (f)
pale green:	leaflet center (b)
green:	peduncle (c), leaflets (d), sepals (e)
pink-purple:	petal lobes (g)
yellow:	spots on abdomen (h), rows of outer forewing spots (i), row of outer hindwing cresents (j)
orange:	row of inner forewing spots (k), row of inner hindwing spots (l), single spot (m), row of spots between j and m
blue:	row of muted areas across hindwing (n)

Flower Pollination by Insects

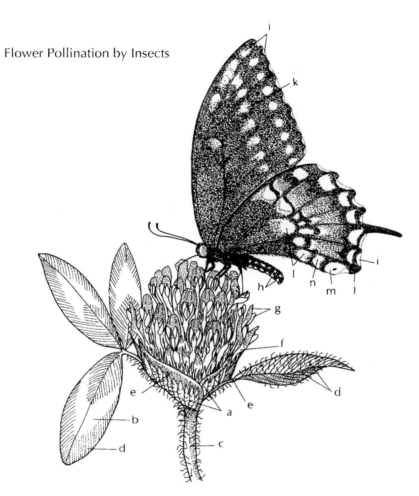

Trifolium pratense
Red Clover
inflorescence with butterfly
× 1⅓

Flower Pollination by Insects
(continued)

Bee-pollinated Flowers

Bees perceive flowers that reflect yellow-green, blue-green, blue-violet, and ultraviolet light. Petals of flowers with colored lines or dark marks guide bees toward nectaries. Rows of dots, stripes, crosses, and checks on flowers attract bees.

The most common type of bee-pollinated flower has a "landing platform" such as the two-lipped (bilabiate) petals found in the flowers of the mint family (see 113, Lamiaceae), or falsely two-lipped petals such as the fused petals of banners and keel found in the pea family (see 100, Fabaceae).

Bee-pollinated flowers have light, sweet, pleasant scents. Bees collect pollen (as a protein source to feed larvae) as well as sugar-rich nectar, which is converted to the honey on which they feed. Honeybees have a so-called pollen basket. Their bristly hairs attract pollen grains that they then brush with their legs into the basket on either side of their bodies. When baskets are tightly packed with pollen, they return to the hive. Of course, not all of the pollen is collected; much is brushed off on the next flower visited insuring pollination.

Bees are the most efficient insect pollinators, as they will visit every flower in a mass of plants until the nectar is exhausted. Some flowers secrete nectar only at a specific time of day when fertilization is to take place; consequently, bees visit these flowers only at that time. Some flowers contain nectar in long spurs and are visited by long-tongued bees (see 115, *Linaria*).

For an illustrated example, see iris family (120, Iridaceae), flag (*Iris*) with a bumblebee (*Bombus affinis*).

Moth-pollinated Flowers

Highly scented and white, green, or yellow flowers are attractive to moths. Night-flying (nocturnal) moths pollinate night-blooming flowers. Nectar at the base of long petal tubes (a) is available only to insects such as moths with very long tubular mouth parts (f). Many moth-pollinated flowers are star-shaped with a central funnel to the nectary.

The illustrated example is a tobacco flower (*Nicotiana*) with a Carolina sphinx moth (*Protoparce sexta*). (Moth adapted with permission from the *National Geographic*).

Fly-pollinated Flowers

Many flowers pollinated by flies have exposed nectar. Some have sweet odors, and other flowers have odor and color (pink, purple, green) of decomposing organic matter, which attracts flies. Usually flies are attracted to white, yellow, and yellow-green flowers.

In the arum family (see 127, Araceae), the odor and color of the flower cluster (spadix), with its bract (spathe, n) attract flies. In some species, a fly visitor slides down the spadix, is trapped in the spathe, and is released within a day, carrying mature pollen.

The illustrated example is a skunk cabbage (*Symplocarpus foetidus*) inflorescence with two syrphid carrion flies (*Allograpta obliqua*).

COLOR CODE

white:	petal tube (a), fly wings (i), spadix flowers (p)
yellow:	stamens (b, o)
green:	calyx (c), peduncle (d), fly thorax (k), lines on spathe (m)
light gray:	moth eyes (e), moth tongue (f), moth thorax (g), moth wings (h)
tan:	fly eyes (j)
orange:	fly abdomen (l)
purple:	spathe (n)

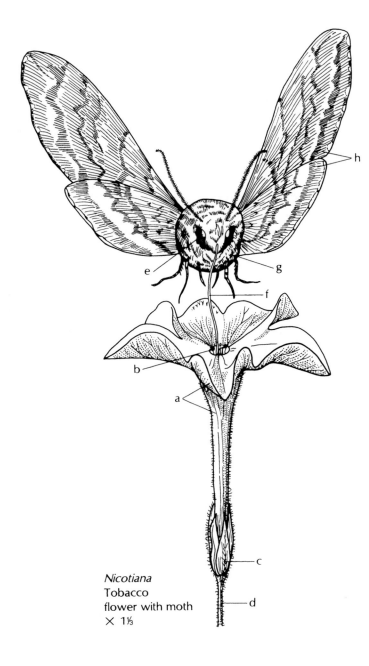

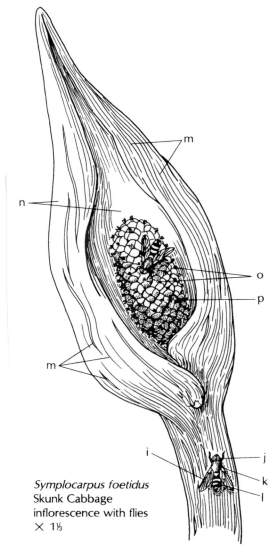

Flower Pollination by Insects

Nicotiana
Tobacco
flower with moth
× 1⅓

Symplocarpus foetidus
Skunk Cabbage
inflorescence with flies
× 1⅓

Flower Pollination by Wind

Wind Pollination

From the fossil record, we learn that wind pollination was a later development in flowering plants. To change from pollination by insects to pollen transferred by wind currents, loss of color, petals, nectaries, scent, and the sticky covering on pollen grains became apparent. With the exception of sexual structures, flower parts developed that were much smaller in size.

The stamens increased in size and number, and some stamens developed very long filaments. Huge amounts of tiny, dry pollen grains developed. Since wind is a chance pollinator, that is, more is better. The stigma of the pistil increased in size and often became feathery (plumose) or divided. The number of carpels in the ovary also increased in some flowers.

Bisexual flowers changed to separate staminate (male) and pistillate (female) unisexual flowers and increased in number. Unisexual flowers tended to aggregate in clusters or to form catkins (aments). In some dicots, staminate and pistillate flowers formed on separate plants (**dioecious**) (see 106, staghorn sumac).

Wind-pollinated Flowers

Some of the monocot herbs that are wind-pollinated are members of the grass (Poaceae, see 123), cattail (Typhaceae), and rush (Juncaceae) families. Bisexual or staminate unisexual flowers are elevated on slender stalks above the vegetative parts of the plant.

Wind pollination is also found in trees, such as elms, ashes, and hackberries of the elm family (see 82, Ulmaceae). Birches, ironwoods, alders, and hazelnuts of the birch family (see 84, Betulaceae), beeches, oaks, and chestnuts of the beech family (see 83, Fagaceae) and poplars of the willow family (see 95, Salicaceae) are also trees with wind pollination.

The notorious ragweed (*Ambrosia*) in the aster family (Asteraceae, see 119) is an obvious herbaceous dicot example. Hayfever sufferers are familiar with its wind-blown pollen.

Willows (*Salix* spp.) have unisexual reduced flowers in catkins but have nectar glands and are insect-pollinated. Staminate flowers, which form in clusters, swing loosely on an axis and open during the windy spring period before the leaves expand. Vestiges of bisexual flowers remain in some wind-pollinated flowers that have reduced, sterile structures of the opposite sex (see 105, maple family, Aceraceae).

The illustrated example is a box elder (*Acer negundo*), which is dioecious. The flower in a cluster of staminate (male) flowers has pendant anthers (b) on a long pedicel (c). On a tree with pistillate (female) flowers, the compound leaves (I) begin to expand as the flower develop. The pistillate flowers has fused, lobed sepals (j), a double ovary (k) and divided stigmas(l).

COLOR CODE

tan:	leaf scars (a, f)
yellow:	anthers (b)
yellow-green:	pedicel (c), shoots (d, g), bud scales (e, h), leaflet (i), sepals (j)
tan-green:	ovaries (k)
white:	stigmas (l)

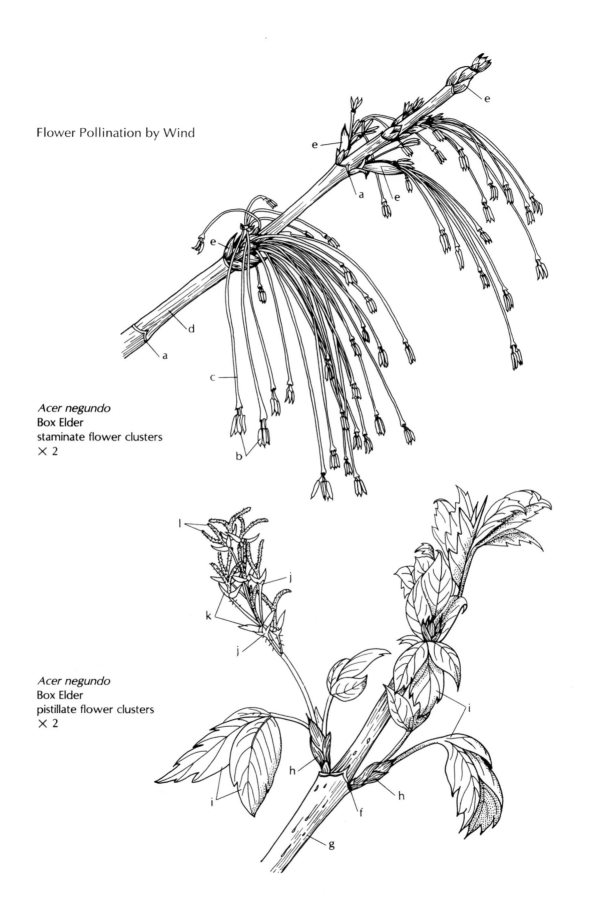

Flower Pollination by Wind

Acer negundo
Box Elder
staminate flower clusters
× 2

Acer negundo
Box Elder
pistillate flower clusters
× 2

Flower Pollination by Birds and Bats

Bird and Bat Pollination

A flower pollinated by birds or bats usually possesses both staminate (male) and pistillate (female) reproductive structures (bisexual). Some have sweet, fleshy bracts surrounding the flower as well as pollen and nectar. Others may have imitation pollen made of proteins and/or starch.

While some flowers attract birds or bats, animals such as slugs, snails, earthworms, mice, rats, squirrels and opossums pollinate other flowers. Gerbils, mice and shrews pollinate various South African *Protea* species. These plants produce low-lying flowers of dull colors.

The attraction for animals is food, i. e., copious amounts of nectar that contains not only sugar, but vitamins, amino acids and other nutrients. While an animal is gathering nectar, pollen grains with hooks and spines and sticky coatings adhere to the animal's coat to be transferred to another flower resulting in pollination.

Bird-pollinated Flowers

Birds pollinate many tropical and subtropical plants. Flowers attractive to birds are usually bright red or orange. Some have striking color contrasts. Hummingbird-pollinated flowers are mainly red, a color not usually perceived by insects. Bird-pollinated flowers have abundant nectar and are without scent. Some are tube flowers with a basal swelling filled with nectar. The ovaries may be inferior or separated from the nectaries,

as in spur flowers, so that damage does not incur while the pollinator is visiting.

Some bird visitors feed on small insects attracted by the nectar. Some bird-pollinated flowers are very large, while others are small. Some plants have sturdy perches near flowers, while others have freely swinging flowers, which attract hovering birds.

Generally, bird-pollinated flower forms are of the two-lipped (bilabiate) type, tubular flowers, flowers with a protruding bottlebrush stamen arrangement, or flowers with nectar in spurs.

The illustration example is fuchsia (*Fuchsia magellanica*) with a Brazilian hermit hummingbird (*Phaethornis eurynoma*).

Bat-pollinated Flowers

Bats pollinate many large tropical herbs and trees. These flowers are usually arranged in a manner easily accessible to flying bats. Flowers attractive to bats bloom at night, have large amounts of nectar and pollen, are usually pale in color, may have a musky or fruity odor, and are strong enough to support a bat. Bat-pollinated flowers are bell-shaped, open saucer-shaped with many stamens, or possess suspended flowers with stamens bunched in a protruding brush.

The illustration example is a saguaro cactus (*Carnegiea gigantea*) flower with a long-nosed bat (*Leptonycteris sanborni*).

COLOR CODE

green:	peduncle (a), ovary (b)
red:	sepals (c), filaments (f), style (h), bat tongue
red-purple:	petals (d)
yellow:	anthers (e), stigma (g), pollen(l), spines (p)
yellow-green:	bracts (o)
light orange:	area below bird's eye (i), breast feathers (j), tips of bracts (n)
white:	unshaded areas of tail, near tail (k), tepals (m)
light brown:	remaining bird feathers, not colored above
brown-orange:	entire bat, except tongue and pollen

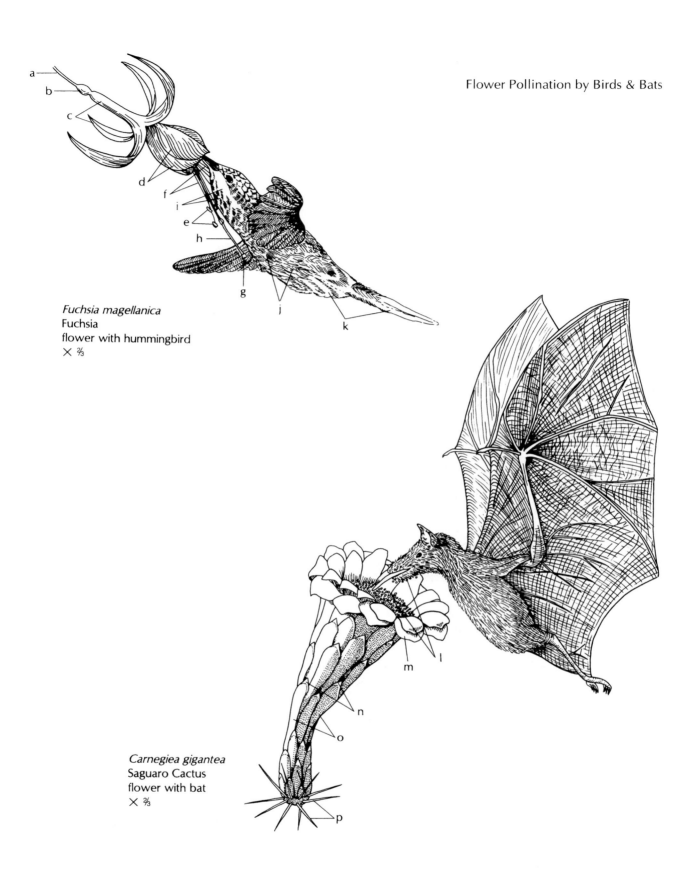

Fuchsia magellanica
Fuchsia
flower with hummingbird
$\times$ ⅔

Carnegiea gigantea
Saguaro Cactus
flower with bat
$\times$ ⅔

Flower—Fertilization and Embryo Development

Fertilization

Pistil. During pollination, pollen (a) from a stamen is transferred to a stigma (b). The stigmatic surface is usually covered with secretory cells (c) possessing copious amounts of sugars needed for growth of the pollen tube (d). The pollen tube grows within the style (e) to the ovule (f) in the ovary (g). There, it enters the embryo sac (h) at the opening (**micropyle**) between the integuments (i). Within the pollen tube are the 1n (haploid) tube nucleus (j) and two 1n (haploid) sperm nuclei (k, l). After entry into the embryo sac, the pollen tube tip bursts and discharges the 2 sperm nuclei. Simultaneously, one or both synergid cells (m) disintegrate (see 32).

Embryo Sac. Within the embryo sac double fertilization occurs. The egg (n) nucleus (o) fuses with one sperm nucleus (k) to form a 2n (diploid) zygote (q). The other sperm nucleus (l) unites with the 2 polar nuclei (p) to form a 3n (triploid) **primary endosperm nucleus** (r).

Embryo Development

Globular Stage Embryo. The primary endosperm nucleus divides repeatedly to form many 3n free, floating nuclei (s). Meanwhile, the zygote starts to divide, forming a basal cell (t), which develops into a **suspensor** (u) to elevate the developing embryo in the sac and to transport nutrients. An apical cell divides to form a globular mass of cells (v).

Heart Stage Embryo. Walls develop around the free endosperm nuclei, forming the cellular endosperm (w). The endosperm cells contain stored food reserves such as hormones, starch, proteins, oils, and fats. In dicots, the embryo becomes heart-shaped with the formation of cotyledon primordia (x).

Torpedo Stage Embryo. As the young embryo elongates, the vascular systems for the young root (y) and shoot (z) are formed.

Mature Seed

In the mature seed, the integuments (i) of the embryo sac have developed into the protective **seed coat** (testa). The embryo illustrated has 2 seed leaves (**cotyledons,** x) of a dicot. (Monocots usually have 1 seed leaf or cotyledon, see 40, corn seed)

COLOR CODE

yellow:	pollen (a), pollen tube (d), tube nucleus (j)
light green:	stigma (b), secretory cells (c), style (e), ovule (f), ovary (g), integuments (i)
gray:	embyo sac (h)
white:	synergids (m), polar nuclei (p), zygote (q), embryo (t, u, v, x, y, z)
tan:	egg (n)
red:	sperm nuclei (k, l), egg nucleus (o)
orange:	endosperm (r, s, w)

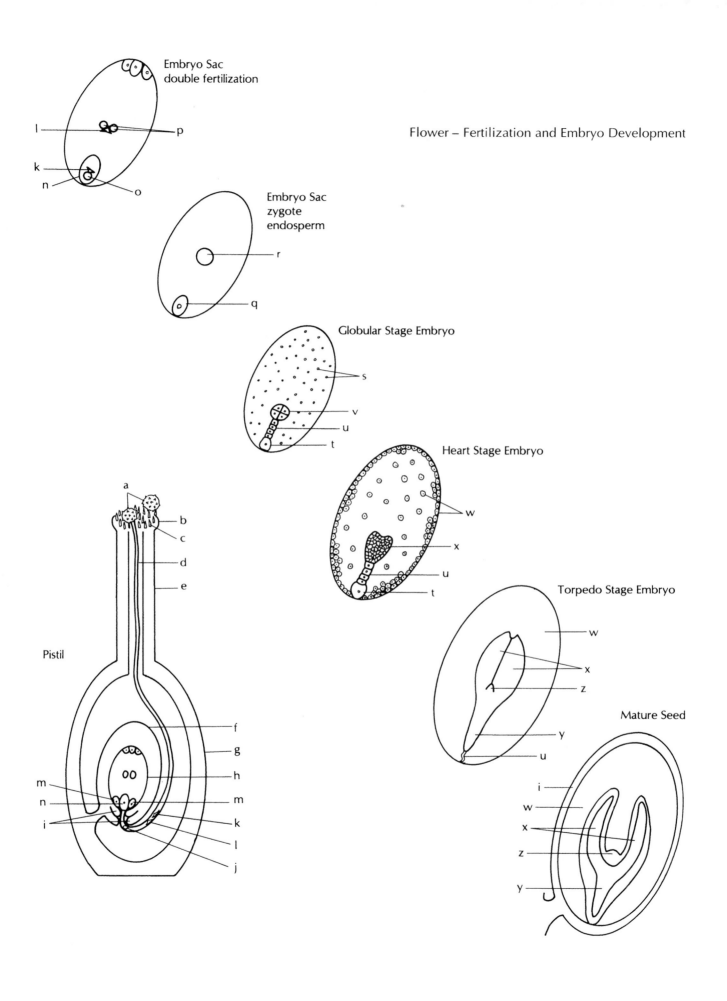

Embryo Sac
double fertilization

Flower – Fertilization and Embryo Development

l
p
k
n
o

Embryo Sac
zygote
endosperm

r
q

Globular Stage Embryo

s
v
u
t

Heart Stage Embryo

w
x
u
t

a
b
c
d
e

Pistil

f
g
h
m
n
m
i
k
l
j

Torpedo Stage Embryo

w
x
z
y
u

Mature Seed

i
w
x
z
y

Fruit—Dry Types

A **fruit** is the seed-bearing structure in flowering plants (conifer seeds are borne in a cone).

Indehiscent, Dry Fruit Types

An **indehiscent, dry fruit** does not open at maturity. The tissues are composed of parenchyma and/or sclerenchyma (see 8) at maturity.

Achene. An achene (a) is a type of one-seeded fruit with a thin wall (pericarp).

Grain (Caryopsis). A grain (c) is a type of achene found in grasses (see 123, grass family, Poaceae), which originates from a superior ovary (see 28).

Cypsela. An achene that originates from an inferior ovary, and commonly has a hairy pappus (d) present, is a cypsela (e).

Nut. An achene-type fruit, the nut (f) is hard and "bony," and may result from a compound ovary.

Samara. The samara is a type of winged achene, which may be single, as in elm (*Ulmus,* h), or double, as in maple (*Acer,* i).

Schizocarp. Carpels of a compound ovary break apart into **mericarps** (j), which function as achenes in the schizocarp type fruit.

Nutlet. A small nut is called a nutlet (k).

Dehiscent, Dry Fruit Types

A dehiscent, dry fruit opens to release seeds at maturity. The fruit splits under tension along definite lines or breakage points.

Follicle. Composed of a single pistil, the follicle (l) splits along one line (see 77, *Magnolia grandiflora*).

Legume. A legume (m) originates from one carpel, which splits along two lines, and is the fruit type of the pea family (see 100, Fabaceae). A type of legume, the seeds of a **loment** (n) are partitioned into sections. The carpel of a **spiral legume** (o) coils.

Capsule. Capsules have more than one carpel and are classified by the mode of opening (dehiscence). The **septicidal** capsule (p) opens in the plane of carpel union. A **loculicidal** capsule (r) opens around a horizontal line. A **poricidal** capsule (t) releases seeds through pores at the top.

Silique. A silique has 2 narrow pieces (**valves**,u) separated by a partition (**replum**, v). A rounded form is called a **silicle** (w, x). These are fruits of the mustard family (see 96, Brassicaceae).

COLOR CODE

tan:	achene (a), cypsela (e), nut (f), schizocarp mericarps (j), nutlets (k), spiral (o), septicidal capsule (p), seeds (s), valve (x)
white:	grain (c), pappus (d), replum (v, w)
red-brown:	bracts (b)
brown:	cup (g), legume (m), loculicidal capsule (q), seeds (y)
yellow-tan:	single samara (h)
green:	double samara (i), circumscissile capsule (r), valve (u)
yellow:	follicle (l), loment (n), poricidal capsule (t)

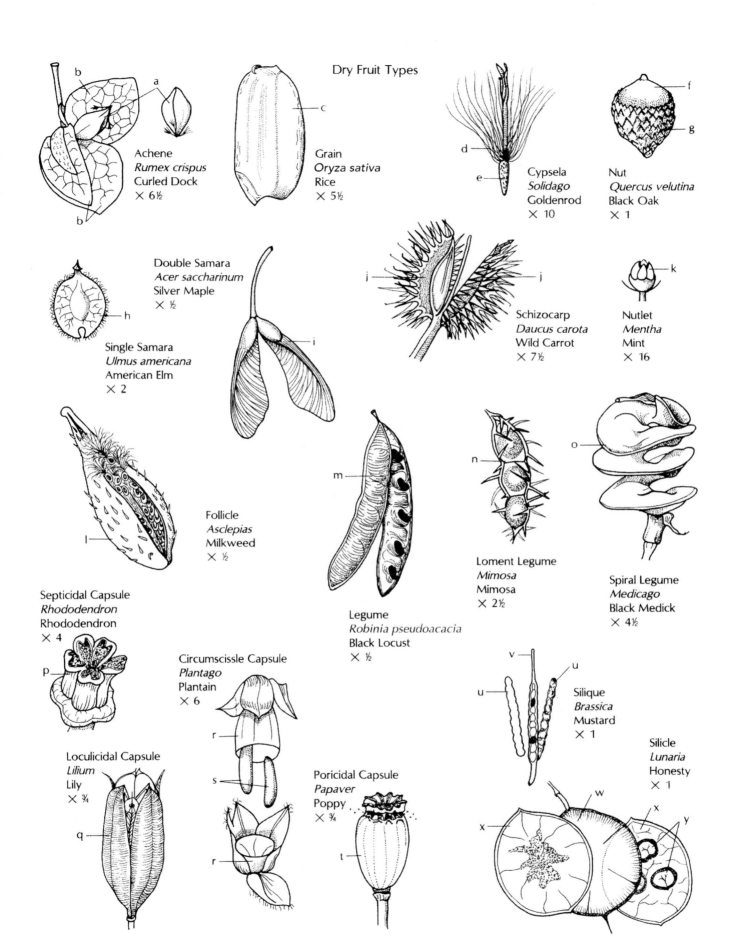

Dry Fruit Types

Achene
Rumex crispus
Curled Dock
× 6½

Grain
Oryza sativa
Rice
× 5½

Cypsela
Solidago
Goldenrod
× 10

Nut
Quercus velutina
Black Oak
× 1

Double Samara
Acer saccharinum
Silver Maple
× ½

Single Samara
Ulmus americana
American Elm
× 2

Schizocarp
Daucus carota
Wild Carrot
× 7½

Nutlet
Mentha
Mint
× 16

Follicle
Asclepias
Milkweed
× ½

Loment Legume
Mimosa
Mimosa
× 2½

Spiral Legume
Medicago
Black Medick
× 4½

Septicidal Capsule
Rhododendron
Rhododendron
× 4

Legume
Robinia pseudoacacia
Black Locust
× ½

Circumscissle Capsule
Plantago
Plantain
× 6

Silique
Brassica
Mustard
× 1

Loculicidal Capsule
Lilium
Lily
× ¾

Poricidal Capsule
Papaver
Poppy
× ¾

Silicle
Lunaria
Honesty
× 1

Fruit—Fleshy Types, Compound

Fleshy Fruit Types

The ovary or fruit wall (**pericarp**) consists of an outer layer (**exocarp**), middle layers of parenchyma (**mesocarp)**, and an inner layer (**endocarp**). The parenchyma tissues are filled with sugars, starch, and/or fats at maturity.

Berry. The entire pericarp (b, c) of a berry is fleshy. There may be one or more carpels and seeds (d).

Pepo. This berry with a hard-rind pericarp composed of exocarp (e) and endocarp is a pepo fruit. It is derived from an inferior ovary (see gourd family, Cucurbitaceae, 94).

Hesperidium. A leathery-rinded berry, the hesperidium exocarp (h) of compact collenchyma contains oil glands. There is a spongy mesocarp (i) and an endocarp (j), which produces juice sacs. This type, commonly called citrus, is found in the rue family, Rutaceae (see 107).

Drupe. The drupe fruit ovary wall develops into a fleshy mesocarp (n) and a stony endocarp (o) over usually one seed.

Pome. A pome has a pericarp of cartilaginous endocarp (r) lining the seed (q) cavities (locules), which are surrounded by fleshy parenchyma exocarp (s). Outside the inferior ovary wall is an expanded floral tube (t) of fleshy parenchyma. (For flower to fruit development, see rose family, Rosaeae, 99).

Compound Fruit

Aggregate Fruits. The separate carpel of one flower stays together as a unit in an aggregate fruit. Examples are an aggregate of drupelets (x), as in black raspberry fruits, an aggregate of achenes, as in strawberry fruits, and an aggregate of follicles (see 77, *Magnolia grandiflora).*

Multiple Fruits. The carpels of separate flowers in a cluster stay together as a unit in multiple fruits as in a multiple of achenes in osage-orange and pineapple.

COLOR CODE

tan:	pedicel (a), peduncle (l, p), seed (d, k, y), endocarp (o)
purple:	exocarp (b)
white:	mesocarp (c, f, i, z), seed (g), exocarp (s), floral tube (t)
green:	exocarp (e), peduncle (v), sepal (w)
yellow:	exocarp (h)
pale yellow:	endocarp (j, r)
red:	exocarp (m), mesocarp (n), epidermis (u)
brown:	seed (q)
dark blue:	drupelet exocarp (x)
yellow-green:	epidermis (y)

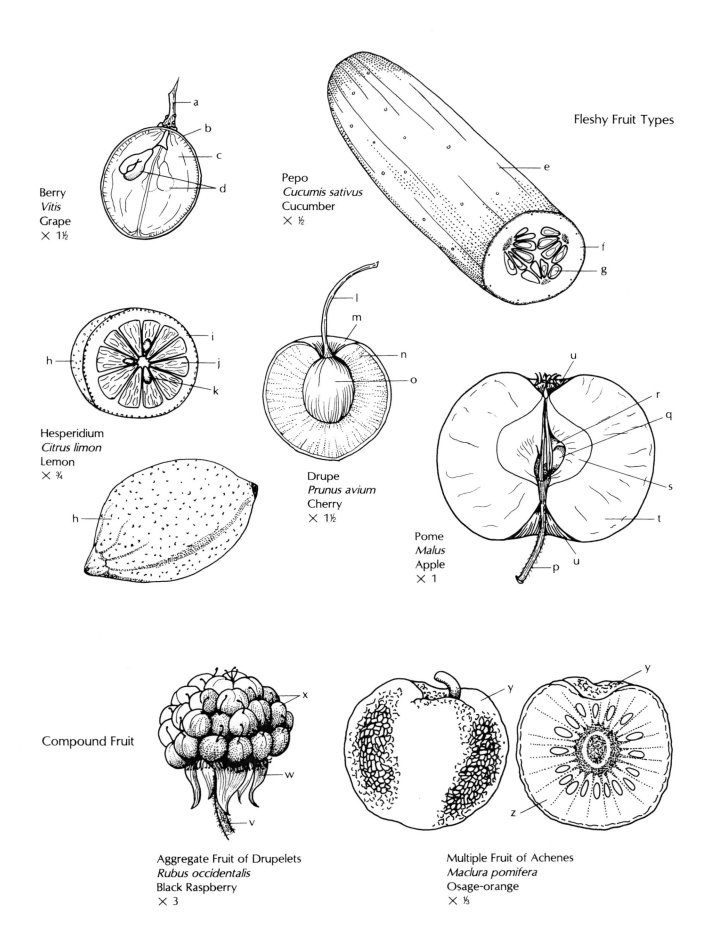

Fleshy Fruit Types

Berry
Vitis
Grape
× 1½

Pepo
Cucumis sativus
Cucumber
× ½

Hesperidium
Citrus limon
Lemon
× ¾

Drupe
Prunus avium
Cherry
× 1½

Pome
Malus
Apple
× 1

Compound Fruit

Aggregate Fruit of Drupelets
Rubus occidentalis
Black Raspberry
× 3

Multiple Fruit of Achenes
Maclura pomifera
Osage-orange
× ⅓

Seed Structure and Germination

A **seed**, which develops from an ovule, consists of a partially developed sporophyte plant (**embryo**) and none or various amounts of food storage tissue (**endosperm**) covered by a protective seed coat (**testa**).

Dicot Seed

The embryo usually has two seed leaves (cotyledons). As the bean seed is split in half, only one cotyledon (a) is shown. In the bean and pea seedlings, two cotyledons (a, b) can be seen. Between the cotyledons is the hypocotyl axis (c) with the shoot meristem (epicotyl) at one end, the root meristem at the other end. An undeveloped root (**radicle**, d) may be present. The **plumule (e)** of the bean seed will develop into stem (f) and leaves (g) in the seedling.

Nutritive/food storage tissue (endosperm) sustains the germinating embryo until roots and leaves are functionally developed. Endosperm cells contain mostly starch with lesser amounts of proteins, oils, and/or fats. In bean seeds, the cotyledons, which occupy most of the space, store the food reserves, with the scant endosperm consisting mostly of protein.

The seed coat (h) is made up of the ovules' fused integuments. An opening (**micropyle**) between the integuments sometimes leaves a pore (i). Where the ovule was attached to the placenta, by a stalk (**funiculus**), there is sometimes a scar (**hilium**, j). Usually the seed coat is very thick and may be colored.

Monocot Seed

Usually monocot seeds have one cotyledon. A grain fruit (caryopsis) is used as a monocot example. The embryo includes one cotyledon called the **scutellum** (k), an outer protective cylindrical leaf (**coleoptile**, l), which encloses young leaves (m) over the shoot apical meristem (n), and a protective cap (**coleorhiza**, o) over the root apex (p).

The outermost layer of the endosperm, the **aleurone** (q), contains most of the protein of the seed, while the remaining endosperm (r) is mostly starch. In grains, the fruit coat (s) is made up of fused integuments and pericarp.

In the monocot grain of rice (*Oryza sativa*), white rice (polished rice) is devoid of aleurone ("protein jacket") and consists of embryo and starchy endosperm only. In contrast, brown rice (not polished) has the aleurone present and is thus much better nutritionally because the aleurone layer contains the bulk of the seed storage proteins (6 to 12 % of dry weight). White rice is primarily starch.

Conifer Seed

The embryo axis (t) consists of cotyledons (u) enclosing the shoot apical meristem at the tip, the hypocotyl, and the root apex (v) at the base. Conifer seeds have no endosperm; the embryo is surrounded by megagametophyte tissue (w) and enclosed by a seed coat (x).

Germination

Seed germination is the growth of the embryo, resulting in emergence of a shoot (**epicotyl**) and root (**radicle**). To germinate, the seed must be alive, not dormant, have a suitable temperature, adequate moisture, an oxygen supply, and for some species, light.

Once the radicle has emerged from the seed coat, the seedling either develops with its cotyledons above ground (**epigeous**), as in bean (a) and pine (u), or below ground (**hypogeous**), as in pea (b) and corn (s). The cotyledons of epigeous seedlings produce chloroplasts and function photosynthetically until true leaves develop. Cotyledons of hypogeous seedlings provide food for the developing plants.

COLOR CODE

Seeds:	white
Seedlings:	
green:	cotyledons (a, b, u), leaves and stipules (g), leaf (m), embryo axis (c, t), stem (f), coleoptile (l)
tan:	seed coat (h, x)
white:	roots
yellow:	fruit coat (s)

Seeds

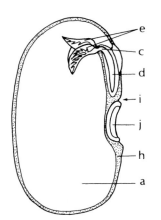
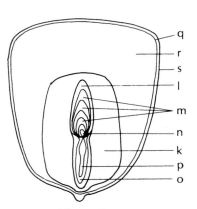
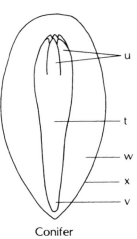

Dicot, ½
Phaseolus
Bean
× 3

Monocot
Zea
Corn
× 5

Conifer
Pinus
Pine
× 8

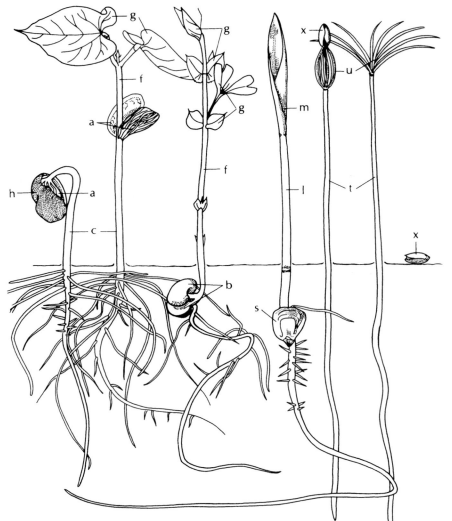

Seedlings

Dicot
Phaseolus
Bean

Dicot
Pisum
Pea

Monocot
Zea
Corn

Conifer
Pinus
Pine

Major Groups; Geologic Time Scale

Learning about the relationships of plants to each other involves the study of plant fossils (paleobotany), physical characteristics (anatomy), form (morphology), inheritance (genetics), chemical characteristics (physiology and plant biochemistry), and plant communities (ecology).

Carolus Linnaeus (1707–1778) organized a system, that has since been expanded, for the classification of plants on the basis of their evolutionary relationships. Some groups of organisms studied by botanists are now no longer recognized as plants, as for example, bacteria, blue-green algae, fungi, and lichens. They are studied by specialists on bacteria and blue-greens (microbiologists), green, red, and brown algae (algologists), fungi (mycologists), and lichens (lichenologists).

The **Bacterial Group** consists of bacteria, blue-greens and other microorganisms. In the past, blue-greens were grouped with algae. Now it is known that they are genetically more like bacteria. The cells of bacteria and blue-greens do not have organelles; instead, the cell membranes carry on many of the biological functions.

There is no membrane-bound nucleus in the cell. This is termed **prokaryotic** (pro = before, karyon = nucleus). Nuclear material consists of strands of DNA (<u>d</u>eoxyribo<u>n</u>ucleic <u>a</u>cid, a molecule that constitutes the genetic information of a cell). Fibrils of DNA in the central region of a bacterial cell are designated by the term **nucleoid**. Asexual reproduction occurs by cell division or budding. Sexual reproduction is a recombination of genetic material. Most are microorganisms, smaller than the **eukaryotic** cells (eu = true, karyon = cell) of the other three kingdoms.

Revision Group. Classification and relationships of molds, mildews, rusts, chytrids and allies, dinoflagellates, algal groups, and stoneworts and others are currently under revision.

A characteristic in common in this group is the presence of nuclei (eukaryotic) in the cells. Many obtain food (organic carbon metabolites) by photosynthesis within cell organelles called **plastids**. A flagellum (whip) for locomotion or feeding is present in many members. Most require oxygen (aerobic). They are aquatic.

Fungi are organisms with nucleated cells (eukaryotic), that form resistant fungal spores, have cell walls composed of chitin, and do not have mobile cells at any stage. Fungi are found from the fossil record 450–500 million years ago (the Ordovician period).

Representative divisions in this book are black bread mold and allies, sac fungi, and club fungi. Lichen classification is a problem because lichens are an association of either a blue-green or a green alga with either a sac fungus or a club fungus. Lichens are included here with the fungi.

Plants represented here include liverworts, hornworts, mosses, whisk ferns, clubmosses, spikemosses and quillworts, horsetails, ferns, and plants bearing seeds. Seed-bearing plants include the gymnosperms (= naked seeds) and the angiosperms (= enclosed seeds).

Living gymnosperm groups are cycads, ginkgo, conifers, and gnetes. The angiosperm group consists of flowering plants. Characteristics of plants are covered in all three sections of the book.

Geologic Time Scale

Eras. The largest units of geologic time are known as eras; each encompasses millions of years. They are basically defined by biological events – the appearance or disappearance of major groups of organisms. These changes in biology are related to major geologic events. However, only the biologic changes are seen worldwide in the fossil record.

Periods. Each era is subdivided into periods, and the periods are broken down into epochs; again their separations are most commonly based on fossils.

Geologic Time Scale

Era	Period	Life
Precambrian		
	Precambrian	The first fossil organisms of bacteria and blue-greens are found in rocks over 3,000 million years old,
		The earliest eukaryotic cells (true nucleated cells) are found in rocks about 1,000 million years old. These fossils are rare.

600 million years ago -

Fossils of invertebrate marine animals are present.

Paleozoic (Old Life)

| | Cambrian | |

500 -

| | Ordovician | The oldest fossils of fungi are from 450 million years ago and algae fossils are common. Dinoflagellates are present. |

430 -

| | Silurian | Oldest fossils of land plants are present. |

400 -

| | Devonian | Land plants became abundant during the Devonian. Fossils of clubmosses, spikemosses, quillworts, giant horsetails, scouring rushes and ferns are found in this period. |

345 -

| | Mississippian | During the Carboniferous (Mississippian and Pennsylvanian periods), |

310 -

| | Pennsylvanian | there were vast forests of trees and swamps preserved now as coal. |

280 -

| | Permian | During the Permian, gnetes were present and true conifers emerged. Ginkgos also originated during this period. |

225 -

Mesozoic (Middle Life)

| | Triassic | The Triassic was the time of reptiles and dinosaurs and the first mammals. There were forests of gymnosperms and ferns. Cycads were also present. |

180 -

| | Jurassic | The Jurassic was the "age of cycads" and it was a time when conifers were distributed worldwide. Birds originated during this period. Reptiles were also dominant. |

136 -

| | Cretaceous | Primitive flowering plants evolved during the Cretaceous. |

65 -

Cenozoic (New Life)

| | Tertiary | During the Tertiary, flowering plants dominated the land. Later, grasslands were apparent. The landscape has been dominated by herbaceous flowering plants ever since. |

2.5 -

| | Quaternary | Humans are present. |

Fossils

Fossil Record and Appearance of Vascular Plants

About 85% of earth's geologic time is within the Precambrian. In the Precambrian, in rocks over 3,000 million years old, the first fossil organisms of bacteria and bluegreens are found. The earliest eukaryotic cells are from rocks about 1,000 million years old. These fossils are rare, and the early evolutionary history of non-vascular plants is obscure.

The division that marks the end of Precambrian time and the beginning of the Paleozoic, about 600 million years ago, is recognized by the worldwide appearance of fossils of invertebrate marine animals. In the late Silurian period, the first vascular plants appear. Vascular plants are so designated because they possess specialized conducting tissues of xylem and phloem (see 9).

Land plants became abundant during the Devonian. During the Carboniferous (Mississippian and Pennsylvanian), there were vast forests of trees and swamps preserved as what we know as coal.

The Jurassic time of the Mesozoic was the "age of cycads" and it was a time when conifers became distributed worldwide.

Primitive flowering plants evolved during the Cretaceous, and by the Tertiary time of the Cenozoic, they dominated the land. Later, grasslands were apparent and the landscape has been dominated by herbaceous flowering plants ever since.

Fossil Evidence. A fossil represents evidence of past life. Fossils are usually found in nature as preserved remains or they may be just an imprint of an organism or part of an organism. Usually, vascular plant fossils occur as fragments. They often show alteration of original structures through damage by transport away from where they lived and when rock deposits were formed.

During the process of fossil formation, plant parts commonly were compressed in soft sediment, which later hardened into rock, leaving an imprint such as a fern leaf. Or the plant was embedded in a soft matrix and later decayed as the rock formed, leaving a mold of its exterior. Or later, a hollow mold may have been filled in with mineral deposits forming a natural cast of the plant structure.

Embedded, **"petrified" plant fossils** are a result of cell contents being replaced by minerals while the cell walls are preserved. Thus, thinly cut sections reveal the microstructure of the fossilized plant parts.

Paleobotany. This is the study of fossilized plants. As parts of plants are found, they are given "form genus" names. For example, a stem section, a leaf, a root, or a reproductive part such as a seed may all have different names, even after it is assumed that all belong to one extinct plant. An illustration of an entire fossil plant is a graphic reconstruction based on fossilized parts found.

Of interest... **organic evolution:** origins and relationships or past and present life forms are better understood from their fossil record. By determining the ages of rock formations, it can be determined when plants first appeared and when they became extinct or yet survive.

Economic: Coal deposits result from peat beds of decaying plants; an example is coal formed during the Carboniferous (Mississippian and Pennsylvanian). The term "carboniferous" is often used to refer to the two Paleozoic periods because of the extensive coal beds formed then. Pressure and heat from overlying rock sedimentation over millions of years compressed and converted the peat into coal. One of the methods for finding oil is the study of samples, obtained from drill holes, of microfossils of specific plants, pollen grains, and animals.

Paleogeography. By studying the distribution of certain fossils over the earth's crust, it is possible to determine ancient relationships of lands and seas.

Paleoecology. Fossil organisms found together indicate ancient relationships, environment and climate.

Cordaites

This conifer-like tree, which reached about 30 meters in height, was prominent during the Carboniferous. Its trunk (a) was woody; the strap-shaped, or sometimes fan-shaped, leaves (b) had parallel veins that branched dichotomously from the base. The leaves were up to a meter long. Reproductive structures (c) that resembled cones on an axis, bore male pollen sacs or female cone scales with ovules that developed into winged seeds after fertilization. (drawing after Grand 'Eury, M. F. Cyrille, 1877)

Williamsonia

Cycadoid plants evolved in the late Carboniferous and became extinct during the late Cretaceous. One of these, *Williamsonia,* reached a height of 2 meters. The plant had a columnar trunk (d) marked with old leaf base scars (e) and whorls of leafy fronds (f) similar to living cycads, although not related. Male pollen sacs and a female seed-bearing organ were borne in a flower-like structure (g). (drawing after Sahni, 1932, Geologic Survey of India)

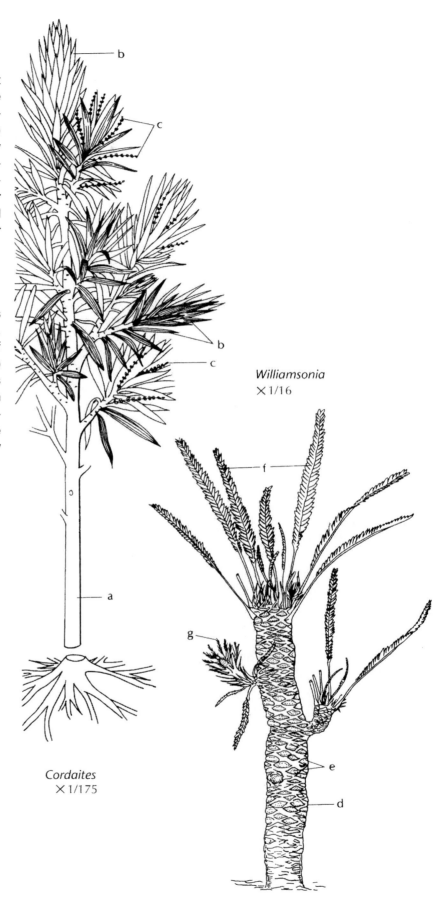

Williamsonia
×1/16

Cordaites
×1/175

COLOR CODE

tan: trunk (a, d)
green: leaves (b, f)
yellow: reproductive structures (c, g)
brown: leaf base scars (e)

Fossils (continued)

Rhynia

This plant lived only during the Devonian. *Psilotum* (see 64) and *Tmesipteris* are living plants that resemble *Rhynia* but are not known in the fossil record. *Rhynia* had simple hair-like rhizoids (a), a creeping underground stem (rhizome, b) that gave rise to upright, branching shoots (c) that reached 50 cm in height. As there were no leaves, it is assumed photosynthesis took place in the green stems. Spore cases (sporangia, d) containing spores were borne at the shoot tips. (drawing after Hirmer, 1927, R. Oldenbourg)

Lepidodendron

This tree reached 30 meters in height and was abundant during the Carboniferous. It is related to living lycopods (see 65). The stem bases (e) and stems (f) were dichotomously branched (fork of two branches of equal length). Diamond-shape leaf scars covered the outer surface of the upright stem (g) and upper branched stems. Extending beyond the leaves (h), cones (i) of sporangia with two kinds of spores (heterosporous condition) grew at the tips of stems. (drawing after Hirmer, 1927, R. Oldenbourg)

Calamites

A tree that had secondary growth and thick bark, *Calamites* reached 10 meters in height. Although much smaller and herbaceous, living *Equisetum* (see 66) plants greatly resemble this ancient relative that lived during the Devonian and Mississipian periods. The upright stem (j) arose from an underground rhizome (k) and possessed roots (l) at the nodes. Stem nodes (m) bore whorls of stems (n). Stem ends had whorls of needle-like leaves (o). (drawing after Hirmer, 1927, R. Oldenbourg)

COLOR CODE

white: rhizoids (a), rhizome (b, k)
green: shoots (c), leaves (h, o), stem (j), whorls of stems (n)
brown: sporangium (d)
tan: stem bases (e), branched stems (f), upright stem (g), cones (i)

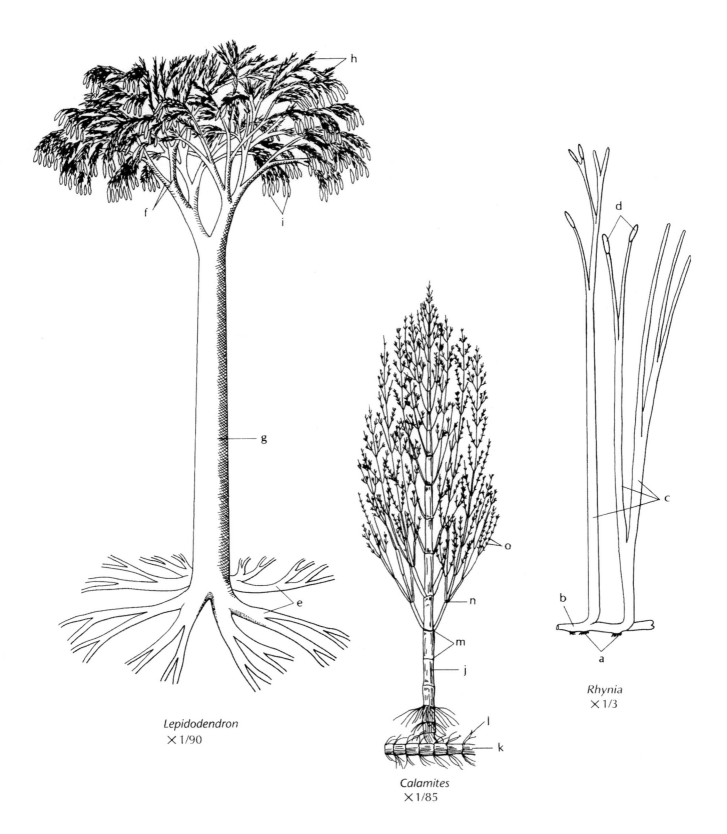

Lepidodendron
×1/90

Calamites
×1/85

Rhynia
×1/3

Blue-greens

Fossil history of the blue-greens extends back to the Precambrian. They are found in water, in and on soil, on rocks, and in the atmosphere. Habitat temperatures range from 0° to 85°C. Some species occur in lichens, liverworts, water ferns, cycads, and flowering plants as symbionts (dissimilar organisms that live together and may, or may not, be beneficial to each other).

Characteristics. The microscopic blue-greens have no membrane-bounded nucleus (prokaryotic, pro = before). [All plants and animals have membrane-bound cell organelles including nuclei (eukaryotic, eu = true).] The cell has a mucilaginous sheath of pectin that surrounds the cell wall and an inner cellulose wall.

Cell contents (protoplasm) include nuclear material with DNA fibrils, chlorophyll a, and accessory pigments of blue (phycocyanins), red (phycoerythrin), orange (carotenes), and xanthophylls (yellow) colors. These pigments are involved in capture of light and subsequent formation of carbohydrates (photosynthesis). Cyanophycean starch is the storage product.

Blue-green forms may be one cell (unicellular), colonial (cells held together by a gelatinous sheath), or filamentous (chains of cells called trichomes). They have no thread-like structures (flagella) for movement.

Reproduction. Blue-greens have no known sexual reproduction. Reproduction of unicellular forms is by cell division in which the cell wall folds in and "pinches" the cell into two cells. Colonial and filamentous forms fragment into separate pieces. Filamentous forms have various specialized cells: an **akinete** is a resistant cell filled with food reserves, which can germinate to form a new filament; a transparent **heterocyst** functions in nitrogen-fixation, and frequently is located where the filament breaks. Some genera have **exospores**. These are cells that pinch off the filament.

Of interest... **ecosystem:** blue-greens are primary colonizers on bare soil and rock; by forming mats that bind to the soil surface, they reduce soil erosion. As **nitrogen-fixers,** they contribute to soil fertility (e.g., growing in rice paddies, *Anabaena,* in association with the floating water fern, *Azolla,* see 70, increases rice production). Blue-greens contribute to the water plankton **food chain; reef building,** in tropical waters blue-greens precipitate calcium carbonate (limestone) out of water and build up rock layers; **toxins:** "blooms" of dense concentrations in the sea kill marine fish, and in reservoirs, they may cause gastrointestinal diseases in cattle and humans.

Synechococcus

Structures of this unicellular spherical (coccoid) blue-green include a mucilaginous sheath (a), cell wall (b), plasma membrane (c) and protoplasm (d). Reproduction is by **binary fission** (splitting into two).

Cylindrospermum

This filamentous blue-green has an akinete (resistant cell, e) and a heterocyst (spore-like cell, f) at the end of the trichome.

Anabaena

A heterocyst (g) is located near the center of the trichome. This filamentous genus is a frequent part of water "blooms." In reservoirs it can cause bad odor and taste to water, killing birds and animals with endotoxin (released on decomposition).

Gloeotrichia

Colonies of trichomes are formed. The tapered trichome has a heterocyst (h) and an adjacent akinete (i).

Hapalosiphon

The trichome is branched, with a heterocyst (j) present. Branches form as the cells divide on a perpendicular plane.

COLOR CODE

white:	sheath (a), heterocyst (f, g, h, j)
tan:	wall (b)
blue-green:	protoplasm (d)
dark blue-green:	akinete (e, i)

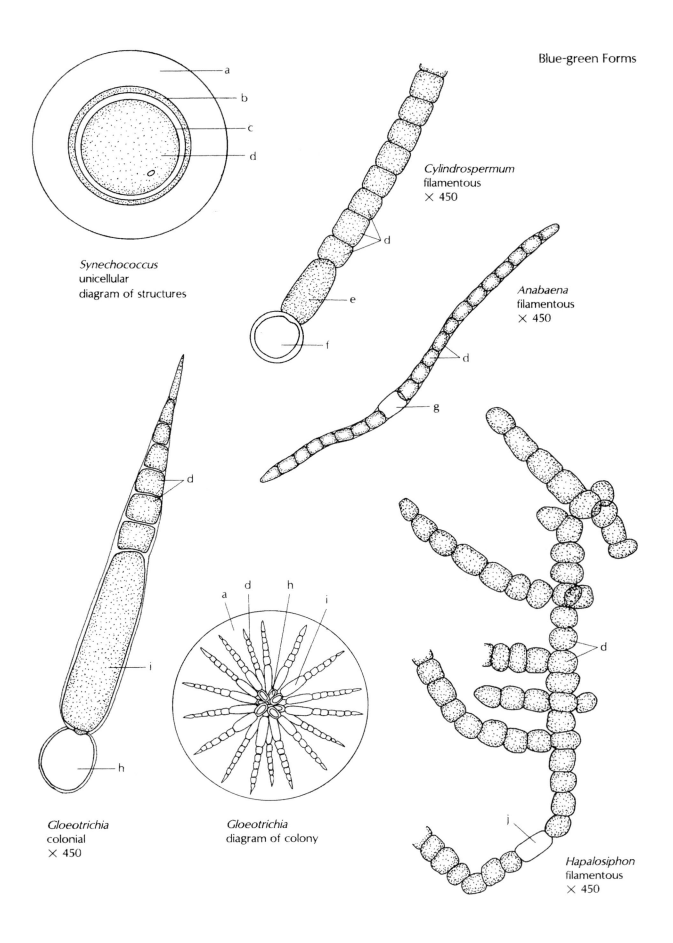

Synechococcus
unicellular
diagram of structures

Cylindrospermum
filamentous
× 450

Anabaena
filamentous
× 450

Gloeotrichia
colonial
× 450

Gloeotrichia
diagram of colony

Hapalosiphon
filamentous
× 450

Slime Molds

Even when organisms were grouped as either plants or animals, no one knew what to do with the slime molds. The creeping acellular phase is animal-like, while the reproductive structures that produce spores are plant-like. Now, true slime molds are recognized as a separate taxonomic group.

Characteristics. There are about 450 species of slime molds distributed worldwide. Most live in shady moist woods on decaying wood, leaves, or moss. Some occur in open, but moist areas, such as lawn grass. They can be found, also, on damp bark mulch used in gardens.

Their diet is bacteria, protozoa, fungal spores, and other minute organisms. The body of a slime mold is called a **plasmodium** (a), consisting of a membrane covering without cell walls, and containing many nuclei. Some plasmodia move in an amoeba-like fashion, flowing over surfaces, engulfing nutrients and other food sources.

When temperature or moisture conditions are unfavorable, the plasmodium may convert to a hardened mass in a resting dormant condition called a **sclerotium** (b). Or with a temporarily dry environment, the plasmodium may mature to the fruiting stage.

In the spring in north temperate regions slime molds can be found on wet rotting logs and examined with a hand lens. Throughout the summer, various forms of fruiting bodies can be discovered, such as those shown here.

Reproduction. The life cycle consists of the plasmodium fusing with other plasmodia, increasing the number of nuclei, which divide simultaneously; engulfing food; and eventually, converting into fruiting bodies called **fructifications**. The three types of fructifications formed by various species are sporangia, aethalia, and plasmodiocarps.

A **sporangium** may have a stalk (h) or not, with a base called a **hypothallus** (i), and a wall of varying lime content, the **peridium** (g), that encloses the spores. Inside the peridium may be a non-living network of hairs, **capillitium** (f), that aid in spore dispersal.

An **aethalium** is cushion-shaped with a peridium (o) enclosing the spores, such as *Lycogala* shown here.

The third type of fructification is a **plasmodiocarp**, which retains the net-shape of the plasmodium.

By reduction division (meiosis), 1n spores are produced. With favorable environmental conditions, the spore germinates into a myxamoeba without **flagella** (flagella, pl.; flagellum, sing. = whip-like structure for movement) or a swarm cell with flagella. These cells function as sex cells (gametes). Two cells fuse to form a 2n zygote nucleus, which divides mitotically to develop into a plasmodium.

Of interest... **blue lawn infestation:** *Physarum cinereum;* **research material:** slime molds provide pure, cell-less protoplasm for genetic, developmental, and cancer studies; **ecosystem:** primary decomposers in the forest food chain.

COLOR CODE

yellow:	plasmodium (a), sclerotia (b)
yellow-brown:	dried plasmodium (c), capillitium (f)
gray:	stalk(d), peridium (e)
tan:	peridium (g), wood (y)
dark brown:	stalk (h), hypothallus (i)
brown:	spores (j)
black:	stalk (k)
white:	lime dots on peridium (l), hypothallus (v), stalk (w)
dark brown:	peridium (z, p), stalk (m, q)
green:	moss (n)
pink:	peridium (o)
orange-brown:	hypothallus (r), stalk (s), peridium(t),capillitium(u)
iridescent (dots of green, red, yellow):	peridium (x)
orange:	plasmodiocarp

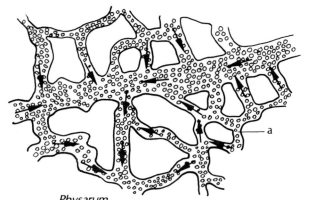

Physarum
plasmodium
arrows indicate protoplasmic flow
magnified

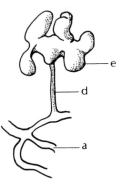

Physarum
sporangium
× 10

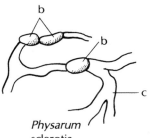

Physarum
sclerotia
× 10

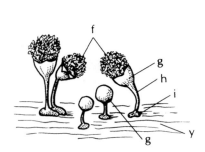

Hemitrichia
sporangia
× 10

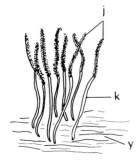

Stemonitis
sporangia
× 3

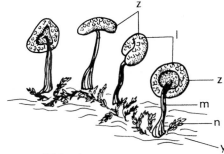

Didymium
sporangia
× 10

Lycogala
aethelia
× 2

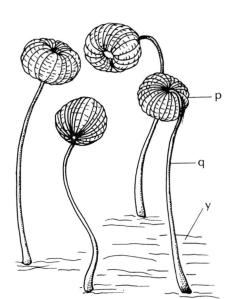

Dictydium
sporangia
× 20

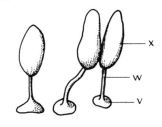

Diachea
sporangia
× 10

Arcyria
sporangia
× 10

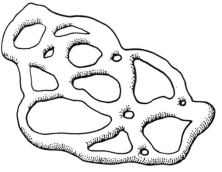

Hemitrichia
plasmodiocarp
× 10

Water Molds, Downy Mildews, White Rusts

In this Oomycete group, organisms are microscopic and found in water and moist soil. The most advanced forms live entirely within a plant or animal host. They range from one cell to copious amounts of threadlike strands. Each strand of a threadlike tubular filament is called a **hypha** (pl. hyphae). A mass of hyphae is called **mycelium**. The cell wall contains cellulose.

Reproduction is by oospores, sexually produced, nonmotile cells; and, asexually, by zoospores with two unlike flagella that are used for motility.

Of interest... *Albugo* spp. (causes white rust of horseradish, cabbage, sweet potato, morning glory, spinach), *Aphanomyces* (causes root disease of sugar beets, peas), *Phytophthora infestans* (causes late blight of potato), *Phytophthora ramorum* (causes sudden oak death), *Plasmopara viticola* (causes downy mildew of grapes). *Pythium debaryanum* (causes damping-off of seedlings), *Saprolegnia* (causes disease of fish eggs and fish).

Achyla

The water mold illustrated here has separate male and female individuals. Eggs (a) are produced in an egg chamber (oogonium, b) on the hypha (c). From a male individual, antheridia (d) with male gametes branch from a hypha (e). After fertilization by a male gamete from the antheridium, an egg develops into a 2n oospore (f). The oospore germinates to form a new body of hyphal stands (g) that produces biflagellated zoospores (h).

Chytrids and Allies

The Chytridiomycetes are simple, microscopic organisms that live in both water and soil. They may be a single cell living within the cell of a host alga or higher plant or have true mycelia and live on the surface of a host. The motile cells have one whiplash flagellum. The cell walls of chytrids are made up of chitin, a tough resistant carbohydrate. Some have cellulose in their cell walls also.

Of interest... Some chytrid parasites destroy algae. Some are parasitic on economic plants that include *Physoderma* (causes brown spot of corn), *Synchytrium* (causes black wart disease of potato tubers), and *Urophlyctis* (causes crown wart of alfalfa).

Allomyces

The 2n mycelium of this chytrid produces thin-walled spore-bearing structures called **sporangia** on hyphal strands (i). From a sporangium (j), 2n spores (k) emerge. Other, thick-walled sporangia (l) produce 1n spores (m) by **meiosis** (nuclear divisions that reduce the number of chromosomes by half). A 1n spore germinates into a 1n hypha (n), which produces 1n sex cells (gametes, o, p) in gametangia (q). Fusion of unlike gametes into a zygote (r) results in the germination of spore-bearing hyphae (i).

COLOR CODE

optional:	egg (a) gamete (o, p), antheridia (d), oospore (f), zygote (r)
gray:	oogonium (b), hypha (c, e, g, i, n)
yellow:	zoospores (h, k, m)
colorless:	sporangium (j, l), gametangia (q)

Achyla
Water Mold
reproduction

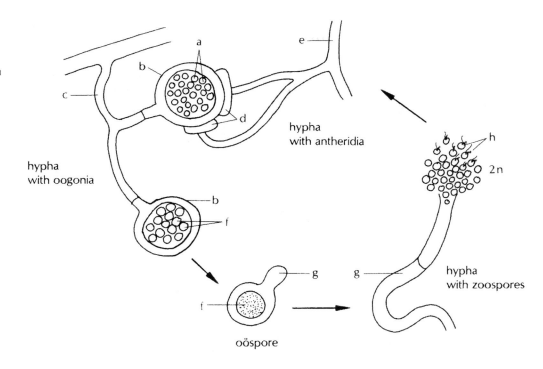

a
b
c
hypha
with oogonia
d
hypha
with antheridia
e
h
2n
b
f
f
oöspore
g
g
hypha
with zoospores

Allomyces
Chytrid
reproduction

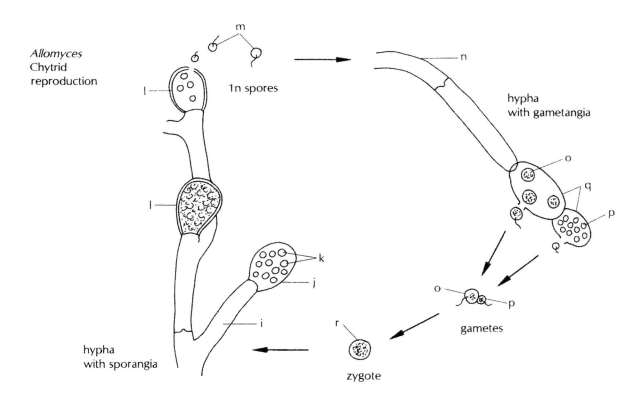

m
l
1n spores
l
k
j
i
hypha
with sporangia
n
hypha
with gametangia
o
q
p
Q
o
p
gametes
r
zygote

Fungi

Characteristics: Fungi are organisms without chlorophyll, and therefore, are dependent on other organisms for nutrition. A **saprophytic** fungus obtains food from dead organic matter. A **parasitic** fungus feeds upon a living organism, the host. **Symbiotic** fungi live in a mutually beneficial relationship with a host. A fungus associated with roots of a higher plant, called a **mycorrhizal** association, may be parasitic or symbiotic. In a symbiotic relationship both the plant and the fungus benefit. The fungal associate on plant roots produces digestive **enzymes** (proteases/peptidases) that release amino acids from proteins and absorb extra amounts of phosphorus from the soil, thence to the plant. The fungus benefits by using carbon—basal metabolites (e.g., sugars, amino acids, vitamins) from the plant roots.

Carl Zimmer reported in 1997 that Suzanne Simard, a forest ecologist, discovered a fungal mycelium connected to several trees and even trees of different species. A forest community of plants can be ecologically interconnected to the point where removal of so-called weed trees disrupts the survival of the entire group of plants. She concluded that the fungus extracted carbon from healthy trees and shared it with trees growing under unfavorable light conditions. It is thus important to maintain diversity in forests for the health of all the plants.

Fungi range in size from one-celled, microscopic organisms to masses of cells strung together in long filamentous strands. Each strand is called a **hypha**. The mass of hyphae that make up a fungal body is called **mycelium**. Cell walls are made up of chitin, which is hard and resists water loss.

Reproduction. All fungi bear spores that germinate into strands of hyphae. The microscopic lower fungi mainly live within a host and reproduce mostly by asexual spores. Higher fungi have elaborate fruiting bodies, composed of hyphae, in which spores are produced. Sexual reproduction, used as the basis of fungal classification, commonly occurs once a year.

Black Bread Mold and Allies

Their sexual resting spores (zygospores) define zygomycete fungi. Another characteristic is the asexual, non-motile, reproductive spore. Of interest . . . *Rhizopus stolonifer* (causes black bread mold, strawberry leak,

and sweet rot of potatoes), *Choanephora cucurbitarium* (causes squash flower and fruit mold); *Absidia* sp., *Mucor* spp., *Rhizopus* sp. (can be fatal to the human nervous system).

Mucor

Hyphae (a) of compatible mating types develop side branches (b) that meet. A swelling or progametangium (c) is produced at each side. The walls (d) between the two break down. The nuclei from each progametangium fuse together and form a zygospore (e) that is held by suspensors (f).

Sac Fungi

Ascomycete fungi sexually produce usually 4 or 8 ascospores (g) within a sac sporangium called an **ascus** (h). Asexual reproduction occurs by breaking off (fragmentation) of mycelium, by splitting of one cell into two cells (fission), by budding of cells, and by formation of conidia. Fungi in the Hemiascomycetidae have little or no mycelia and the asci are not produced in a spore-bearing structure or fruiting body. Yeasts have evolved from several evolutionary lines, but are grouped together for convenience. They are found in sugary substances such as flower nectar and on the surface of fruits, in soil, in animal wastes, in milk, and in other substances. They have the ability to ferment carbohydrates, producing alcohol and carbon dioxide, as in wine and beer making. The leaf curl disease-causing fungi are classified with the yeasts, as both may form buds on the **ascospores**. Of interest . . . *Saccharomyces cerevisiae* (baker's and brewer's yeast), *Taphrina* spp. (causes leaf curl or peach, chokecherry, oak, witch's broom of cherries).

Saccharomyces cerevisiae Yeast

Yeasts are microscopic, unicellular organisms (i). Asexual reproduction occurs by cell division and budding (j). Sexual reproduction, while rare, is by fusion of two cells to form an ascus (k) with 4 ascospores (l).

Taphrina deformans Peach Leaf Curl

Budding spores infect the peach leaf (m) producing more buds or mycelia, which penetrate the leaf tissue. The host tissue reacts to this infection by forming swollen, tumor-like masses (n) and by tortuous curling of the leaves.

Lower Fungi   Sac Fungi

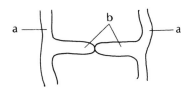
a — b — a

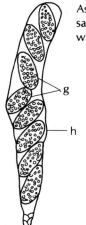
Ascus
sac sporangium
with 8 ascospores

g

h

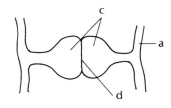
c
a
d

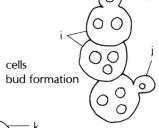
j
i
j
cells
bud formation

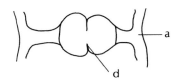
a
d

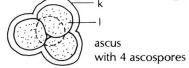
k
l
ascus
with 4 ascospores

Saccharomyces cerevisae
Yeast

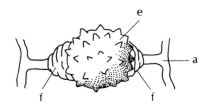
e
a
f   f

Mucor
zygospore formation

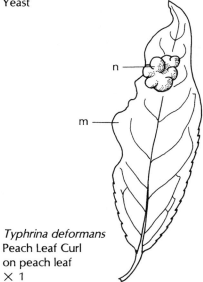
n
m

Typhrina deformans
Peach Leaf Curl
on peach leaf
× 1

COLOR CODE

gray: hypha (a), tips (b), progametangium
             (c), zygospore (e), suspensors (f)
yellow: ascospore (g, l)
beige: cell (i, j)
colorless: ascus (h, k)
brown: masses (n)
green: leaf (m)

Molds, Mildews, Morels (Sac Fungi)

Usually members of this Euascomycetidae group have asci (sac sporangia) enclosed in a fruiting body called an **ascocarp**. A cleistothecium, a perithecium, and an apothecium are types of ascocarps. A **cleistothecium** (c) is an enclosed ascocarp composed of loose masses of hyphae with asci irregularly arranged within. A **perithecium** (h) is a vase-shaped ascocarp with a pore (ostiole, i) at the top. It may be embedded in a compacted mycelial structure (stroma, j). An **apothecium** is a cup-shaped ascocarp with asci lining the open surface (see 54, *Xanthora*, j).

Of interest ... **molds:** *Aspergillus* spp. (mold on bread, leather; causes human skin and respiratory disease), *Aspergillus oryzae* is used to make sake from rice in Japan, *Claviceps purpurea* (ergot of rye), *Neurospora* (pink bread mold), *Penicillium italicum* (blue mold on citrus fruits and preserves), *P. digitatum* (green mold on citrus fruits), *P. camemberti* (flavoring in Camembert cheese), *P. roqueforti* (for flavor in production of Roquefort cheese), *P. notatum* and *P. chrysogenum* (source of the antibiotic, penicillin); *Rhizopus nigricans* (black bread mold); **wilts:** *Ceratocystis* spp. (oak wilt, Dutch elm disease); **blight:** *Endothia parasitica* (chestnut blight); **powdery mildews:** *Chaetomium* (of clothes), *Podosphaera leucotricha* (of apple), *Sphaerotheca pannosa* (of roses); **rot:** *Monilinia fructicola* (brown rot of stone fruits such as peach, plum, cherry, apricot); **wild:** *Helvella* and *Gyromitra* (false morels), *Morchella* (morel), *Tuber maganatum* (white truffle), *T. melanosporum* (black Perigord truffle), *Verpa* (bell morel).

Truffles form a mycorrhizal association with roots of oak, hazel, poplar and many other trees. Of the 200 species, two edible species are found in France and Italy. Hunters using trained pigs and dogs gather them. Small mammals (chipmunks, voles and squirrels) feed on truffles and disperse fungal spores in their fecal matter.

Penicillium

Asexual reproduction occurs by microscopic spores called **conidia** (a) formed on the side of a hyphal strand (b). By subjecting conidia of *Penicillium chrysogenum* to ultra-violet irradiation, surviving colonies have a high yield for penicillin production. It is a naturally occurring antibiotic. Discovery came when it was observed that in Petri dishes, bacterial colonies died when "contaminated" with *Penicillium*.

Microsphaera alni   Powdery Mildew of Lilac

The microscopic fruiting body is a cleistothecium. The wall (c) has been broken to reveal 2 asci (d) with ascospores (e). Appendages (f) aid in spore dispersal.

Diatrypella

In this fruiting body, asci (g) line perithecia (h) with terminal pores (i). They are embedded in a hyphal stroma (j).

Morchella   Morel

Easily recognized, this edible mushroom is prized for its distinctive flavor. Apothecia with asci line the cavities (k) of the spongy-looking fruiting body. In spring, morels are a mushroom hunter's favorite find in northern deciduous wooded areas.

Claviceps purpurea   Ergot of Rye

An ascospore germinates to form a mat of mycelium on a rye flower (see 123). Instead of a rye grain (m) developing, the fungal mycelium forms an overwintering asexual mass—a dark, hardened **sclerotium** (n). This sclerotium contains ergot alkaloids that poison cattle when grazing and humans when eating contaminated rye flour products. Also in humans, ergot causes gangrene.

Xylaria   Dead Man' Finger

Perithecia are embedded in the outer wall of the stroma (o) or "finger," which develops on rotten, burned wood. Mycelial filaments (p) are shown.

Daldinia concentrica

A vertical section reveals perithecia (q) lining the outer stromatic surface (r) of this cushion-shaped fruiting body.

Geoglossum difforme, Microglossum rufum Earth Tongues

Apothecia of asci form on the surface of the club-shaped fruiting bodies (s, t).

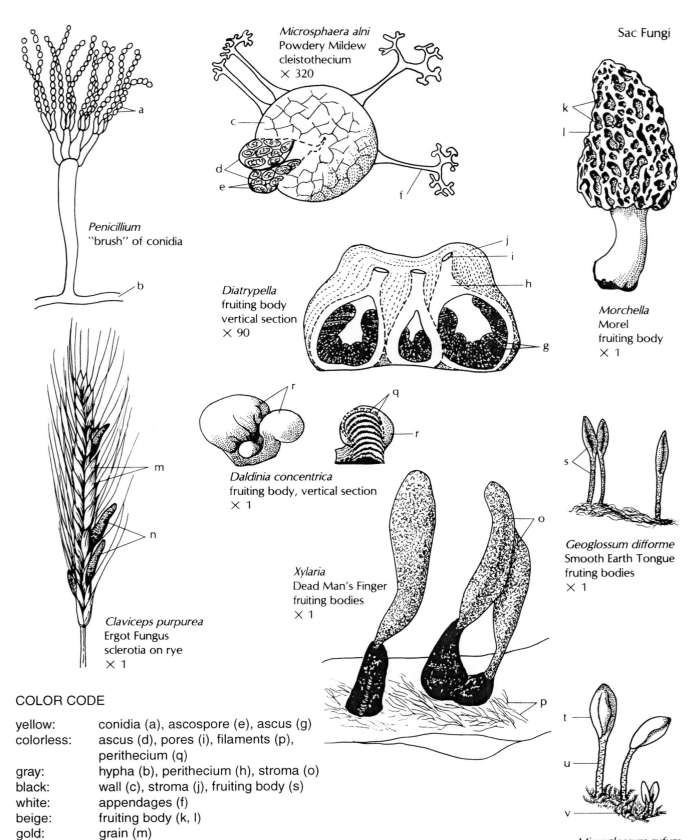

Penicillium
"brush" of conidia

Microsphaera alni
Powdery Mildew
cleistothecium
× 320

Sac Fungi

Diatrypella
fruiting body
vertical section
× 90

Morchella
Morel
fruiting body
× 1

Daldinia concentrica
fruiting body, vertical section
× 1

Geoglossum difforme
Smooth Earth Tongue
fruiting bodies
× 1

Xylaria
Dead Man's Finger
fruiting bodies
× 1

Claviceps purpurea
Ergot Fungus
sclerotia on rye
× 1

Microglossum rufum
Earth Tongue
fruiting bodies
× 1

COLOR CODE

yellow:	conidia (a), ascospore (e), ascus (g)
colorless:	ascus (d), pores (i), filaments (p), perithecium (q)
gray:	hypha (b), perithecium (h), stroma (o)
black:	wall (c), stroma (j), fruiting body (s)
white:	appendages (f)
beige:	fruiting body (k, l)
gold:	grain (m)
dark brown:	sclerotium (n), stalk (u)
purple:	stroma (r)
orange-red:	fruiting body t)
green:	moss (v)

Rusts, Smuts, Jelly Fungi
(Club Fungi)

For club fungi sexual reproduction results in usually 4 basidiospores (a) produced on the outside of a microscopic club-shaped sporangium called a **basidium** (b). When basidia are borne in a fruiting body, it is called a **basidiocarp**. Familiar forms are mushrooms, shelf fungi and puffballs. Also included are rusts, smuts, jellies, stickhorns, bird's-nest fungi, coral fungi and earth stars. There are many edible species; however, there are no rules, only myths, for distinguishing poisonous from edible fungi. There are about 22,000 species of basidiomycetes.

Rusts (Uredinales). Fungi in this order are **obligate parasites** (require living hosts) on plants. No fruiting bodies are produced. Classification is based on the presence of teleutospores, representing the overwintering, resting stage. Many rusts attack only specific plants. Some require two hosts to complete their life cycles. For example, wheat rust alternates between wheat (*Triticum*) and barberry (*Berberis*)—unrelated hosts. This is called **heteroecism** (different homes).

Of interest . . . *Cronarium ribicola* (white pine blister rust with currants and gooseberries as the alternate hosts), *Gymnosporangium* spp. (rusts of juniper, apple, hawthorn, and pear), *Piccinia* spp. (rusts of wheat, barley, oats, rye, and hollyhock), *Uredinopsis osmundae* on alternate hosts of balsam fir (*Abies balsamea*) and cinnamon fern *(Osmunda cinnamomea).*

Gymnosporangium juniperi-virginianae

The life cycle of this fungus is completed by alternation of the fungus between apple *(Malus)* and red cedar *(Juniperus virginiana)* host plants. The cedar apple gall (c) is a mass of mycelium on cedars (d). During spring, gelatinous structures, telia (e), expand with water uptake (e.g., from rainfall) and protrude from the surface. Inside a telium are hundreds of teleutospores, basidial structures that produce basidiospores after germination on apple trees.

Smuts (Ustilaginales). Smuts form masses of black, sooty-looking teleutospores. Smuts are similar to rusts but do not need living host material and can be grown in the laboratory. Of interest . . . *Tilletia foetia* (bunt or stinking smut of wheat), *Ustilago* spp. (smut of corn, oats), *Urocystis cepulae* (onion smut). *Ustilago maydis* (corn

smut) is sold as a gourmet food called "corn mushroom." Economically, smuts cause extensive damage to crops and consequent reductions in yields.

Urocystis cepulae   Onion Smut

Black teleutospores (f) form an infestation on bulbs of onions (g). This is commonly seen on commercially sold onions.

Jelly Fungi (Auriculariales). Fruiting bodies of gelatinous material are produced. When wet, basidia are formed. Most genera are saprobic, living on dead organic matter. A few genera are parasitic on mosses and flowering plants. Of interest . . . *Herpobasidium deformans* (blight of honeysuckle).

Auricularia auricula   Ear Fungus

The gelatinous fruiting body (h) resembles a human ear.

Jelly Fungi (Tremellales). Fruiting bodies are gelatinous with leaf-like folds that vary in color with species from white, pink, yellow, and orange to red.

Tremella mesenterica

This jelly fungus (i) is found on hardwood tree bark (j). Some species are edible.

Jelly Fungi (Dacrymycetales). Waxy, bright-colored gelatinous fruiting bodies are produced. The yellow coloring is due to β-carotene pigment. They grow on living or dead trees.

Dacryopinax spathularia

This jelly fungus (k) is found on rotting wood (l).

COLOR CODE

yellow:	basidiospore (a), onion (g)
tan:	basidium (b), fruiting body (h)
orange:	telium (e), fruiting body (k)
brown:	gall (c), wood (l)
green:	cedar (d)
pale yellow:	fruiting body (i)
gray:	bark (j)

Basidiocarp
gill section

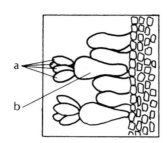

a
b

Basidia
with basidiospores
× 450

Gymnosporangium
gall on cedar
× 1

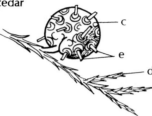

c
e
d

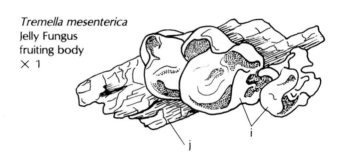

h

Auricularia auricula
Ear Fungus
fruiting body
× 1½

Tremella mesenterica
Jelly Fungus
fruiting body
× 1

i
j

f
g

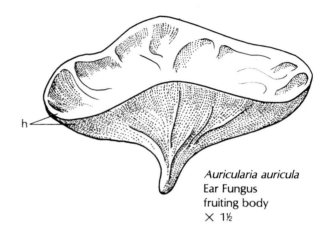

k
l

Urocystis cepulae
Onion Smut
on onion
× 1

Dacryopinax spathularia
Jelly Fungus
fruiting body
× 1

Gill Fungi

Gill, Pore, Coral, and Toothed Fungi (Agaricales). In this order, a basidium usually produces 4 basidiospores, which are forcibly discharged from the fruiting body. Fruiting bodies may occur in forms commonly known as mushrooms, toadstools, shelves, corals, and toothed fungi.

A "spore print" (basidiospore collection) can be made by placing a fruiting body cap with gills or pores half on white paper, half on black paper. If dark spores are discovered, use an entire cap on white paper and if light spores use black paper. Spore color is useful for identification. If can be white, pink, yellow, brown, purple, or black.

Of interest... **food:** *Agaricus campestris* (commercial mushroom), *Calvatia gigantea* (giant puffball), *Cantharellus cibarius* (chanterelle), *Lentinus edodes* (shiitake, high in vitamin B$_{12}$, cobalamin; grown commercially on white oak logs to produce fruiting bodies for up to 10 years after inoculating the mycelium in holes drilled in the logs. Logs are stacked crisscrossed like Lincoln logs in the shade of deciduous forests and watered frequently with sprinklers. This technique is widely used in Korea and now in the U.S. by Korean entrepreneurs.), *Pleurotus ostreatus* (oyster mushroom, this mushroom is grown commercially on vertically stacked sections of American elm, *Ulmus americana,* trunks covered with black plastic film.);

wild: *Amanita verna* (destroying angel, contains a deadly poison that destroys liver and kidney cells), *Armillaria mellea* (honey mushroom, causes root rot of trees, exhibits bioluminescence or "fox-fire" in mycelia-penetrated organic matter), *Boletus edulis* (porous mushroom), *Ganoderma applanatum* (shelf or bracket fungus, a type of artist's fungus, so called because marks made on the smooth white underside will remain when the mushroom dries), *Marasmius oreades* (fairy ring mushroom), *Polyporus sulphureus* (sulfur mushroom, causes wood rot of trees), *P. squamosus* (heart rot of trees), *Psilocybe* spp. (hallucinogenic mushrooms). Psilocybe mushrooms in Mexico are collected, dried, and served in pairs by women in all-night religious ceremonies of Amerinds, as documented by the famous Harvard University botanist, Charles Evans Schultes.

Amanita muscaria   Fly Amanita, Fly Agaric

The fruiting body of a gill mushroom begins as a mass of hyphal cells and enlarges into a "button." At this stage, the fruiting body is surrounded by a **universal veil** (d). As expansion of the gills (**lamellae**) takes place, the veil is broken. The veil may remain as patches or scales (b) on the cap (**pileus**) and as veil remnants (c) or as a cup (**volva**) at the base of the stalk (**stipe**).

Some species of agarics also have an inner or **partial veil** (d) between the developing stalk (e) and gills (f). When the partial veil is broken upon expansion of the gills, it may remain attached to the stalk as a ring (**annulus**, g) or remain attached to the margin of the cap, hanging down as a curtain, as in *Cortinarius*.

The microscopic basidia are borne on the inner surface of the gill. Individual cap (h) color in this species varies from white, yellow, and orange to red. This is a poisonous mushroom, found mostly in mycorrhizal association (see 12) with roots of aspen and conifer trees.

Cortinarius

Remains of the partial veil, which hangs down from the cap as a curtain (cortina), form web-like lines (i) on the stalk (j). As the cap expands, gills (k) are exposed. This fruiting body was found in a beech-maple forest.

COLOR CODE

white:	scales (b), veil remnants (c), partial veil (d), stalk (e), gills (f), annulus (g)
orange-red:	cap (h)
light tan:	stalk (j), gills (k)
yellow:	cap margin (l)
yellow-brown:	cap center (m)

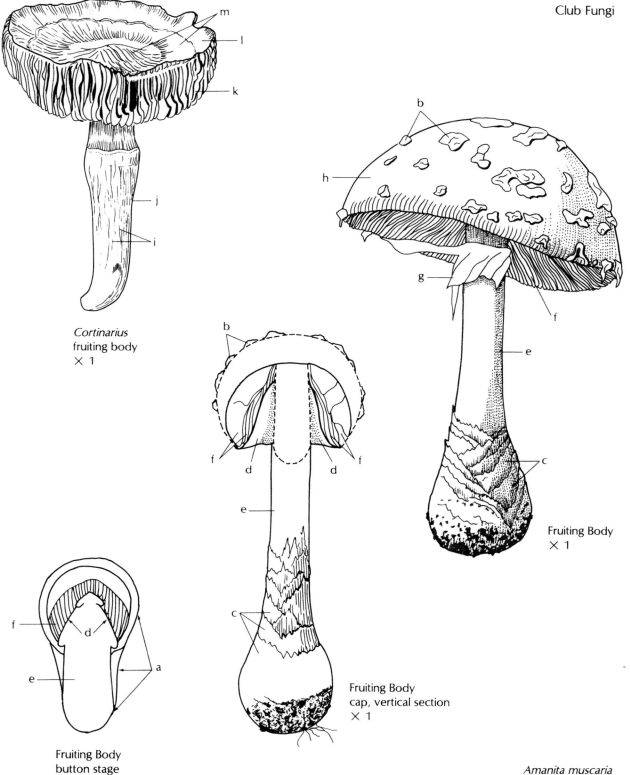

Cortinarius
fruiting body
× 1

Fruiting Body
button stage
vertical section
× 1

Fruiting Body
cap, vertical section
× 1

Fruiting Body
× 1

Amanita muscaria
Fly Amanita

Gill and Pore Fungi

Agaricales (continued)

Coprinus comatus   **Shaggy Mane**

As the white fruiting body matures, the gills undergo self-digestion with black basidiospores released in an inky black fluid. This mushroom is found in lawns, golf courses and fields and is edible and preferred in the button stage before dripping black fluid. Remnant of a veil (a) remains on top. Caution: when eating this mushroom, alcoholic drinks should not be taken; otherwise, you can become quite ill!

Chantharellus cinnabarinus   **Cinnabar Chanterelle**

Basidiospores are produced on blunt gills (b). With age, the cap (c) becomes funnel-shaped on the fruiting body stalk (d). This mushroom is edible and found in forests throughout the United States and Canada. It is considered a gourmet treat, sold in grocery markets and used in restaurant offerings.

Russula emetica   **Emetic Russula**

The fruiting body has a stout stalk (e) supporting a bright red cap (f) with white, brittle gills (g). This fragile fruiting body is probably poisonous.

Suillus americanus

Instead of gills, this mushroom has fleshy pores (h) and is therefore designated as a **bolete**. Basidia line the inner surfaces of layers of tubes with pore openings. The fruiting body is fleshy and decays quickly. Brown scales (i) dot the yellow cap (j). It is edible, but little flesh remains after the tubes have been removed.

COLOR CODE

light tan:	veil remnant (a)
dark orange:	gills (b), cap (c), stalk (d)
pale pink:	stalk (e)
red:	cap (f)
white:	gills (g)
yellow:	pores (h), cap (j), stalk (k)
brown:	scales (i)
green:	moss (l)

Club Fungi

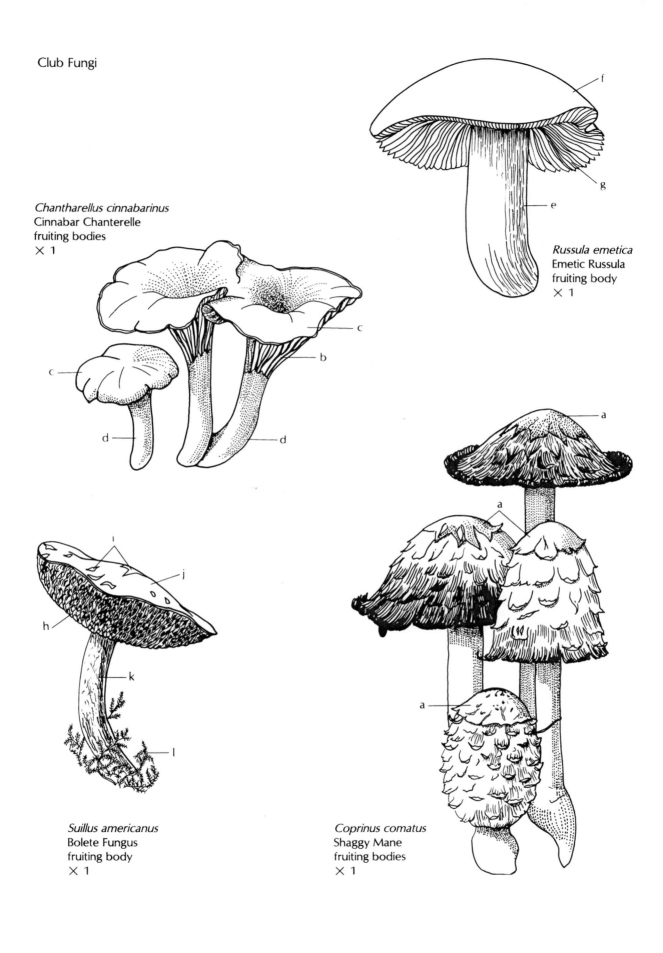

Chantharellus cinnabarinus
Cinnabar Chanterelle
fruiting bodies
× 1

Russula emetica
Emetic Russula
fruiting body
× 1

Suillus americanus
Bolete Fungus
fruiting body
× 1

Coprinus comatus
Shaggy Mane
fruiting bodies
× 1

Pore, Coral, and Toothed Fungi

Agaricales (continued)

Coriolus (Polyporus, Trametes) versicolor
Turkey-tail Fungus

In polypore (= many pores) fungi, basidia are formed on the inner surface of leathery pores (a) on the lower portion of the fruiting body (b). This fan-shaped bracket fungus is found on rotting wood. The bands of tan, brown, white are like the colors of a turkey's tail feathers, and thus account for the common name.

Fomes formentarius   **Rusty-hoof Fungus**

This perennial polypore adds new growth (c) yearly to the woody bracket-type fruiting body (d). It is parasitic on birch (*Betula*) and beech (*Fagus,* e) and some other hardwood deciduous trees. It is shaped like a horse's hoof and is common in northern temperate zones.

Irpex lacteus   **Crust Polypore**

A crust-like fruiting body (f) of this polypore grows on the lower surface of dead tree limbs (g).

Clavulinopsis fusiformis   **Coral Fungus**

The fruiting body (h) in this coral fungus is simple and unbranched.

Ramaria stricta   **Straight Coral Fungus**

This coral fungus has a more complexly branched fruiting body (i). It is usually found on rotting beech (*Fagus*) wood.

Hericium   **Toothed Fungus**

Basidia are produced on downward-pointed, tooth-like projections on the fruiting body (j). It is found on dead deciduous hardwood trees.

COLOR CODE

tan-brown-white bands:	fruiting body (a, b)
rust (yellow-brown-red):	new growth (c)
dark and light gray bands:	fruiting body (d)
brown:	bark (e)
yellow-tan:	fruiting body (f, i)
gray:	bark (g)
orange:	fruiting body (h)
white:	fruiting body (j)

Club Fungi

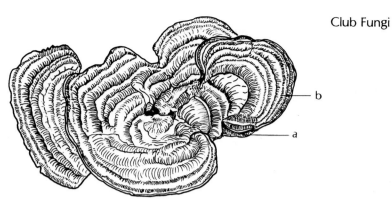

Coriolus versicolor
Turkey-tail Fungus
fruiting body
× 1

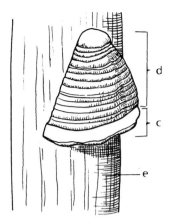

Fomes fomentarius
Rusty-hoof Fungus
fruiting body
× ½

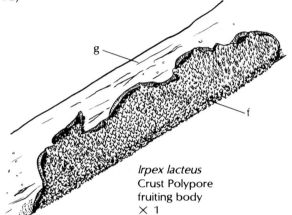

Irpex lacteus
Crust Polypore
fruiting body
× 1

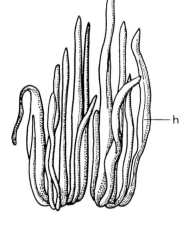

Clavulinopsis fusiformis
Coral Fungus
fruiting body
× 1

Ramaria stricta
Straight Coral Fungus
fruiting body
× 1

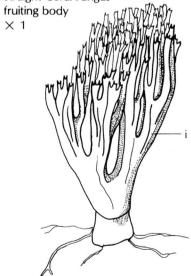

Hericium
Toothed Fungus
fruiting body
× 1

Puffballs, Stinkhorns, Bird's-nest Fungi

Puffball Fungi (Lycoperdales). Spores are dispersed by wind from puffball fruiting bodies. An inner fertile portion, called the **gleba**, is composed of basidiospores and sterile thread-like structures (**capillitia**). Powdery masses of spores are exposed when the puffball wall (**peridium**) disintegrates, or emerge in a cloud from an opening (**ostiole**) when the flexible peridium is disturbed.

Of interest ... *Calvatia gigantea* (giant puffball) found in mostly open or sometimes wooded areas and after rains in late summer or early autumn and is edible when pure white inside, *Geastrum* and *Astraeus* (earthstars), *Lycoperdon* (stalked puffball).

Geastrum   Earthstar

This type of puffball has a star shape when the rigid outer wall or peridium (a) becomes wet and splits open. An inner flexible peridium (b) is exposed. A single pore (ostiole, c) for spore dispersal occurs at the center.

Lycoperdon   Stalked Puffball

At maturity, a powdery gleba of spores (d) puffs out of an ostiole (e) in the peridium (f) due to rain drops or other physical disturbances.

Stinkhorn Fungi (Phallales). Basidiospores are formed within a gleba. A receptacle, bearing the gleba, emerges from a protective enclosure, the peridium. As the gleba breaks down, the basidiospores are exposed in a gelatinous mass at the top of the stinkhorn. The foul smell of the mass attracts flies, which disperse the spores.

Of interest ... *Dictyophora* (collared stinkhorn), *Clathrus* (net stinkhorn), *Mutinus* (dog stinkhorn), *Phallus* (common stinkhorn)

Dictyophora duplicata Collared Stinkhorn

At the egg stage, the peridium (g) encloses the gleba (h) where basidia and basiospores are formed. The receptacle (i) and stalk (k) will enlarge and emerge from the peridium, which remains at the base.

Mutinus caninus   Dog Stinkhorn

From the broken peridium (k), a pink stalk (l) emerges. The dark pink receptacle (m) at the top of the fruiting body is covered by a green, slimy mass (gleba, n) of spores.

Bird's-nest Fungi (Nidulariales). The fruiting body appears as a hollow nest filled with egg-like peridioles, which contain basidiospores. The nest acts as a splashcup. Raindrops cause the periodioles to bounce out of the fruiting body. The periodiole's sticky covering adheres to nearby surfaces where the spores are then released. This type of fungus is often found on rotting bark mulch used in gardening and on deciduous forest duff litter.

Of interest ... *Nidularia* (white bird's-nest fungus), *Crucibulum* (common bird's nest fungus), *Cyathus* (striate bird's-nest fungus).

Crucibulum vulgare   Common Bird's-nest Fungus

This genus is often found on wood chip (o) paths and on the forest "floor." Small peridioles (p) are formed in the nest-like fruiting body (q).

Of interest ... many of these fungi are important in forest tree communities in causing the breakdown of leaf/twig litter and dead tree branches to form organic humus (like compost) over the forest floor. This forms a great substratum for forest tree seed germination and subsequent development of seedlings.

COLOR CODE

tan:	peridium (a, b, f, g, q)
green-brown:	gleba (h, n)
white:	receptacle (i), stalk (j), peridium (k), periodiole (p)
light pink:	stalk (l)
dark pink:	receptacle (m)
brown:	wood chips (o)

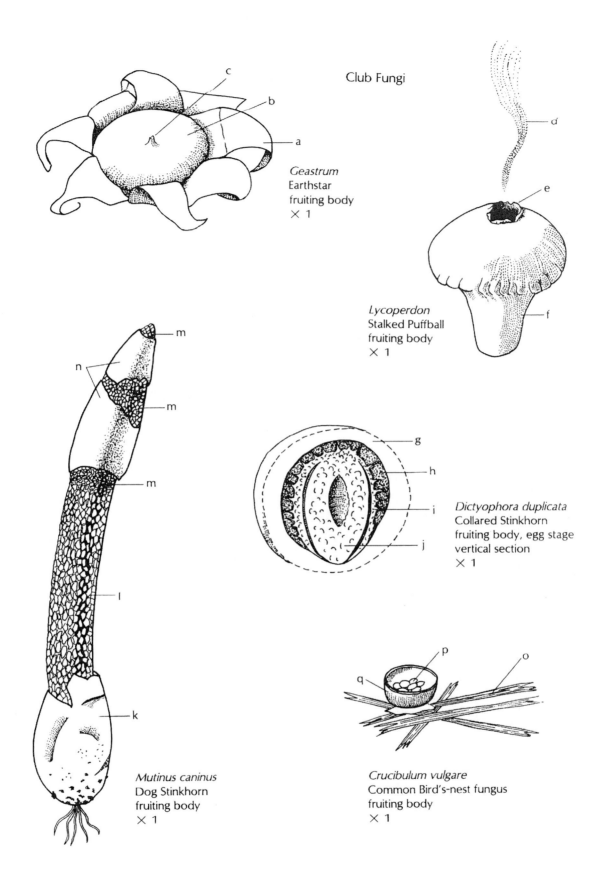

Club Fungi

Geastrum
Earthstar
fruiting body
× 1

Lycoperdon
Stalked Puffball
fruiting body
× 1

Dictyophora duplicata
Collared Stinkhorn
fruiting body, egg stage
vertical section
× 1

Mutinus caninus
Dog Stinkhorn
fruiting body
× 1

Crucibulum vulgare
Common Bird's-nest fungus
fruiting body
× 1

Lichens

A **lichen** is an association of a fungus (mycobiont, mykes = fungus + bios = life) and an alga (phycobiont, phykos = alga + bios = life). The fungus is an ascomycete or a basidiomycete. The algal partner is either a blue-green or a green alga. The association can be **parasitic**, in which the fungus eventually kills the algal cells; or it can be **mutualistic**, in which both fungus and algal cells benefit. Mutual benefits include algal nutrition that benefits the fungus because of the presence of chlorophyll for photosynthesis in the alga. And protection for the smaller alga is provided by the fungus. Lichens can tolerate harsh conditions such as dryness, arctic cold, and bare rock habitats. They possibly live for thousands of years.

Characteristics. The upper cortex (a) is the outer protective layer with fungal hyphae gelatinized and cemented together. Next is the algal layer (b). Most lichens are associated with the unicellular green alga, *Trebouxia.* So far, over 30 genera of algae have been found associated with lichens. Most of the lichen body (thallus) is made up of medulla (c) with hyphae loosely interwoven. The medulla retains moisture and stores food.

The thallus may take different forms. **Crustose lichens** are extremely flat and only ascocarps (fruiting bodies of the fungus, see 48) may be visible. (See drawing opposite of *Graphis scripta*). They are found on rocks and trees. The medulla may be embedded in rock surfaces. **Foliose lichens** have a leafy thallus and may have a compacted lower cortex (d) and rootlike anchoring devices called **rhizines** (e). (See drawings opposite of *Xanthoria* and *Physcia aipolia*.) **Fruticose lichens** with three-dimensional projections, called podetia, arise from a scale-like, loosely-attached (squamulose) base. (See drawings opposite of *Cladonia cristatella* and *Cladonia pyxidata*.)

Reproduction. Various vegetative structures may be found on lichens, such as soredia and isidia. **Soredia** originate in the algal layer as masses of algae cells with a few gelatinized hyphae erupting through cracks in the surface cortex. By a process of breaking off, new lichen thalli are formed. An **isidium** is a projection on a foliose or fruticose lichen that serves as a vegetative propagule when broken off. Depending on the type of fungus, fruiting bodies with ascospores or basidiospores are produced.

Lichens also have asexual structures called pycnidia, flask-shaped structures that produce microconidia. The fungal spore or microconidium must form an association with an appropriate alga in order to live.

Of interest . . . **arctic animal food:** *Cetrarias, Cladonia* spp. (reindeer mosses), *Usneas;* **human skin rash:** *Evernias, Usneas;* **dye:** *Letharia, Ochrolechia, Parmelia, Roccella;* **ecosystem succession role:** lichens are often the first colonizers of bare substrate, breaking up rocks, due to their secretions of lichen acids; thalli form an anchor base for later colonizing plants.

Sac Fungi: Euascomycetidae (Lecanorales)

Graphis scripta   Script Lichen

Found on hardwood trees, this crustose lichen thallus (f) has dark eruptions of apothecia (g).

Xanthoria

Found on rotten wood, the thallus (h) of this foliose lichen has apothecia composed of gray-green receptables (i) with brightly colored ascopsore layers (hymenia, j).

Physcia aipolia   Blister Lichen

This foliose lichen has a white-spotted thallus (k) with apothecia emerging from the surface cortex. The fertile layer of asci (hymenium, l) is gray with a gray-green receptacle (m).

Cladonia cristatella   British Soldiers

The fruticose podetium (n) arises from a squamulose thallus (o). Red apothecia (p) dot the top of the podetia.

Cladonia pyxidata   Pyxie Cup

The fruiticose podetia (q) resemble pyxie cups arising from a squamulose thallus (r). Brown apothecia (s) dot the cup rims.

COLOR CODE

gray:	cortex (a, d), thallus (f), hymenium (l)
green:	algae (b)
white:	medulla (c)
black:	rhizines (e), apothecia (g)
gray-green:	thallus (h, k, o, r), receptacle (i, m), podetium (n, q)
yellow-orange:	hymenium (j)
red:	apothecium (p)
brown:	apothecium (s)

Lichens

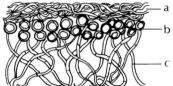

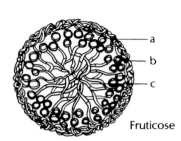

Fruticose

Lichen Structure
diagrams

Crustose, Squamulose

Foliose

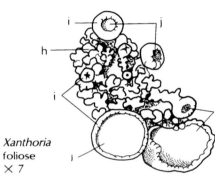

Xanthoria
foliose
× 7

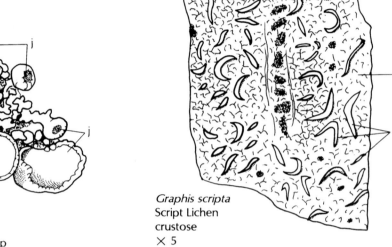

Graphis scripta
Script Lichen
crustose
× 5

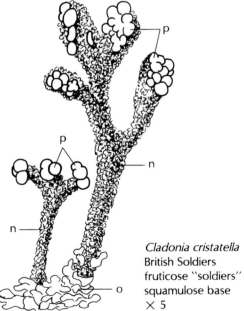

Cladonia cristatella
British Soldiers
fruticose "soldiers"
squamulose base
× 5

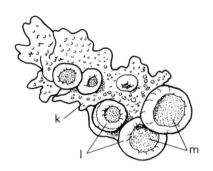

Physcia aipolia
Blister Lichen
foliose
× 7

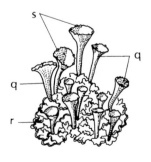

Cladonia pyxidata
Pyxie Cup Lichen
fruticose "cups"
squamulose base
× 1

Dinoflagellates

Fossils of dinoflagellates extend back to the Ordovician time of the Paleozoic. Most are marine.

Characteristics. These microscopic organisms are unicellular and are usually motile due to the presence of a pair of unequal flagella. There are diverse forms. Nutrition varies from self-feeding (**autotrophic**) to mutually beneficial (**mutualistic**) to dependency upon a host (**parasitic**). Pigments are chlorophylls a and c, carotene, and the xanthophyll, peridinin, which gives a gold-brown color to the organisms. Starch is the primary stored food.

The nucleus is large and the chromosomes remain visible at all stages (see 6), while the nuclear membrane remains intact. Some dinoflagellates have light-sensitive eyespots. Other marine species are bioluminescent: when cells are vigorously agitated, they give off flashes of light.

Reproduction. Asexual reproduction occurs by cell division and cyst formation. Sexual reproduction is by sex cells (gametes) that look alike (**isogamy**) or unlike (**anisogamy**).

Of interest . . . **"blooms"** or **"red tides:"** under optimal conditions rapid reproduction produces heavy concentrations of dinoflagellates that color the ocean red, red-brown, or yellow; **toxins:** these "blooms" can be toxic to other organisms in different ways: killing only fish, killing mainly invertebrates, or not killing but concentrating their toxins in bivalve mollusks. The toxins produced by species of *Gonyaulax* are thousands of times more potent than cocaine, and may cause human death within 12 hours after consumption of affected bivalves, but in low concentrations, their toxin usually produce severe digestive upsets.

Dinophyceae. This class of dinoflagellates is made up of biflagellated motile cells. One flagellum is coiled around the transverse girdle groove (cingulum), providing a rotating movement. The other flagellum is located in a longitudinal groove (sulcus); it functions as a propellant from the posterior end.

Gymnodinium

The name means "naked whorl," as it is without armored plates. Structures shown are the girdle groove (a) with flagellum (b) and the sulcus groove (c) with flagellum (d). Many chloroplasts (e) are visible.

Ceratium

The name means "horned." Armored plates (f) ornamented with pits cover the organism. Structures shown are the girdle groove (g) with flagellum (h) and the sulcus groove (i) with flagellum (j).

COLOR CODE

gold-brown: chloroplasts (e), plates (f)

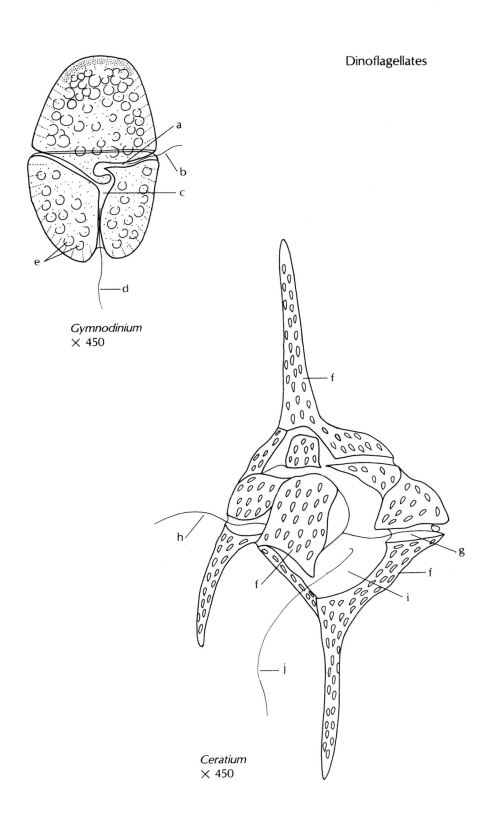

Gymnodinium
× 450

Ceratium
× 450

Golden Algae, Yellow-green Algae, Diatoms

This diverse group of algae has several common features: a complex carbohydrate food reserve called chrysolaminaran (leucosin); more carotene pigments than chlorophylls, thus the gold and yellow-green colors; and chlorophylls a and c. The pigment, fucoxanthin, occurs in some of the divisions. Forms range from amoeboid cells, flagellated cells, filaments, to parenchymatous thalloid types.

Golden Algae (Chrysophyceae). Members in this group live mainly in fresh water. There are a few marine forms. Some have cell coverings that may contain silica. Two-part resting spores (statocysts) are formed when unfavorable conditions occur. Asexual reproduction is by fragmentation and production of an asexual reproductive cell (zoospore). Sexual reproduction is not well-known.

Dinobryon

This fresh water alga forms free-swimming colonies of biflagellated cells. The protoplast (a) is connected by a thin strand to its protective covering: a transparent, open-ended, vase-shaped lorica (b). The protoplast has 2 flagella (c) of unequal length, an eyespot (d) and a chloroplast (e).

Yellow-green Algae (Xanthophyceae). Fresh water habitats are typical for this group. They form a scum in standing water, and some occur on the surface of moist mud and on tree trunks. There are less than 100 known species.

Vaucheria

This alga is a branching filament (f) containing many chloroplasts (g). It may form mats under moist, greenhouse clay pots. The filament has no cross-walls except where reproductive chambers are formed. Reproduction is by asexual zoospores and by the sexual union of a small motile sperm with a large non-motile egg (**oogamy**). An antheridial branch (h) containing sperm cells (i) forms laterally on a filament. A biflagellated sperm cell swims to a pore (j) in the oogonium (k) to fertilize the egg (l). The fertilized egg develops into an oospore (m), which remains in the oogonium as a resting spore until germination occurs.

Diatoms (Bacillariophyceae). Two hundred surviving genera of unicellular diatoms are known, with as many as 100,000 species. Diatom walls (**frustules**) have the structure of a narrow box with a lid. The lid, called the epivalve, fits over the box or hypovalve. The overlapping area is the girdle region. Reproduction is by cell division. It involves separation of the epivalve and hypovalve, with new, smaller container walls being formed. When an ultimate, minimum size reduction is reached, new cells are formed by the fusion of the protoplasm of two cells. Diatoms also have sexual reproduction.

Of interest ... **ecosystem:** diatoms, the major component of plankton—which form the base of aquatic food chains, are the single most important group of algae. In relation to other families, diatoms also supply the major percentage of the world's oxygen supply. They are affected by temperature, pollution, and light. Some characteristic species of diatoms and other algae are indicators of polluted water, while other species are indicative of clean, non-polluted water; **fossils: diatomaceous earth** is represented by the silica remains of diatoms; it is used for filters, abrasives, insulation, in pavement paints, and as an indicator of oil- and gas-bearing formations.

Pinnularia

The epivaluve (n) portion fits over the hypovalve (o) as shown in this diatom diagram.

Asterionella

This diatom forms a star-shaped colony of cells (p).

Coscinodiscus

Surface striations (q) of puncture holes mark the wall (r).

COLOR CODE

red:	eyespot (d)
yellow:	protoplast (a), diatoms (n, o, p, q, r)
yellow-brown:	chloroplast (e)
yellow-green:	filament (f), oospore (m)
tan:	sperm cells (i), egg (l)

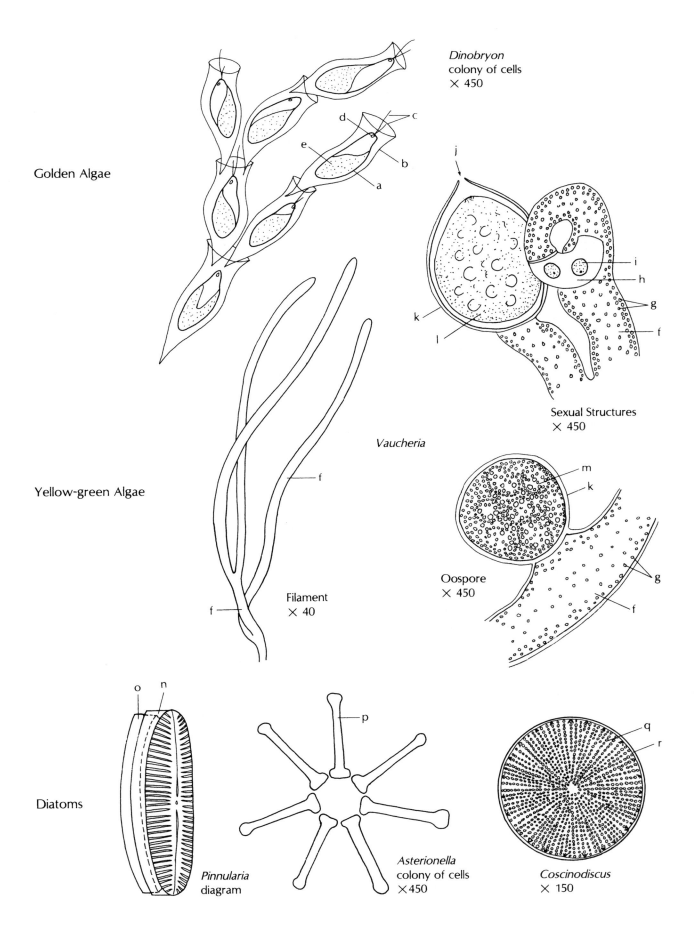

Golden Algae

Dinobryon
colony of cells
× 450

Sexual Structures
× 450

Yellow-green Algae

Vaucheria

Filament
× 40

Oospore
× 450

Diatoms

Pinnularia
diagram

Asterionella
colony of cells
× 450

Coscinodiscus
× 150

Red Algae

The red algae, with about 400 genera, are mostly marine with only a few known freshwater genera. Most are found in tropical areas where they occur at great depths or along intertidal regions.

Characteristics. Their color varies greatly depending on the amount of pigments and can range from shades of green, red-brown, bright red, blue, and purple-blue to black. Usually, the green pigments, chlorophylls a and d, are masked by the phycobilin pigment (red phycoerthyrin and blue phycocyanin). Other pigments include carotenes and xanthophylls (see 4).

Red algae are relatively simple vegetative organisms but have complex reproductive systems. The primary food reserve is floridean starch (a polysaccharide = carbohydrate made up of many glucose units). There are a few microscopic unicellular forms, but most have branched filaments or flat leaf-like forms.

Reproduction. Asexual reproduction occurs by fragmentation and production of vegetative spores. Red algae are divided into two subclasses. In the more primitive Bangiophycidae, sexual reproduction is mostly absent. In the more advanced Florideophycidae, sexual reproduction is oogamous: non-flagellated male gametes (**spermatia**) are carried by water to a stationary female gamete-bearing structure called the **carpogonium**.

Of interest ... **colloids:** *Chondrus crispus* (**carrageenan**, used as a stabilizing agent in milk products, jellied foods, cosmetics, insect sprays, and water-based paints), *Gloiopeltis* (**funori**, used as a water-soluble sizing in water-based paints and as a flexible starch in laundering); **agar:** *Gelidium* (agar, used as a culture medium in microbiology, as non-irritant bulk in human intestinal disorders, and to make pill capsules); **coral reefs:** *Corallinna, Lithothamnium,* and *Galaxaura* accumulate calcium carbonate, which contributes to reef-building; **ecosystem:** red algae provide food for marine animals; **human food:** *Chrondrus crispus* (Irish moss), *Porphyra* (nori), *Palmaria* (dulse).

Porphyra   Nori

This genus, in the subclass Bangiophycidae, has a leafy thallus (a) with rhizoidal cells at the base for attachment to rocks.

Polysiphonia

Placed in the subclass Florideophycidae, this genus has a filamentous thallus (b) constructed of many (poly) tubes (siphons, c); hence its name. Male gametophyte individuals produce special branchlets called **trichoblasts** (d), that bear spermatia (e). Female gametophyte individuals produce urn-shaped structures called **cystocarps** (f), which contain egg cases (**carpogonia**). After fertilization and maturation, carpospores (g) are released. Carpospores develop into a tetrasporophyte generation, producing tetraspores that germinate into male or female gametophytes (gamete-bearing). Both the tetrasporophyte and the gametophyte generations have the same external filamentous appearance.

Of interest ... to preserve filamentous marine plant specimens, place the plant in a large shallow pan of water and allow it to float naturally. Then slide art paper, bond paper, or, ideally, rag paper under the specimen. Slowly lift the paper with the plant on top and drain excess water off one end. Blot the bottom of the paper on newsprint, then place plant-on-paper in a single folded sheet of newspaper. Stack with blotters and newspaper and press with weights or place in a plant press until dry.

COLOR CODE

red-purple:	thallus (a)
red:	thallus (b), tubes (c), trichoblast (d, e), cystocarp (f), carpospores (g)

Red Algae

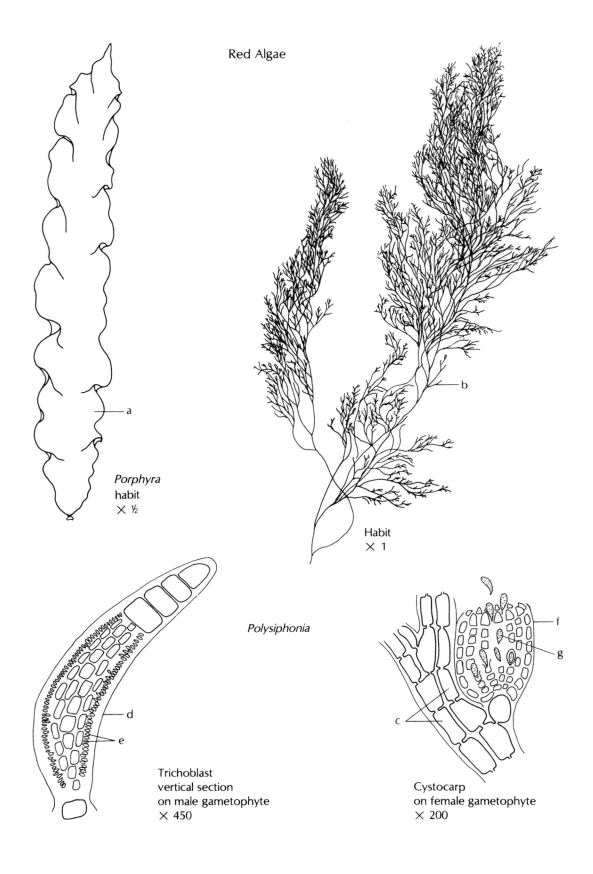

Porphyra
habit
× ½

a

Habit
× 1

b

Polysiphonia

Trichoblast
vertical section
on male gametophyte
× 450

d

e

Cystocarp
on female gametophyte
× 200

c

f

g

Green Algae

Green algae constitute a very ancient group of plants with fossils occurring back to early periods of the Paleozoic era. They are mainly freshwater plants, but also occur in saltwater, in soil, on tree bark, and on snow.

Characteristics. Green algae forms are unicellular, colonial, filamentous, membranous (resembling) leaves), and tubular. The cells have many of the organelles found in cells of higher plants, including a rigid cell wall; a nucleus with one or more nucleoli; a large vacuole; well-organized chloroplasts; pigments of chlorophylls a and b, carotenes, and xanthophylls; self-feeding nutrition (autotropic); and pyrenoids associated with starch formation. Motile forms usually have two whiplash flagella of equal length. There are many varying types of asexual and sexual reproduction.

Of interest . . . **phylogeny:** green algae are considered to be on the direct evolutionary line to higher vascular plants; **ecosystem:** they contribute to the freshwater plankton of the food chain.

Note: Of the 450 genera, the few illustrated here are microscopic, with the exception of *Acetabularia,* which is macroscopic.

Chlamydomonas

This unicellular alga is motile by means of two flagella (a). The cup-shape chloroplast (b), that contains a red eyespot (c) and a pyrenoid (d), surrounds the nucleus (e).

Eudorina

Sixteen cells make up this **coenobium** (a colony of a fixed number of cells). A gelatinous sheath (f) surrounds the flagellated (g) cells. The chloroplast (h) contains an eyespot (i).

Cladophora

This alga has branched filaments whose cells contain numerous pyrenoids (j) in reticulate chloroplasts (k).

Acetabularia

This marine alga, known as "mermaid's wineglass," is one immense cell with a giant nucleus (l) in the basal rhizoid (m). The nucleus migrates to the cap (n) where reproductive cysts are formed.

Scenedesmus

This is a coenobium composed of four cells that are joined laterally. The laminate chloroplast (o) contains a pyrenoid (p). There is one nucleus (q) per cell.

Micrasterias

The cell wall, enclosed by a gelatinous sheath (r), is divided into two halves. Each half of this unicellular alga has a chloroplast (s) with pyrenoids (t). The nucleus is located at the contraction (isthmus, u) between the cell halves. For asexual propagation, the nucleus divides in half, the cell splits into two halves and separates. Then each half grows a new partner to form a cell as illustrated. This plant occurs in freshwater ponds and lakes.

Spirogyra

Spirogyra is an unbranched filamentous freshwater alga. One cell in a filament is shown. It has two spiral chloroplasts (v) with numerous pyrenoids (w). The nucleus (x) is suspended in the center by threads of cytoplasm.

COLOR CODE

colorless:	flagellum (a, g), sheath (f, r), rhizoid (m)
white:	pyrenoid (d, j, p, t, w)
gray:	nucleus (e, l, q, u, x)
red:	eyespot (c, e, i)
green:	chloroplast (b, h, k, o, s, v), cap (n)

Green Algae

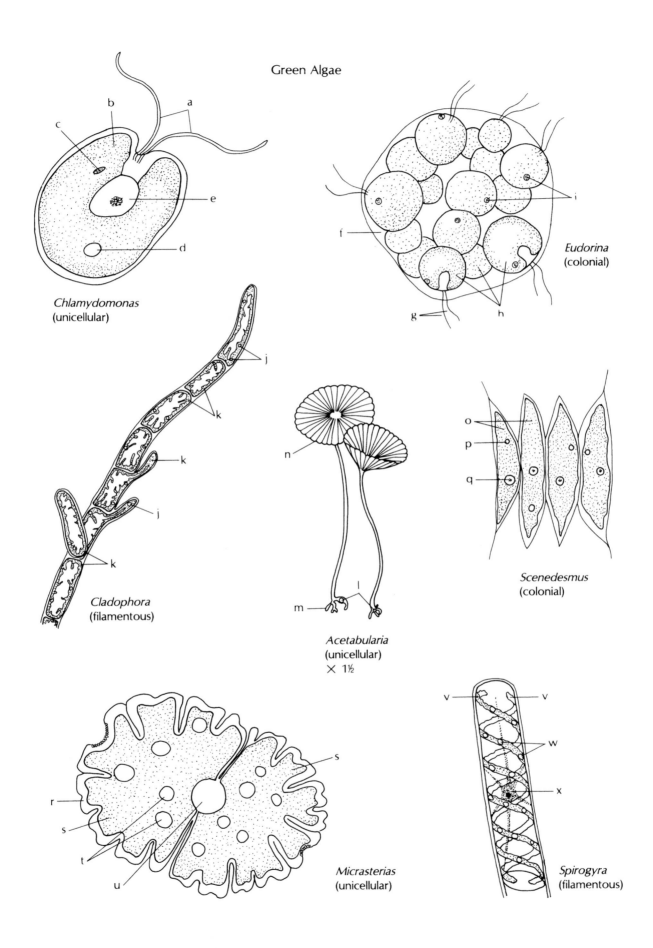

Chlamydomonas
(unicellular)

Eudorina
(colonial)

Cladophora
(filamentous)

Acetabularia
(unicellular)
× 1½

Scenedesmus
(colonial)

Micrasterias
(unicellular)

Spirogyra
(filamentous)

Brown Algae

The algae in this division are mainly marine. They live along cooler coastal, intertidal areas, forming large beds of brown seaweed. Common features among the different genera are the pigments of chlorophylls a and c, β-carotene, violaxanthin and fucoxanthin, source of the brown color; a sugar food reserve (**laminarin**); and cell walls made up of an inner cellulose layer and an outer mucilaginous layer. Diverse size ranges from microscopic to enormous kelps up to 60 meters in length. Of the 250 known genera there are more than 900 species.

Asexual reproduction occurs by fragmentation and production of sporangia with zoospores. When a section of a living filament breaks off, the fragment becomes a new individual. A sporophyte individual produces sporangia with zoospores, that when released, develop into more sporophyte individuals.

Sexual reproduction is by sex cells (gametes) that look alike (isogamy), or are unlike (anisogamy), or by formation of motile sperm with a stationary egg (oogamy).

Of interest . . . **ecosystem:** dense beds of kelp along sea coasts provide a substrate for small animals, protection, and foods for swimming animals and fish, and they contribute to phytoplankton communities; **industry: algin** (a cell wall constituent) contains sodium, ammonium and potassium salts (these salt compounds are used in fireproofing fabrics, as laundry starch substitutes, as ice cream stabilizers, as binders in printer's ink, in soaps and shampoos, in photographic film coatings, in paints and varnishes, and in leather finishes; insecticides; toothpaste, shaving cream, and lipstick); **fertilizer:** brown algal organisms contain more potassium, less phosphorus, and about the same proportion of nitrogen as animal manure; **food:** species of kelps are used as food (**lombu**) in Japan.

Ectocarpales. Of the 13 orders, the most primitive brown algae are found in this order. The body (thallus) consists of erect filaments of a single row of cells (uniseriate) with basal branching systems (heterotrichy). Their size ranges from microscopic to forms up to 25 cm. long.

Ectocarpus

This alga is found on most seashores growing upon other plants (**epiphytic**). The filaments (a, b) of the alga have a filmy appearance. The cells have net-like (reticulate) chlorophasts (c). Alternate generations are represented by sporophytes and male and female gametophytes. A **sporophyte** is an individual that produces sporangia with 2n spores, while a **gametophyte** produces 1n gametes (sex cells).

The sporophyte produces two types of reproductive structures: (1) a muticellular (plurilocular) sporangium (d) produces 2n zoospores that germinate into another sporophyte plant; (2) a unicellular (unilocular) sporangium (e) produces 1n zoospores that germinate into gametophytes. Gametophytes develop plurilocular gametangia (f) that produce motile gametes of the same sex as that of the parent. Fusion of male and female gametes results in development of a 2n (diploid) sporophyte.

If a gamete does not fuse with another gamete, it may be by parthenogenic germination produce a 1n (haploid) sporophyte. **Parthenogenesis** is the development of a new individual from an unfertilized egg. There are also other more complicated variations in the life cycle of *Ectocarpus.*

COLOR CODE

tan-green:	sporophyte (a), filament (b), chloroplasts (c)
tan:	sporangium (d, e), gametangium (f)

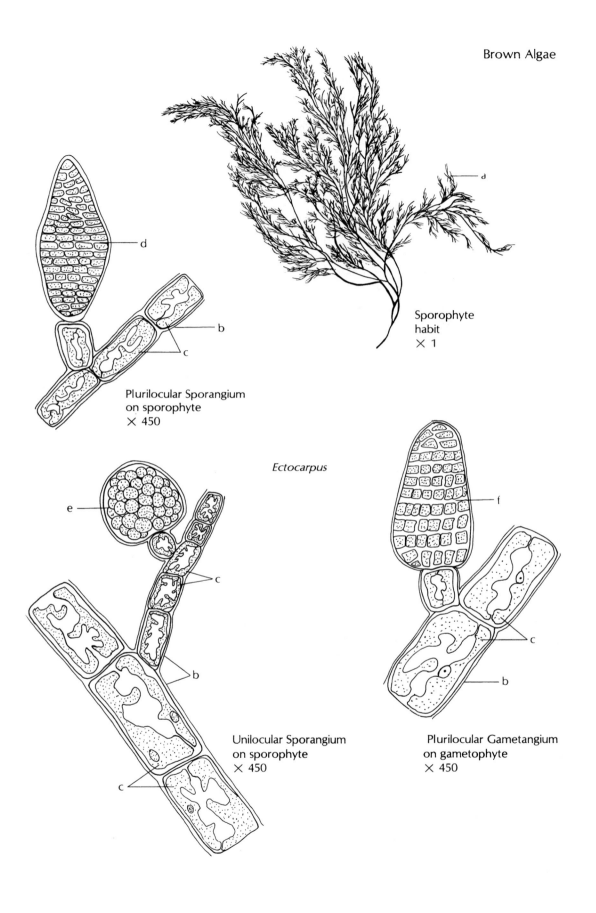

Brown Algae

Sporophyte
habit
× 1

d

Plurilocular Sporangium
on sporophyte
× 450

b
c
d

Ectocarpus

Unilocular Sporangium
on sporophyte
× 450

e
c
b
c

Plurilocular Gametangium
on gametophyte
× 450

f
c
b

Brown Algae (continued)

Laminariales (Kelps). This order consists of the kelps, the largest of all algae. Some reach 100 meters or more in length. The microscopic gametophytes have male and female gametes on separate individuals. The characteristic tissue of the sporophyte is composed of randomly arranged cells (parenchyma, for photosynthesis and food storage), rather than a continuous filament of cells.

Well-developed meristems generate new cells that contribute to growth of the sporophyte thallus in length and width. (**Meristems** are areas of undifferentiated cells that can develop into specialized tissues and organs.) The macroscopic sporophyte structure consists of a blade, stalk (stipe), and holdfast part that is attached to the rocky substrate.

Laminaria

Most species in this genus have perennial basal structures consisting of a stipe (a) and a root-like holdfast (b). There is a loss, but later replacement, of the sporopyte blade (c) from the stipe's meristematic region.

The reproductive structures of the sporophyte are asexual unilocular sporangia (d) containing zoospores (e). Between the sporangia are thread-like structures termed **paraphyses** (f), with distended tips that provide protection for the zoospores.

Nereocystis leutkeana   Ribbon Kelp

This annual kelp has blades that develop in the same manner as *Laminaria*. Due to progressive splitting, the blade (g) appears ribbon-like. The top of the stipe (h) bloats into a gas-filled **pneumatocyst** (i), which serves as a float to maintain the buoyancy of the blades at the water's surface as the stipe elongates.

Sexual reproduction is oogamous as follows: the microscopic male gametophyte produces clusters of antheridia (k) containing biflagellated sperm at the filament tips. The microscopic female gametophyte produces a non-motile egg in the oogonium (l). After fertilization, a zygote is formed which emerges from the oogonium and begins sporophyte development (m). (A **zygote** is the product of fertilization and is potentially a new individual.)

COLOR CODE

tan-green:	stipe (a), blade (c)
tan:	holdfast (b), sporangium (d), zoospores (e), antheridium (k)
white:	paraphysis (f)
yellow-brown:	blade (g), stipe (h), pneumatocyst (i), sporophyte (m)
brown:	holdfast (j)
transparent:	empty oogonium (l)

Brown Algae

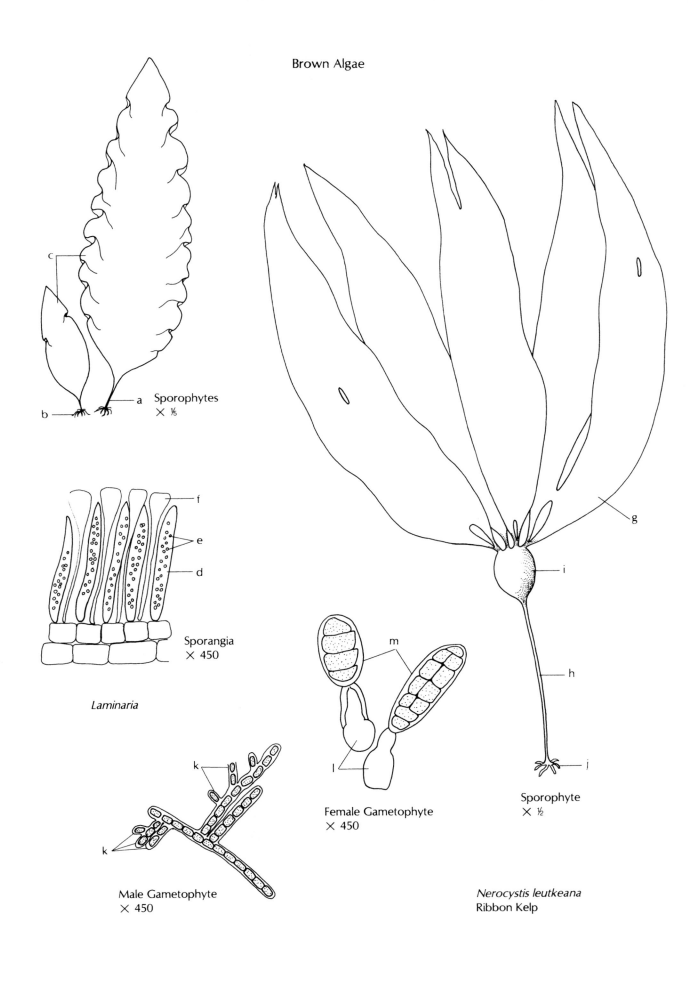

a
b
c
Sporophytes
× ⅕

f
e
d
Sporangia
× 450

Laminaria

k
k
Male Gametophyte
× 450

m
l
Female Gametophyte
× 450

g
i
h
j
Sporophyte
× ½

Nerocystis leutkeana
Ribbon Kelp

Brown Algae (continued)

Fucales. The algae in this order are unique in that the gametophyte generation has been suppressed. Thus, there is only a 2n (diploid) organism. It has oogamous sexual reproduction by 1n (haploid) gametes (see below), which arise by meiosis. **Meiosis** is a process of sexual reproduction in which a 2n (diploid) organism produces 1n (haploid) sex cells (gametes) by reduction division from one 2n cell to form four 1n cells.

Fucus vesiculosus

The thallus of this alga consists of a holdfast, stipe (a), and flat, bilaterally branched blades (b) with midrib (c) and paired air bladders (pneumatocysts, d). The separate male and female plants have swollen fertile areas on the blade tips called **receptacles** (e). They produce either male or female **conceptacles** (f).

A conceptacle has a pore (g) that opens from the surface of the receptacle to surrounding water. Also present in the conceptacle are sterile, hair-like threads called **paraphyses** (h), and either oogonia (i) that produce eggs, or antheridia (j), that produce biflagellated sperm. When eggs are released from the oogonium into the water, they produce **fucoserraten**, a sex attractant (pheromone), that causes the sperm to swarm around them. Fertilized eggs develop into 2n (diploid) individuals.

COLOR CODE

brown-red:	stipe (a), blade (b, c)
tan:	air bladders (d)
brown:	receptacle (e), oogonium (i), antheridium (j)
colorless:	conceptacle wall (f), paraphysis (h)

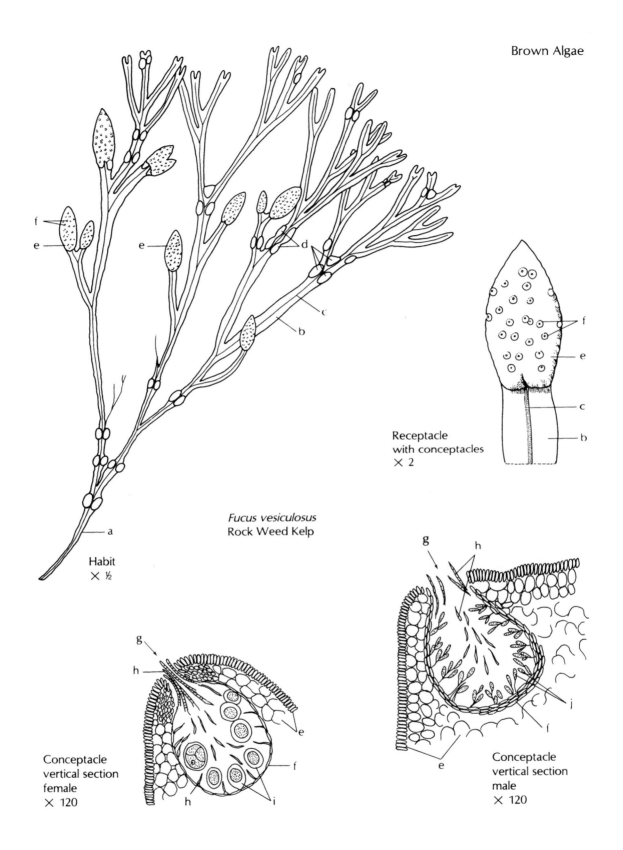

Receptacle
with conceptacles
× 2

Fucus vesiculosus
Rock Weed Kelp

Habit
× ½

Conceptacle
vertical section
female
× 120

Conceptacle
vertical section
male
× 120

Stoneworts

According to the fossil record, this ancient group of 6 living genera lies developmentally between the green algae and the liverworts and mosses (bryophytes). Fossils of stoneworts are found from the late Silurian period, over 400 million years ago. Stoneworts grow submerged in fresh water or sometimes in brackish (salty) water with rhizoidal attachment to the substrate of either soil or rock. With a preference for hard water (high pH), stoneworts have incrustations of calcium carbonate that account for the "stonewort" name. An obnoxious odor is characteristic of the plants.

Stoneworts grow continuously by means of a single dome-shaped apical cell that cuts off by mitotic divisions new node and internodal cells.

Their pigments are chlorophylls a and b, carotenes, and xanthophylls. Sexual reproduction is oogamous, which means a motile sperm unites with a larger, stationary egg.

Chara   Stonewort

The base of this plant, the **protonema** (a), has substrate-anchoring rhizoids (b), while the upper portion of the stem (c) has whorls of branches (d) at the nodes. The stem is a long axis of cells divided into small regions (nodes and internodes). The node divides producing whorls of branches. The internode is made up of one cell containing a single nucleus. The reproductive structures, male **globules** and female **nucules** (e), arise at nodes on the branches.

The male globule (antheridium) has an outer layer of sterile shield cells (f). Small cells (g) arise on the inner surface which produce antheridial filaments (h). Each cell in the filament produces a single sperm with two flagella for motility. The shield cells break open, permitting sperm cells to swim free.

The female nucule (oogonium) is attached to a node by a short stalk-like cell (i). Sterile sheath cells (j) spiral around the single egg (k). At the top are small **coronal cells** (l). At maturity, slits that form under the coronal cells allow the sperm to enter the oogonium.

After fertilization, the sheath cells thicken around the zygote (oospore), which overwinters as a dormant structure. After germination in the spring, the zygote wall splits and the primary protonema (m) emerges to form a new sporophyte plant.

Of interest . . . **gravisensors:** barium sulfate crystals in the tips of the rhizoids sediment in the gravitational field, thus enabling the rhizoids to respond to gravity and bend downward towards the substratum (starch-filled amyloplasts in root caps of higher plants serve the same function). The large cells of stoneworts have proven to be highly suitable for **experimental study** of patterns of cell growth, orientation of cell wall microfibrils, transport of ions into and out of the cells, and cytoplasmic streaming (**cyclosis**).

COLOR CODE

gray-green:	protonema (a, m), stem internode (c), branches (d)
white:	rhizoid (b)
red:	nucules (e), sheath cells (j)
tan:	egg (k), coronal cells (l)
colorless:	shield cells (f), filaments (h)
red-brown:	cells (g)
green:	stalk cell (i)

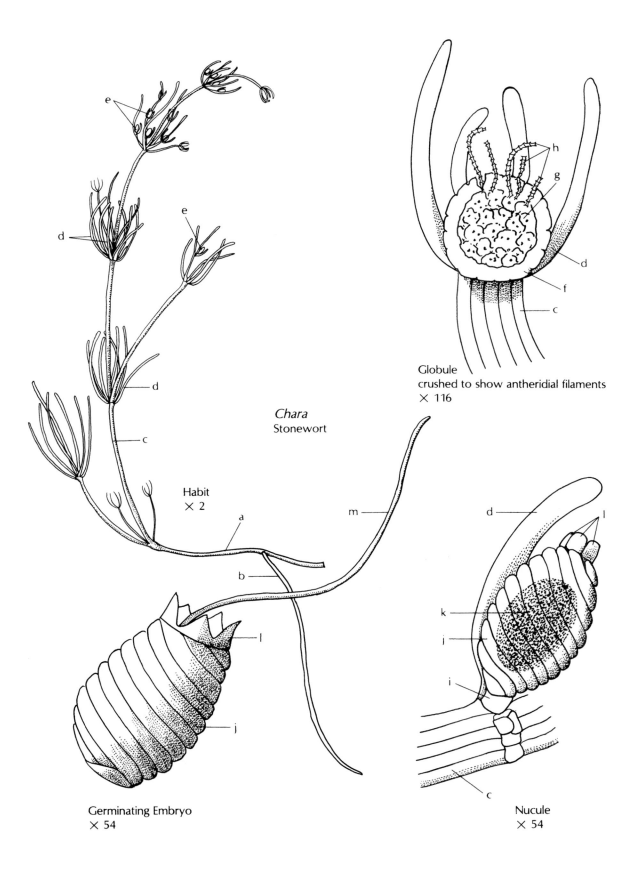

Chara
Stonewort

Globule
crushed to show antheridial filaments
× 116

Habit
× 2

Germinating Embryo
× 54

Nucule
× 54

Liverworts, Hornworts, Mosses

This most primitive group of green land plants are called **bryophytes.** Mostly perennials, the plants are small in size and have no vascular tissues. Humid habitats provide a water medium for the motile (flagellated) sperm to reach the egg. Bryophytes cannot tolerate salt water, excessive heat, or pollution. Mosses can dry down during drought, then recover with rainfall. They are also tolerant of freezing. Liverworts, on the other hand, cannot tolerate dryness, but can survive freezing.

Bryophytes have two generations. The dominant plant (gametophyte generation) develops from a 1n (haploid) spore and produces sperms and eggs. A fertilized egg forms a 2n zygote, which develops into the sporophyte. The smaller sporophyte generation produces 1n spores by meiosis that are dispersed and develop into gametophytes. In some bryophytes, asexual structures called **gemmae cups** (a) contain gemmae (b) that develop into gametophytes.

Liverworts (Hepaticae). The liverworts are primitive bryophytes. The gametophyte is flat and strap-shaped, as in *Marchantia,* or leafy as in *Bazzania.* The gametophyte body (thallus, c) branches equally (**dichotomously**).

Marchantia polymorpha

This species has unisexual thalli (**dioecious**). The male gametophyte develops antheridia-bearing stalks (**antheridiophores, d**), which produce motile sperm in its disc-shaped top. The female gametophyte develops an **archegoniophore** (e), which produces archegonia, flask-shaped structures containing eggs, under its umbrella-shaped top. After fertilization, a 2n zygote develops into a sporophyte, which in turn produces spores that develop into gametophytes.

Hornworts (Anthocerotae)

Anthoceros punctata   Common Hornwort

The gametophyte thallus (f) forms a flat rosette. Male and female structures occur in the upper layers of the thallus. The union of sperm and egg produces the sporophyte. A foot (g) secures the sporophyte (h), to the gametophyte thallus. The sporophyte splits at the top into 2 valves to expose the spores.

Mosses (Musci). All moss gametophytes have 3- to 5-ranked leaves. Most mosses develop from a spore

into a thread-like structure (**protenema**) with rhizoids, then into a leafy gametophyte. The young sporophyte is green and photosynthetic. Its firm tissue eventually turns brown and may persist on the gametophyte for years. Unlike the liverworts and hornworts, mosses show a great deal of variation in gametophytes and sporophytes. Mosses form large colonies by fragmentation and gemmae.

Sphagnum centrale   Peat Moss

The peat mosses (*Sphagnum,* i) grow in acid bogs. They secrete hydrogen ions, contributing to the acidity. This acid environment delays the vegetation decay process. Thus, sphagnum mosses contribute to mat formation at the open water margin of the bog. This provides a habitat for other types of plants. Undisturbed, a bog can eventually develop into a forest through the process of **succession.** Due to its great water-holding capacity *Sphagnum* peat, derived from old bogs, is added to garden soil to acidify it and hold moisture.

Andrea rupestris   Rock-loving Moss

This plant is found mainly on rocky outcroppings. The leafy gametophyte (j) is branched. The sporophyte is raised on a stalk (pseudopodium, k). When the capsule (l) dries the inner walls contract into 4 slits, releasing the spores.

Tetraphis pellucida   Four-tooth Moss

Moist rotting logs in coniferous forests provide the main habitat for this plant. Leafy gametophytes (m), attached by rhizoids (n), produce gemmae cups (o). The sporophyte is borne on a long stalk (seta, p), that twists near the capsule (q). A pleated covering (**calyptra,** r) protects the capsule; it falls off at maturity. A lid (**operculum**) opens to expose 4 teeth (**peristome,** s). When wind or animals shake the capsules, spores are released.

COLOR CODE

green:	gemma cups (a, o), gemmae (b), thallus (c, j, m)
tan:	antheridiophore (d), archegoniophore (e), foot (g), pseudopodium (k), capsule (l, q), seta (p), calptra (r), teeth (s)
black:	sporophyte (h)
gray-green:	thallus (i)
white:	rhizoids (n)

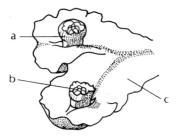

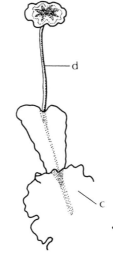

Gametophyte
with gemmae
× 2½

Marchantia polymorpha
Common Liverwort

Male Gametophyte
× 2½

Female Gametophyte
× 2½

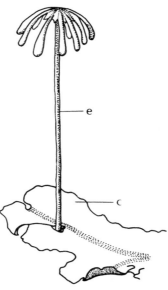

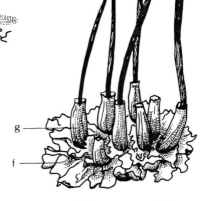

Anthoceros punctata
Common Hornwort
gametophyte at base
sporophytes above
× 3

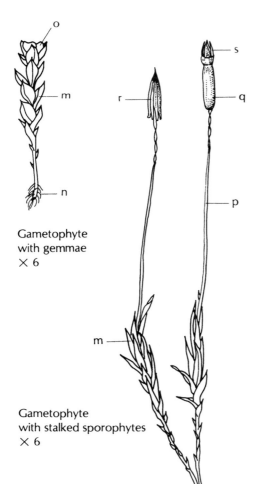

Gametophyte
with gemmae
× 6

Gametophyte
with stalked sporophytes
× 6

Tetraphis pellucida
Four-tooth Moss

Sphagnum centrale
Peat Moss
gametophyte
× 1⅓

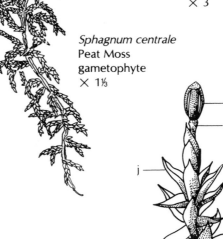

Andreaea rupestris
Moss
gametophyte
sporophytes at top
× 13

Whisk Ferns

The whisk ferns are believed to be among the most ancient of living vascular plants. There is no known fossil record to determine relationships with extinct forms. There are two psilophyte genera: *Psilotum* and *Tmesipteris*.

Characteristics. These are rootless, simple plants with rhizoids, as anchoring structures, and hairless (**glabrous**) upright stems with simple sterile appendages (**enations**). The large spore cases (sporangia) are fused into two's or three's.

Reproduction. In whisk ferns, an alternation of generations results in a dominant sporophyte plant producing alike spores (**homosporous**), which germinate to form 1n (haploid) gametophytes. The underground gametophyte, which forms mycorrhizal associations with fungi for nutrition, develops archegonia, each with 1 egg, and antheridia with flagellated sperm.

After fertilization, a 2n (dipoid) zygote forms, which eventually develops into a sporophyte plant. Asexual reproductive bodies (gemmae) may be produced on the sporophyte.

Of interest . . . Whisk ferns are dominant components of the tropical ecosystems of Hawaii's island archipelago. In fact, they are considered to be weeds in Hawaii due to their propensity to colonize disturbed areas, (W. H. Wagner, Jr., personal communication).

Psilotum nudum   Whisk Fern

This plant may be terrestrial or grow upon other plants (**epiphytic**) in the tropics and subtropics. The shoot is upright with stems (a) that branch into two equal parts (dichotomous branching). Simple leaf-like appendages, called enations (b), emerge from the stem. The three-lobed spore case (c) has a small, forked, leaf-like scale (d) below.

Tmsipteris tannensis var. *lanceolata*

This plant is found only in the southwestern Pacific Ocean area. It has a simple stem (e) with flat, single-veined, leaf-like appendages (f). The sporangium (g) is two-lobed.

COLOR CODE

light green:	stem (a), enations (b), scale (d)
yellow:	sporangium (c)
green:	stem (e), enations (f)
tan:	sporangium (g)

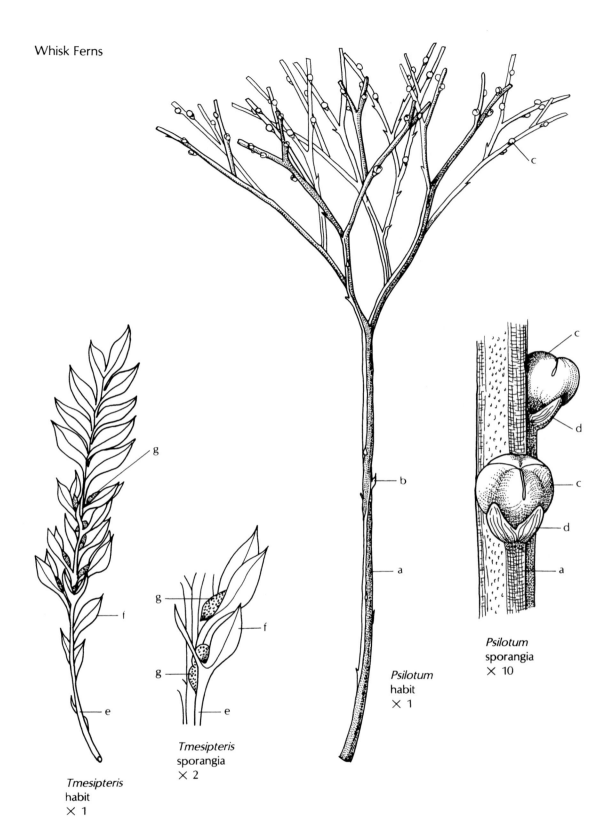

Tmesipteris
habit
× 1

Tmesipteris
sporangia
× 2

Psilotum
habit
× 1

Psilotum
sporangia
× 10

Clubmosses, Spikemosses, Quillworts

This distinct group of ancient plants has five living genera. Common features are single-veined leafy appendages (**microphylls**), two-part branching which may be equal or unequal, axillary sporecases (sporangia), and flagellated sperm.

Clubmosses (Lycopodiales). *Lycopodium*, with 700 species, is found worldwide, with most species occuring in the tropics. Plants are herbaceous, perennial, and usually evergreen. No leaf outgrowths (ligules) are present. Some species bear sporangia in terminal spikes (**strobili**).

Lycopodium annotinum   Stiff Clubmoss

The creeping horizontal stem (rhizome) of this plant has upright shoots with annual constrictions (a). The spirally arranged leaves (b) are narrow and spread in different directions. In the alternation of generations, the spore germinates into a plant (gametophyte) bearing male and female sex cells (gametes). Then, after fertilization, an embryo develops into a sporophyte, the dominant plant.

A "cone" (strobilus, c) of specialized leaves (**sporophylls**), bearing sporangia in the axils, terminates an upright shoot. The sporangia contain one type of spore (**homosporous**).

Spikemosses (Selaginellales). *Selaginella*, the single genus, has about 600 species. Differing from *Lycopodium*, there are **ligules**, root-like organs (**rhizophores**) occurring at dichotomies, and sporangia always in a "cone," bearing spores of two sizes (**heterosporous**). Reproduction is the same as shown below for *Isoetes*.

While most species prefer moist conditions, there is an unusual species, *Selaginella lepidophylla* which is called resurrection plant. It is native to dry regions of the southwestern United States and Mexico. Dry specimens of this plant are occasionally found in plant stores as a curiosity because it appears as a dried leafy mass until "revived" with water that causes it to turn green and flatten into a rosette. It can be dried and revived repeatedly.

Selaginella rupestris   Rock Spikemoss

The creeping stem has spirally arranged leaves (d), root-like appendages (rhizophores, e), and 4-sided fertile spikes (f).

Quillworts (Isoetales). These plants are considered by some botanists to be the last remnant of the fossil tree lycopods due to their peculiar secondary growth, heterospory and presence of ligules. *Isoetes*, with 60 species, grows mostly submerged in lakes, ponds, and streams.

Isoetes echinospora   Braun's Quillwort

The sporophyte has an enlarged perennial underground stem, the **corm** (g). The roots (h) branch dichotomously, and annual strap-shaped leaves (i) grow in a spiral cluster (**rosette**). The leaf has a ligule (j), 4 vertical air chambers (k), and an expanded fertile base covered by a flap (**velum**, l).

At the bases heterosporous leaves (**mega-** and **microsporophylls**) bear megasporangia (m) containing large spores (**megaspores**, n) or microsporangia (o) with small spores (**microspores**). The megaspore wall is variously ornamented according to species. The megaspores develop into female gametophytes, while the microspores produce male gametophytes.

Stylites andicola   Andes Quillwort

This plant, forming cushions in marshes, was discovered in the Andes Mountains of Peru in 1957. The stem (p) branches dichotomously with the leaves (q) atop each section. The roots (r) develop on the sides of the stem.

COLOR CODE

dark green:	leaves (b)
green:	spike (c), leaves (d, q), ligule (j)
white:	rhizophores (e), roots (h, r), megasporangium (m), microsporangium (o)
tan:	velum (l), stem (p)
light green:	spike (f), leaves (i)
brown:	corm (g)
yellow:	megaspore (n)

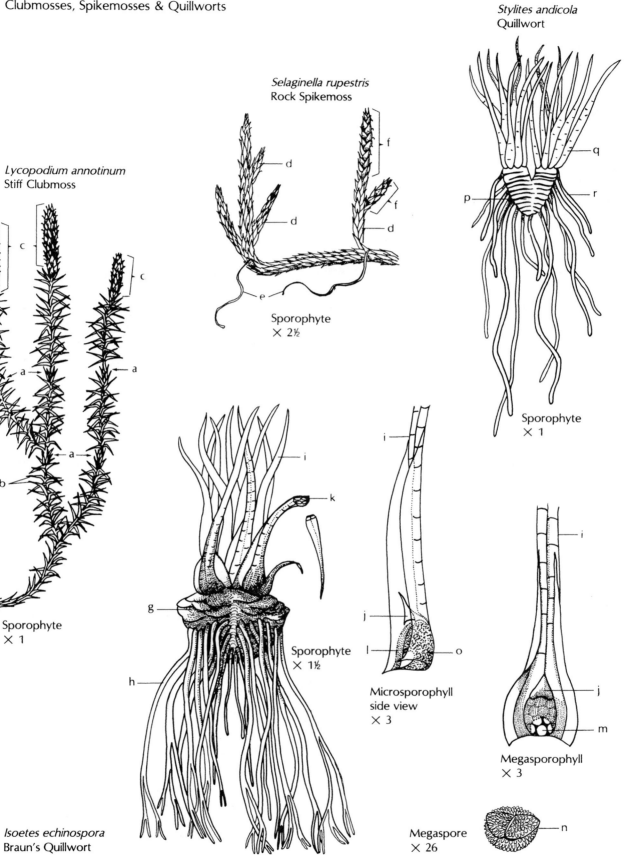

Stylites andicola
Quillwort

Sporophyte
× 1

Selaginella rupestris
Rock Spikemoss

Sporophyte
× 2½

Lycopodium annotinum
Stiff Clubmoss

Sporophyte
× 1

Sporophyte
× 1½

Isoetes echinospora
Braun's Quillwort

Microsporophyll
side view
× 3

Megasporophyll
× 3

Megaspore
× 26

Horsetails

The single genus, *Equisetum,* with 15 species, is the only present-day member of plants in this family, Equisetaceae. It is related to the fossil plant, *Calamites* (see 43). Species of *Equisetum* are found worldwide except in Australia and New Zealand.

Characteristics. These annual or perennial terrestrial plants are less than 1 meter in height, with the exception of the Costa Rican tropical *E. giganteum,* which grows to 8 meters in height. The plant consists of a jointed underground perennial rhizome and annual upright jointed shoots.

Species with branched shoots are commonly called "**horsetails**" and unbranched species are known as "**scouring rushes**" because of their earlier use as pot scrubbers (effective because of the presence of amorphous silica, $SiO_2 \cdot nH_2O$, in the stem surface).

The photosynthetic stem is round and grooved on the outside and hollow in the center. At the stem node is a sheath of greatly reduced leaves of various teeth-like shapes, depending on the species. In branched species, whorls of branches develop at the base of the leaf sheath. Internodes of the rhizomes and aerial shoots elongate by means of basally localized intercalary meristems (see 15).

Reproduction. The plant has an alternation of a sporophyte generation with an either male or bisexual gametophyte generation as follows: The sporophyte produces a terminal "cone" (**strobilus**) with whorls of hexagonal scales (**sporophylls**) on stalks. Attached to each scale are 5 to 10 spore-bearing sacs (**sporangia**) containing alike spores (**homosporous**). The spores contain chlorophyll. They germinate within a few days, developing into tiny green thalloid gametophytes that produce antheridia with sperm or both antheridia and archegonia. Fertilization of the egg in an archegonium by a sperm results in a zygote that develops into a new sporophyte plant.

Equisetum variegatum   Variegated Scouring Rush

Equisetum stems (a) are grooved and possess hollow air channels within. The air channels include a central canal (b), **carinal canals** (c), and **vallecular canals** (d) at the bottom of each groove. Size and shape of the air canals vary with species.

Equisetum arvense   Field Horsetail

This species has two forms (**dimorphic**). In the spring, an unbranched fertile shoot (e) emerges from the rhizome (f). It is short-lived and dies back when the spores are shed from the strobilus (g). The other form of stem (h) is smooth and branched (i). By summer, the branches have elongated. The leaf sheaths (j) in this species have dark, lance-shaped teeth (k).

Equisetum laevigatum   Smooth Scouring Rush

The scouring rush stem (l) is unbranched. The leaf sheath (m) is flared and has deciduous teeth (n). The strobilus (o) shows the hexagonal surface of the sporophylls (p).

Equisetum hyemale   Common Scouring Rush

A sporophyll removed from the stobilus exposes the sporangia (q) containing the spores.

Equisetum scirpoides   Dwarf Scouring Rush

A spore (r) surface splits into 4 strips of water-absorbing (hydroscopic) **elators** (s). When dry, the elators (t) expand and aid in spore dispersal.

COLOR CODE

green:	stem (a, h), branches (i), sheath (j), sporangium (q)
colorless:	canals (b, c, d)
tan:	shoot (e), strobilus (g)
dark brown:	rhizome (f)
yellow-green:	stem (l), sheath (m), surface of sporophylls (p), spores (r), elators (s, t)

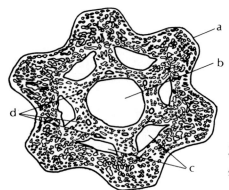

Equisetum variegatum
Variegated Scouring Rush
stem cross section
× 34

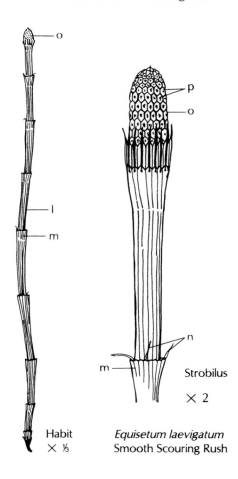

Habit
× ⅓

Equisetum laevigatum
Smooth Scouring Rush

Strobilus
× 2

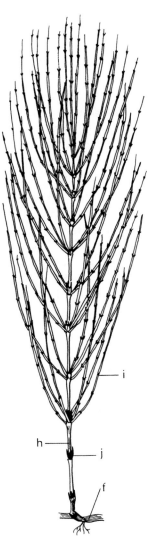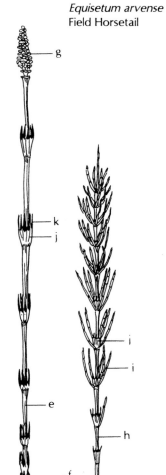

Equisetum arvense
Field Horsetail

Spring Habit
× ½

Summer Habit
× ½

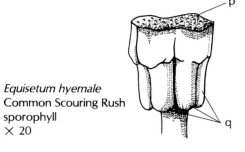

Equisetum hyemale
Common Scouring Rush
sporophyll
× 20

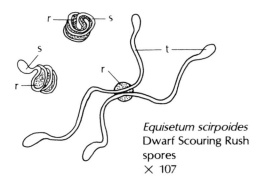

Equisetum scirpoides
Dwarf Scouring Rush
spores
× 107

Ferns

Characteristics. Although fern origins are uncertain, fossils have been found dating back to the Devonian. Most fern species have leafy fronds that unroll as the leaf expands (circinate development, see 69). Stems lack secondary growth, have roots that are usually creeping or below ground, and may have protective hairs or scales. The leaves may be simple, feather-like (**pinnate**), lobed, with veins arising from one point (**palmate**), or with veins which appear to branch. Plant size ranges from a few millimeters to 15 meters.

Ferns are perennials and may take the form of twining vines, floating plants, trees, epiphytes, or, most commonly, terrestrial herbs. While most species of ferns are found in the moist tropics, some grow in bare rocky habitats, or temperate swamps, fields and forests.

Reproduction. Asexual reproduction occurs by offshoots from a rhizome and by spores. The sporophyte generation is dominant and produces 1n spores in spore cases (**sporangia**), usually on the lower leaf surface. The spore cases are massive in the primitive orders, Ophioglossales (Adder's-tongue Ferns) and Marattiales.

In the more advanced orders, the spore cases are minute, nearly microscopic, stalked, and in a "fruit dot" cluster called a **sorus**. The sculptured, thick-walled spores are dispersed by wind and germinate into the sexual gametophyte generation. The tiny, rarely seen gametophyte produces sperm in antheridia and eggs in archegonia. For fertilization to take place, the sperm must swim in water to the archegonium, a vase-shaped structure with one egg. After fertilization, a 2n zygote is formed and develops into a sporophyte on the gametophyte. The gametophyte then dries up and dies.

Of interest ... **ornamentals:** *Adiantum pedatum* (maidenhair fern), *Athyrium filix-femina* (lady fern), *A. nipponicum* 'Pictum' (Japanese painted fern), *Asplenium nidus* (bird's-nest fern), *Camptosorus* (walking fern), *Cyrtomium* (holly fern), *Dryopteris erythosora* (autumn fern), *D. filix-mas* (robust male fern), *D. marginalis* (marginal wood fern) *Nephrolepis* (Boston fern), *Onoclea sensibilis* (sensitive fern), *Osmunda cinnamonea* (cinnamon fern), *O. regalis* (royal fern), *Platycerium* (staghorn fern), *Polystichum acrostichoides* (Christmas fern), *P. polyblepharum* (tassel fern); **poisonous:** *Pteridium* (bracken fern); **horticulture:** *Osmunda* (dried roots used as medium for growing orchids); **fertilizer:** *Azolla* (mosquito fern, see 70).

Ophioglossales. Plants in this order are mostly small and succulent with the leaf divided into sterile and fertile, sporangial-bearing parts. There are four genera: *Botrychium* (grape ferns), *Ophioglossum* (adder's-tongue ferns), *Cheiroglossa,* and *Helminthostachys.*

Botrychium dissectum   Dissected Grape Fern

The single succulent compound leaf (a) is divided into **pinnae** (b), a primary subdivision of the leaf and **pinnules** (c), a further subdivision of the pinnae. The fertile stalk (d) is divided into many branches where sporangia (e) are borne in grape-like clusters. The roots (f) are coarse. This terrestrial fern is found on the forest floor of woods and in swamps.

Helminthostachys zeylanica

This fern is found in Southeast Asian and Polynesian swamps. The compound leaf is divided into 3 leaflets (h). The fertile spike (i) has very short branches bearing sporangia (j).

Ophioglossum vulgatum   Adder's-tongue Fern

The single stalk (k) has one simple blade (l) without a petiole (sessile), and sunken sporangia (m) borne terminally in a spike. The roots (n) are smooth and fleshy. Adder's-tongue fern grows in damp areas in fields and woods in the temperate region.

COLOR CODE

green:	leaf (a, b, c, h), stalk (d, g), spike (i)
orange:	sporangia (e, j)
brown:	roots (f)
light green:	stalk (k), leaf (l)
yellow:	sporangia (m)
white:	roots (n)

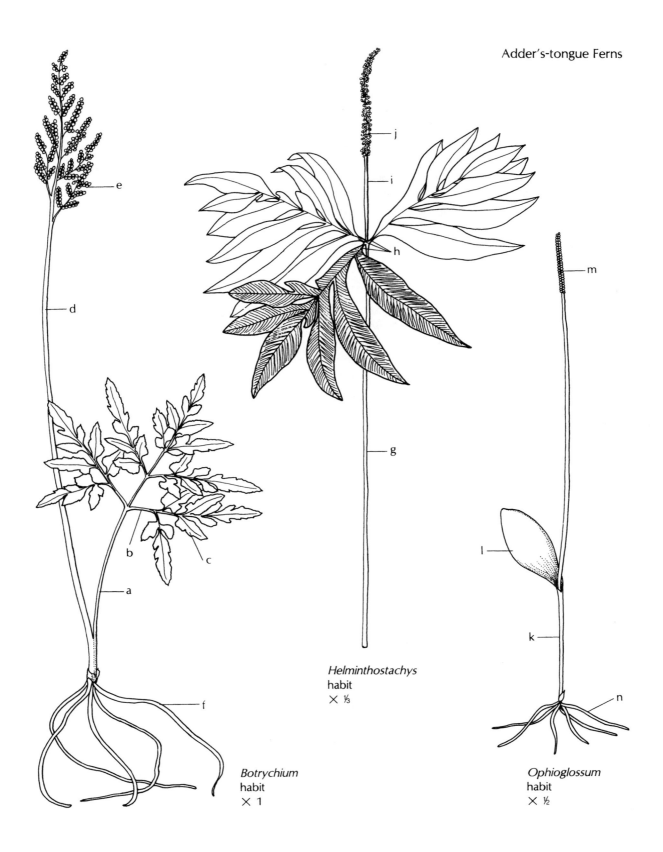

Helminthostachys
habit
× ⅓

Botrychium
habit
× 1

Ophioglossum
habit
× ½

Common Ferns

Common Ferns (Polypodiales). Most common ferns are in this order. They are usually of medium size among the ferns and are found on most parts of the earth. Sporangia are single, or more often, there are many in a cluster called a **sorus**. In some genera, the sorus has a protective covering (**indusium**). Some ferns in this order produce sporangia on separate, fertile stalks, but most have the sporangia on the underside of the frond (leaf) in sori, or on the entire surface, or in rows along the veins, or under the curled edge of the leaflets. There are over 8,000 species.

Osmunda regalis   Royal Fern

There are three species of *Osmunda*: *O. cinnamomea* (cinnamon fern), *O. claytonia* (interrupted fern), *and* *O. regalis* (royal fern). *O. cinnamomea has* two types of stalks (**dimorphic**): a sterile frond (leaf) and a separate fertile stalk with cinnamon-colored sporangia. It grows in shady, wet areas with acid soil. The fertile fronds of *O. claytonia* have a fertile central area of sporangia on the stalk with pinnae above and below. It prefers a dry type of soil compared to the other two species of *Osmunda*.

Osmunda regalis is shown here. The fern leaf (a) is twice divided into pinnae (b) and bears sporangial branches (c) terminally. The round single spore cases (sporangia, d) open in halves and occur in clusters (e). The deciduous leaves arise from a perennial rhizome with black wiry roots. Royal fern grows in very wet habitats with acid soil.

Dryopteris carthusiana (D. spinulosa) Toothed Wood Fern

The usually evergreen leaf of this fern is three-times dissected, forming delicate **pinnules** (f) in the pinna (g). Scales (h) cover the petiole (i). Sori are produced on the leaf underside. The sorus (j) is covered by a kidney-shaped **indusium** (k) that protects the sporangia (l). The preferred habitat is moist shade to partial sun. There are about 150 species of *Dryopteris* (wood ferns).

COLOR CODE

red-brown:	petiole (a), sporangium (d, l), sporangial clusters (e)
green:	pinnae (b), pinnules (f)
tan:	scales (h), petiole (i), indusium (k)

Common Ferns

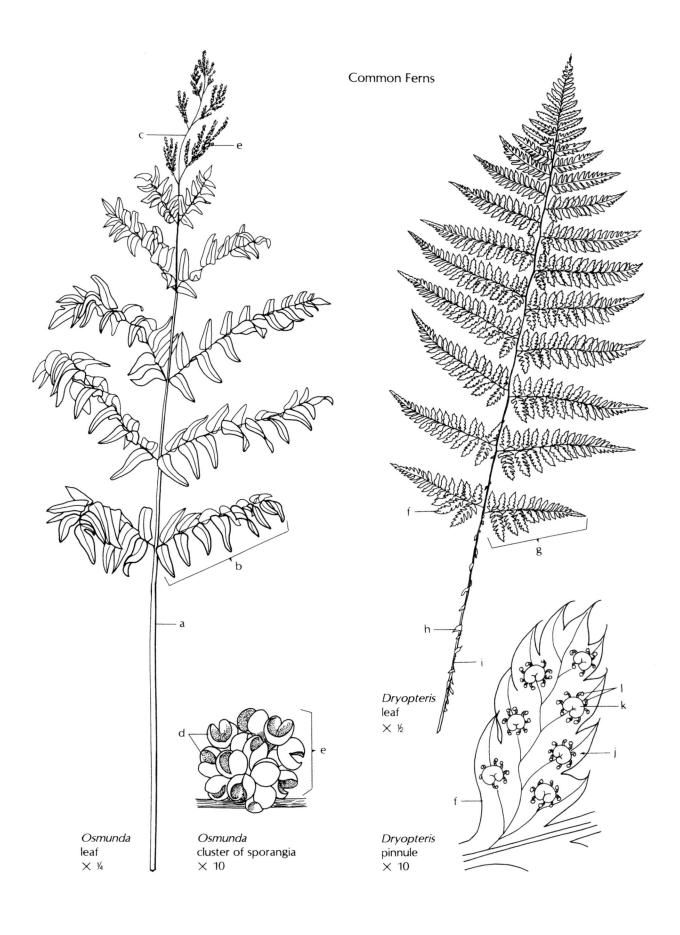

Osmunda
leaf
× ¼

Osmunda
cluster of sporangia
× 10

Dryopteris
leaf
× ½

Dryopteris
pinnule
× 10

Fern Leaf Development

Common Ferns (continued)

Rumohra adiantiformis   **Leatherleaf Fern**

Florists frequently use the sturdy fronds of this tropical fern. The young leaf meristem is protected by scales (a, b). As in other common ferns, the fiddlehead (**crozier, c, d**) expands by unrolling (**circinate development**). Ferns are the only plants with this type of leaf development.

As the stalk (petiole) elongates, the blade midrib develops leaflets (pinnae, e). In this genus, the leaflets (pinnae) are further divided into smaller leaflet sections called pinnules (f, g).

(This illustration was made from a series of drawings from one leaf as it developed.)

COLOR CODE

brown:	scales (a, b)
green:	crozier (c, d)
dark green:	leaf (e, f, g)

Fern Leaf Development

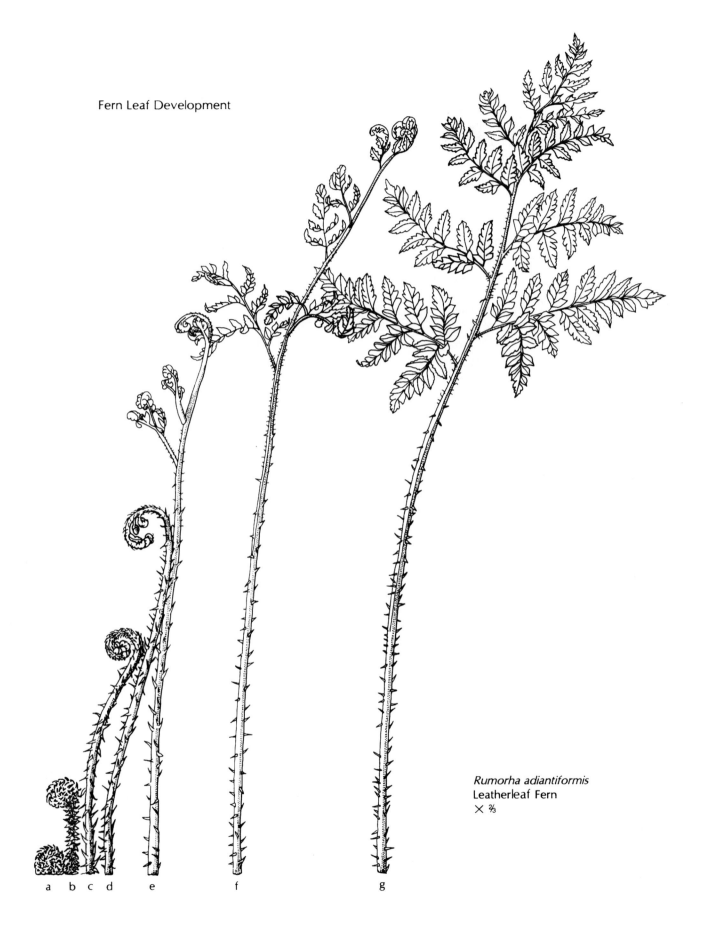

a   b c d   e   f   g

Rumorha adiantiformis
Leatherleaf Fern
× ⅔

Water Ferns

Water Ferns (Marsileales). Water ferns in this order usually have rooted rhizomes. The leaf may have 0, 2, or 4 leaflets on a long petiole. Spore-bearing receptacles (**sporocarps**) are formed along the petiole. A sporocarp develops both small microspores and larger megaspores (a **heterosporous** condition) on one plant (**monoecious**).

Marsilea quadrifolia   Water Clover

This pond plant is rooted (a) in mud with leaves floating on or near the water surface. The creeping rhizome (b) develops new leaves (c), which have 2 pairs of opposite leaflets (d).

Salviniales (Floating Ferns). The floating ferns include *Salvinia* and *Azolla*. The plant consists of a branched stem and leaves. Sporocarps containing two types of spores (heterosporous) are borne on the leaves.

Salvinia   Water Spangles

The leaves occur in whorls of 3 with two keeled, green floating leaves (f) and a colorless, finely divided, submerged leaf (g). The upper surface of the floating leaves is covered with rows of 4 rib-topped, transparent hairs (h). The hairs cause water droplets to maintain surface tension and roll off the leaf surface.

The submerged leaf filaments are covered with brown hairs, which function as roots for water and nutrient absorption. Sporocarps (i) are borne on submerged leaf filaments.

Azolla   Mosquito Fern

This floating fern has simple roots (j) that hang down in the water. The leaves are alternate on the stem (k) and two-lobed, having a green, fleshy lobe (l) above the water and a colorless flat lobe (m) on the water surface.

Some *Azolla* species have a symbiotic relationship with a blue-green, *Anabaena* (see 44), which lives in cavities of the fern. *Anabaena* converts nitrogen from the air into a reduced form (ammonia - NH_3) that can be used by the fern and surrounding water plants such as rice *(Oryza sativa)*. Because the presence of *Azolla* with *Anabaena* increases rice production, this tiny plant is probably the economically most important fern.

COLOR CODE

brown:	roots (a), leaf hairs (g)
white:	rhizome (b), root (j), stem (k)
green:	leaf (c), leaflets (d), petioles (e)
light green:	leaf (f), leaf lobe (l)
tan:	sporocarps (i)

Water Ferns

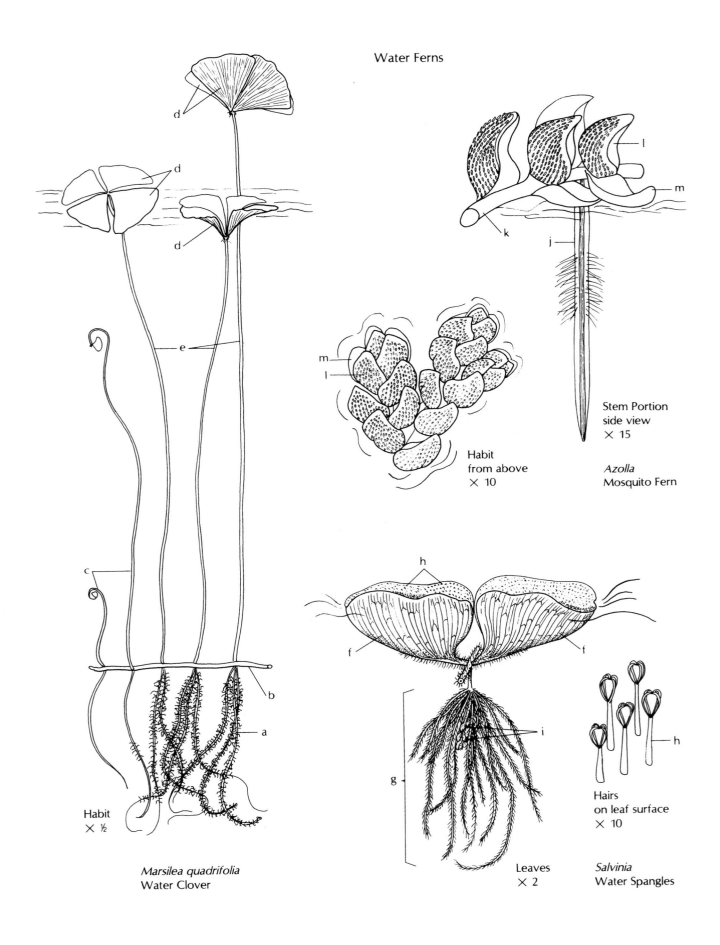

d
d
d

e

c

b

a

Habit
× ½

Marsilea quadrifolia
Water Clover

l
m
k
j

Stem Portion
side view
× 15

Azolla
Mosquito Fern

m
l

Habit
from above
× 10

h
f
f
i
g

Leaves
× 2

h
h

Hairs
on leaf surface
× 10

Salvinia
Water Spangles

Cycads

The fossil history of the cycads extends back 200 million years to the Mesozoic era. Cycads are now confined to the tropics and subtropics as living relics of once worldwide distribution. They are the most primitive living seed plants.

Characteristics. Cycads look like sturdy ferns or palm trees. They have thick upright stems with a primary thickening meristem and not much secondary growth. The leaves are large, compound fronds with leaflets. Leaflets are thick and tough, usually with equally branching (**dichotomous**) veins.

Typically, cycads reach a maximum height of 2 to 3 meters, but some species grow to 20 meters high. Cycad plants are slow-growing and long-lived, some as old as 1,000 years.

Reproduction. Unisexual cones (**strobili**) are borne on separate plants (**dioecious**). The male cone has spirally arranged cone scales (**microsporophylls**) with spore cases (**sporangia**) on the lower surface. The spore case contains microspores that develop into pollen grains that are dispersed to the female reproductive cones. There, the pollen grain develops mutiflagellated sperm. The cycads and *Ginkgo* are the only living seed plants having flagellated sperm.

The female cone scales (**megasporophylls**) produce ovules in which eggs develop. After fertilization, brightly-colored, usually red, seeds are formed. Cycads, *Ginkgo*, conifers and gnetes are referred to as **gymnosperms**, meaning "naked seed, " because the seeds are borne in an exposed position on the sporophyll.

Of interest . . . **houseplants or landscape plants** for tropical areas include *Bowenia serrulata, B. spectabilis, Ceatozamia latifolia, C. robusta, Cycas circinalis* (fern palm), *C. revoluta* (sago palm), *Dioön edule, Encephalartos,* and *Zamia.*

Zamia furfurscea

Leaflets (a) have dichotomous venation but do not have midribs. They leave prominent leaflet scars (b) when they are shed (abscise). The male cone (d) has stalked cone scales (e) with clusters of spore cases (f) on the under surfaces. The female cone scale (g) has a hexagonal surface (h), and below, are 2 ovules (j) in which the eggs develop.

Dioön edule

The habit illustration of this cycad shows the prominent stem (l) covered with old leaf bases (m). A spiral pattern of leaves (n) emerges at the apex. Leaflets (o) are arranged in two rows.

COLOR CODE

green:	leaflets (a, o)
tan:	leaf stalk (c), spore case (f), cone stalk (k), cone scale stalk (i)
brown:	cone (d), cone scale stalk, surface (e)
dark green:	cone scale (g, h)
white:	ovules (j)
red-brown:	stem, leaf (l, m)

Cycads

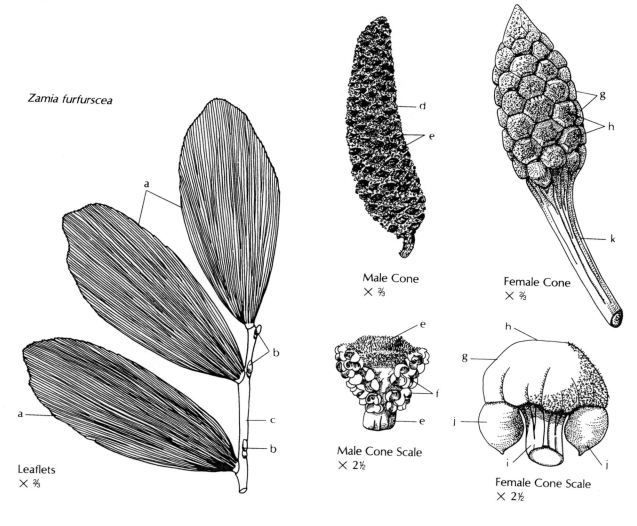

Zamia furfurscea

Leaflets
× ⅔

Male Cone
× ⅔

Female Cone
× ⅔

Male Cone Scale
× 2½

Female Cone Scale
× 2½

Dioön
habit
× ¹⁄₁₀

Ginkgo

This division has only one living member: *Ginkgo biloba* (Maidenhair Tree). Fossil records show it probably originated in the Permian about 250 million years ago. Members of the division were globally distributed, but became extinct except for this single species in southeastern China. *Ginkgo* is similar to cycads in reproductive structures and gametophyte development, but the leaves are distinctively different. The stem, with extensive wood, small pith and cortex, and cell wall pitting, is similar to that of the conifers (see 15).

Ginkgo biloba　Maidenhair Tree

This plant was cultivated in temple and monastery gardens of China and Japan for centuries. The Chinese name was mistransliterated and should be *Ginkyo*. The common name is derived from a resemblance of the leaves to the leaflets of the maidenhair fern *(Adiantum pedatum).*

Characteristics. A hardy tree with a slow maturation rate, *Ginkgo* can grow to 30 to 40 meters (60 to 80 feet) in height and over 1 meter (3.3 feet) in diameter. It can live to 1,000 years. After the first 10 to 20 years of vertical growth, laterally spreading branches develop.

The leaf blades (a) are fan-shaped, lobed, and have **dichotomous venation** (each vein branches into two smaller veins). The alternate leaves are mainly two-lobed (biloba, b) on long shoots (c) and wavy-margined on short spur shoots (d) where they appear in a whorled arrangement. In autumn, the deciduous leaves fade to yellow (chlorophylls disappear) and all are shed from a tree within a few hours, especially after a hard frost.

Reproduction. Unisexual structures are borne on separate trees (**dioecious**). The male structures are catkin-like strobili (e) of spore cases that develop in leaf axils. Two male spore cases (f) develop on a stalk (g). The female strobilus has two terminal, "naked" ovules (h) on a stalk (i) and thus, *Ginkgo* is referred to as a **gymnosperm** (= naked seed).

In the spring, wind carries pollen to an opening (**micropyle**) in the ovule. The seed (j) has a fleshy outer layer with the odor of rancid butter when mature.

An embryo with two primary leaves (cotyledons) develops after the seed has been shed.

Of interest . . . this "living fossil" may be the oldest surviving seed plant. It is cultivated as an ornamental shade tree because of its resistance to pathogens and pollutants, and its drought and low-temperature tolerance. The tree needs sun, but grows in almost any soil.

Because of the odoriferous seeds from female trees, male trees are preferred. There are various cultivars selected for particular shapes. 'Autumn Gold' is a broad-shaped cultivar. The 'Fairmount' cultivars are cone-shaped. 'Sentry' *(Ginkgo biloba* cv. *fastigata)* cultivars are narrow, column shaped.

COLOR CODE

yellow:	long shoot leaf (b), male strobili (e), spore cases (f)
green:	leaves (a), stalks (g, i)
tan:	long shoot (c)
brown:	spur shoots (d)
light green:	ovules (h)
yellow-orange:	seed (j)

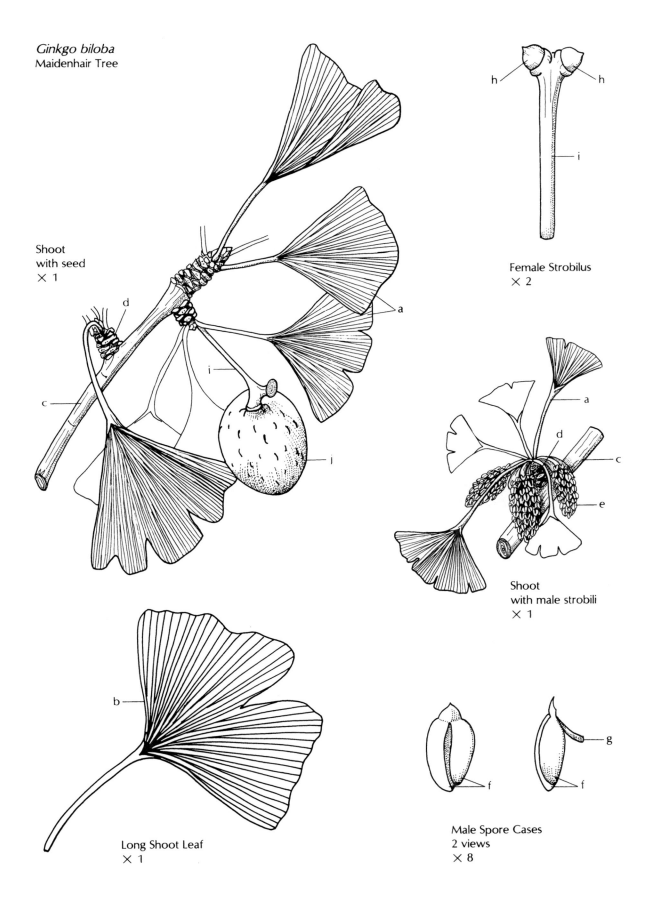

Ginkgo biloba
Maidenhair Tree

Shoot
with seed
× 1

d

c

i

a

j

Female Strobilus
× 2

h   h

i

Shoot
with male strobili
× 1

a

d

c

e

Long Shoot Leaf
× 1

b

Male Spore Cases
2 views
× 8

f   g

f

Conifers

Conifers have been present on the earth since the Carboniferous. Present-day conifers have fossil records dating back to the Jurassic time of the Mesozoic. Conifers, as well as cycads, Ginkgo and gnetes, are referred to as **gymnosperms** (gymno = naked, sperm = seed) because of their naked seeds. There are 550 species of 50 genera of conifers.

Characteristics. Almost all conifers are woody trees, having one central trunk from which branches extend. Most have narrow, evergreen leaves commonly called **needles** (see 21, 22). Leaves are borne singly or in fascicles. They may be arranged in a spiral, opposite, or whorled manner, or occur on short spur shoots. *Larix* (larch) and *Taxodium* (bald cypress) have deciduous leaves that are shed in autumn.

Conifers prefer cooler climates forming climax forests at high altitudes, as well as in the temperate regions. Conifers are the tallest known trees, with giant sequoia (*Sequoiadendron giganteum)* and redwood *(Sequoia sempervirens),* which may reach 115 meters and live thousands of years. There are seven families: pine family (Pinaceae), araucaria family (Araucariaceae), sequoia family (Taxodiaceae), cypress family (Cupressaceae), podocarpus family (Podocarpaceae), plum-yew family (Cephalotaxaceae) and yew family (Taxaceae).

Reproduction. Conifers bear pollen sacs and ovules in separate structures (stobili = cones) on the same plant (monoecious). Wind currents carry pollen to female strobili, where the naked seeds are borne. Or the strobilus may be an open fleshy aril that surrounds a single seed, as in *Taxus* (yew).

Of interest ... **timber** and **paper pulp:** *Abies* (fir), *Chamaecyparis* spp. (false-cypress), *Picea* spp. (spruces), *Pinus* spp. (pines), *Pseudotsuga menziesii* (Douglas fir), *Sequoia* (redwood), *Taxodium* (bald cypress); **naval stores** (turpentine, wood oil, wood tars, rosin): *Picea* spp. (spruces), *Pinus* (pines); **food:** *Juniperus* (juniper "berry" flavoring), *Pinus* spp. (seeds of pinyon pines); **ornamentals:** *Abies* (fir), *Araucaria araucana/A. imbricata* (monkey puzzle tree), *A. excelsa* (Norfolk Island pine), *Cupressus* (cypress), *Juniperus* spp. (juniper, red cedar), *Larix decidua* (European larch), *L. laricina* (tamarack, eastern larch), *L. occidentalis* (western larch), *Picea* spp. (spruces), *Taxus* (yew), *Thuga* (arborvitae, white cedar), *Tsuga*

spp. (hemlocks); **poisonous:** *Taxus* (yew: wood, bark, leaves, seeds); **medicinal:** *Taxus brevifolia* (Pacific yew) bark contains taxol, used to treat breast and ovarian cancers; *Abies balsamina* (balsam fir), pitch from blister/bubbles on trunks was used to treat burns by Ojibwa Indians. It prevents infections, hastens healing, and kills the pain; **tallest tree on earth:** *Sequoia sempervirens* can be up to 25 feet (7.6 meters) in diameter and up to 350–375 feet (106.2–113.75 meters) high. It is only found in Northern California where 96% of virgin redwood forests have been cut down.

Picea glauca   White Spruce

As a member of the pine family (Pinaceae), this tree has needle-like, evergreen leaves (a). The male cone is composed of spirally arranged cone scales (b) with pollen sacs on the sides. In the spring, after clouds of yellow pollen (c) are dispersed, the cone disintegrates.

The young female cone has woody, spirally arranged cone scales (d) with outer bracts (e). On the inner surface of each scale, two ovules (f), that after fertilization, become embryos that develop into seeds. At maturity, the cone (g) opens and winged seeds (h) are released.

Taxus   Yew

This yew family (Taxaceae) shrub has flat, evergreen leaves (i) and bears male and female structures on separate plants (dioecious). The female reproductive structure has scales (j) at the base, and a band of tissue (k) that matures into a fleshy aril (l) that envelops the seed (m). The red aril attracts and is eaten by birds that discard the poisonous seed.

COLOR CODE

green:	leaves (a, i)
red:	cone scales (b), aril (l)
yellow:	pollen (c), scales (j)
tan:	cone scale (d), bract (e), ovule (f), seed (h)
brown:	cone (g)
light green:	tissue (k)
brown-green:	seed (m)

Conifers

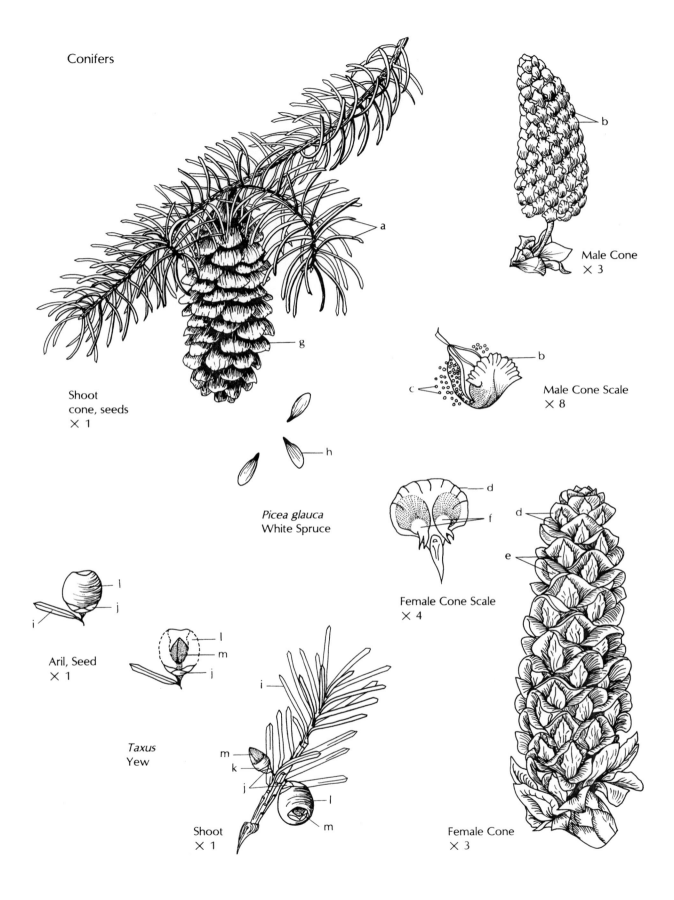

Shoot
cone, seeds
× 1

Male Cone
× 3

Male Cone Scale
× 8

Picea glauca
White Spruce

Female Cone Scale
× 4

Aril, Seed
× 1

Taxus
Yew

Shoot
× 1

Female Cone
× 3

Gnetes

There are three genera in this division: *Ephedra, Welwitschia,* and *Gnetum*. Pollen grains of *Ephedra* have been found from the Permian. Even though origins and relationships of these plants with others are not known, they are termed gymnosperms because they bear exposed ovules. Most Gnetes are **dioecious** (two households) with male and female cones on separate plants.

Ephedra

These plants grow in warm, dry temperate areas. They are perennial shrubs. Oppositely branching stems (a) bear whorls of scale-like evergreen leaves (b) and male structures (**strobili**, c) at the nodes. Sporangia (d) are protected by bracts (e) and bracteoles (f). Stalked female structures (strobili, g) are borne at nodes and protected by bracts (h). Two seeds (i) develop after fertilization.

An alkaloid, ephedrine, occurs in these plants. In ancient China, ephedrine was used to relieve hay fever and asthma symptoms. Its action stimulates the heart and nervous system, by constricting blood vessels and expanding bronchial tubes. This raises blood pressure and can cause heart attack, tachycardia, palpitation, psychosis and death.

Welwitschia mirabilis

This unusual plant, discovered in 1859 by Friedrich Welwitsch, is native to a small area on the dry, southwest coast of Africa. Plants may live for over 2,000 years. Its woody stem (j) protrudes slightly out of the sandy soil and has only 2 strap-shaped leaves (k) produced from long-lived intercalary meristems (l).

Unisexual cones, borne on stalks that arise from the apical meristemic region (see 15), are produced on separate plants (dioecious). The male cone is composed of **bracts** (cone scales) with **bracteoles** (m) covering a smaller **strobilus** on each bract. With the bracteoles removed, six spore cases (**sporangia**, n), with three lobes each, may be seen. They surround a central, sterile ovule (o). Pollen is formed in the sporangia. The four-sided female cone has overlapping scales (p). An **integument** (q), consisting of an outer layer of cells that covers the ovule, protrudes from each scale.

Gnetum

A plant of the tropics, most species are vines (lianas) with a few shrubs and trees. Opposite leaves (r) resemble those of dicot flowering plants. Unisexual strobili are produced on separate plants (dioecious).

COLOR CODE

green:	stems (a), leaves (b, r), stobilus (c, g), bracts (e, h), bracteoles (f)
yellow:	sporangia (d,)
tan:	seeds (i), stem (j), bracteoles (m), ovule (o), scales (p)
light green:	apical meristem (l)
gray-green:	leaves (k)

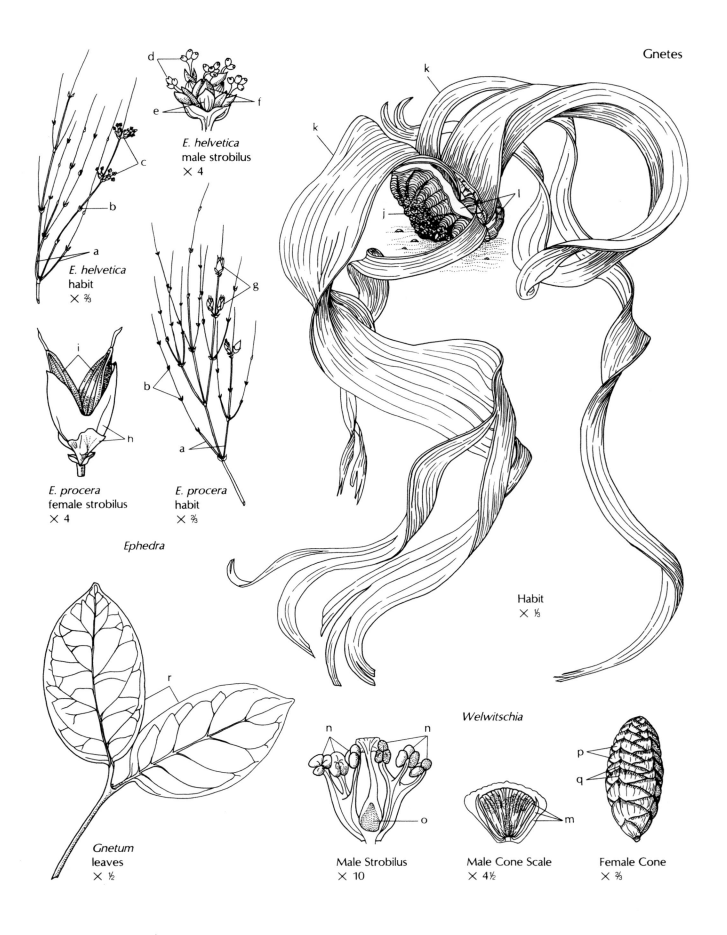

Gnetes

E. helvetica
male strobilus
× 4

E. helvetica
habit
× ⅔

E. procera
female strobilus
× 4

E. procera
habit
× ⅔

Ephedra

Habit
× ⅓

Welwitschia

Gnetum
leaves
× ½

Male Strobilus
× 10

Male Cone Scale
× 4½

Female Cone
× ⅔

Flowering Plant Classification

The dominant plants of the earth today are the flowering plants. While their fossil record dates from the Cretaceous when they diversified rapidly, their ancestral origin is not known (see 41). Sizes range from a one-millimeter duckweed (*Wolffia*) to the 100-meter tall Australian gum tree *(Eucalpytus).*

Flowering plants are called **angiosperms** (angio = covered, sperm = seed), plants that have seeds enclosed in a fruit that develops from an ovary. There are two classes: Magnoliopsida, the **dicots** (embryos with two seed leaves—dicotyledons), and Liliopsida, the **monocots** (embryos with one seed leaf - monocotyledons). There are 354 families in the two classes. Flower characteristics and differences in chloroplast DNA separate subclasses and families.

Dicot (Magnoiliopsida) Subclasses

Magnoliidae (36 families). These flowering plants are among the most primitive. Their flowers usually have an indefinite number of parts with many petals, many stamens, and many separate carpels. Pollen grains have one pore (called **unicolpate** condition, see 31).

Hamamelidae (23 families). The flowers are small, usually unisexual, and are adapted towards wind pollination.

Caryophyllidae (14 families). This group alone has the water-soluble betalain pigments (see 4) of betacyanins (beet red-purple) and betaxanthins (yellow, orange, orange-red). The flower ovary usually has free-central to basal placentation (see 28) and the ovule integuments are twisted and bent.

Dilleniidae (69 families). The flower has many stamens, which mature in a sequence from inside to the outside. Parietal placentation of the ovules is common.

Rosidae (108 families). Structurally, this group has no one common feature, but developmentally, the plants are placed between the primitive Magnoliidae and the advanced Asteridae.

Asteridae (43 families). In 70% of these plants, the flower petals are fused, at least at the base, and the stamens arise form the petals.

Monocot (Liliopsida) Subclasses

Alismatidae (14 families). These primitive monocots are mostly aquatic plants. The flower carpels are usually separate and the pollen grains possess three pores (**tricolpate condition**, see 31).

Commelinidae (16 families). Plants in this subclass are mostly terrestrial. The flower carpels are usually joined and the seed usually has a starchy endosperm (see 40). The sepals and petals are separate or reduced to bristles or absent.

Zingiberidae (9 families). This subclass is separated from Commelinidae by the presence of septal nectaries, and vessels (see 8) are found only in the roots.

Arecidae (5 families). The flower are usually small and numerous in a cluster, subtended by a bract (**spathe**), and often aggregated into a fleshy spike (**spadix**).

Liliidae (17 families). Tepals (sepals and petals that look alike) and well-developed nectaries are usually present in the flower. This subclass includes the most specialized monocot flowers.

Evolutionary Floral Trends

Evolutionary changes from primitive to more advanced forms may include the following. **Reduction**, where flower structures are less in size, number, and kinds of parts. Petals are most frequently lost; stamens or carpels may be lost (bisexual to unisexual flower, see 28), with, usually, sepals the last to be lost because they protect the young flower. There may be **fusion** of parts, such as petals forming a tube instead of being separate. Instead of a superior position, the **ovary** is in an **inferior position** (see 28). The lower, less-exposed position provides more protection of the ovules. The shift from **insect to wind pollination** is regarded as an advanced evolutionary change. A change from **radial to bilateral symmetry** is an adaptation for specific insect pollinators. In radial (actinomorphic) symmetry, similar parts are regularly arranged around a central axis. With bilateral (zygomorphic symmetry), flowers can be divided into equal halves in one plane only. Some advanced flowers are irregular (asymmetrical), incapable of being divided into equal halves in any plane. **False flowers** (pseudanthia) are composed of a head (capitulum) of small flowers that look like one flower. In the center of the head are fertile flowers surrounded on the outside by petaloid structures or large-petalled sterile flowers.

Classification of Flowering Plants

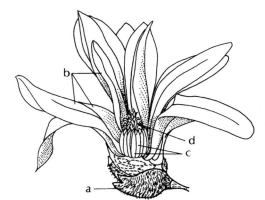

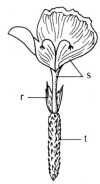

Magnoliidae
Magnolia Family (Magnoliaceae)
Magnolia
many, separate parts
bisexual
superior ovaries
insect pollinated
radial symmetry
woody plant
× 1

Hamamelidae
Birch Family (Betulaceae)
Betula Birch
few, separate parts
unisexual
inferior ovary
wind pollinated
irregular flowers
clustered in catkins
woody plant
× ½

Asteridae
Aster Family (Asteraceae)
Tagetes Marigold
few, fused parts
unisexual, bisexual
inferior ovary
insect pollinated
irregular flowers
clustered in false flower
herbaceous plant
× 2

Arecidae
Arum Family (Araceae)
Anthurium Flamingo Flower
few, fused parts
bisexual
superior ovary
insect pollinated
radial symmetry
spadix of flowers
herbaceous plant
× ⅓

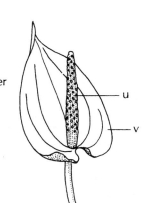

Caryophyllidae
Pink Family (Caryophyllaceae)
Dianthus Pink
some fused parts
bisexual
superior ovary
insect pollinated
radial symmetry
herbaceous plant
× 1

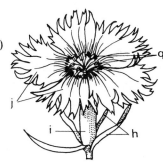

Liliidae
Orchid Family (Orchidaceae)
Phalaenopsis Moth Orchid
few, fused parts
bisexual
inferior ovary
insect pollinated
bilateral symmetry
herbaceous plant
× ⅔

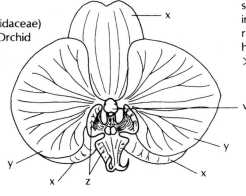

Dilleniidae
Mustard Family (Brassicaceae)
Lobularia Alyssum
few, separate parts
bisexual
superior ovary
insect pollinated
radial symmetry
herbaceous plant
× 6½

COLOR CODE

green:	bract (a, h, r), leaf (e), female catkin (g), sepals (i, k, o)
white:	perianth (b), spadix of flowers (u), male-female column (w)
yellow:	stamens (c, m, q)
light green:	pistils (d, n), ovary (t)
pink:	petals (p, y), sepals (x)
tan:	male catkin (f)
red:	petals (j), bract (v)
purple:	petals (l)
orange:	petal (s)
dark pink:	lip petal (z)

Rosidae
Rose Family (Rosaceae)
Malus Apple
some fused parts
bisexual
inferior ovary
insect pollinated
radial symmetry
woody plant
× 1

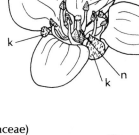

Major Land Plant Communities

Biomes are characterized by their natural vegetation of dominant plants determined by climate and position of the continents. In earth's history, biomes changed as the climate cycles cooled or warmed and continents drifted.

Arctic Tundra (a). This biome is wet, arctic grassland that supports lichens, grasses, sedges, and dwarf woody plants and is cold most of the year. Below a few centimeters, the soil is permanently frozen (**permafrost**).

Northern Conifer Forests (b). Other names are boreal forest, spruce-fir forest and taiga. Low temperatures prevail for at least half the year. The plants include spruces, firs, and pines.

Temperate Deciduous and **Rain Forests** (c). In the Temperate Zone, deciduous plants have an even distribution of 30 to 60 in. (76 to 152 cm) of rainfall annually and moderate seasonal temperatures. Redwoods, Douglas fir, western hemlock and western red cedar (see 73) dominate temperate rain forests along the West Coast of the United States. Rain forests of the subtropics with high moisture and even temperatures have broad-leaved evergreens.

Temperate Grassland (d). These areas have a low annual rainfall of 10 to 30 in. (25 to 76 cm) which effectively prevents forest formation. Temperatures range from very cold winters to very hot summers.

Tropical Savanna (e). Warm regions with 40 to 60 in. (102 to 152 cm) of rainfall and with a prolonged dry season have drought- and fire-resistant grasses with scattered trees.

Desert (f). With less than 10 in. (25 cm) of annual rainfall in desert regions, plants include rain-season annuals, succulents, and shrubs which can remain dormant for long periods. Plants have light-reflecting surfaces created by white hairs or waxy cuticle. Small stomata (see 10) are sunken or closed much of the time, opening at night. They also have extensive root systems.

Chaparral (g). A mild climate with abundant winter rains and dry summers supports a community of trees or shrubs with thick evergreen leaves. Fire is an important maintenance factor. Native plants are fire-adapted.

Tropical Rain Forest (h). Equatorial lowlands have 80 to 90 in. (203 to 230 cm) of annual rainfall. Broad-leaved angiosperms with evergreen leaves include trees, vines, and epiphytes. Without freezing temperatures, there are no deciduous-leafed plants. Photosynthesis is continuous throughout the year.

Tropical Scrub (i) and **Deciduous** (j) **Forests.** These are areas without an even distribution of rainfall. They are composed of thorn forests of distorted, small hardwood trees.

Mountains (k). Many irregular bands of different types of plant communities are present. **Alpine tundra** is above the treeline (too high for trees to survive). It occurs in mountainous areas much south of the **arctic tundra.** Coniferous forest in temperate zones, such as the Appalachian Mountain region in Eastern United States, has dominant spruces, pines, firs and tamarack. Highland rain forests in the tropics can support trees.

Of interest . . . Forests of the earth absorb carbon dioxide and release oxygen during photosynthesis (see 24). As forests are cut down and burned, potential oxygen is lost and carbon dioxide stored in the trees is released into the atmosphere. The rise of atmospheric carbon dioxide causes a rise in temperature. Along with gas-emitting power plants, automobiles and factories, the temperature rose 1° Fahrenheit during the last century. As temperatures continue to rise, land habitats disappear. Satellite imaging can monitor deforestation. Brazil and Central Africa have the largest tropical rain forests, although deforestation by logging and mining is destroying this biome. From 1978 to 1996 burning ("slash and burn") and logging to create pastures for livestock and fields for agriculture destroyed 12.5% of the Amazon's rain forest. Destruction continues unabated.

COLOR CODE

tan:	tundra (a)
dark green:	northern conifer forest (b)
red:	temperate deciduous and rain forest (c)
yellow:	grassland (d)
light green:	tropical savanna (e)
orange:	desert (f)
pink:	chaparral (g)
purple:	tropical rain forest (h)
brown:	tropical scrub forest (i)
gray:	tropical deciduous forest (j)
blue:	mountains (k)

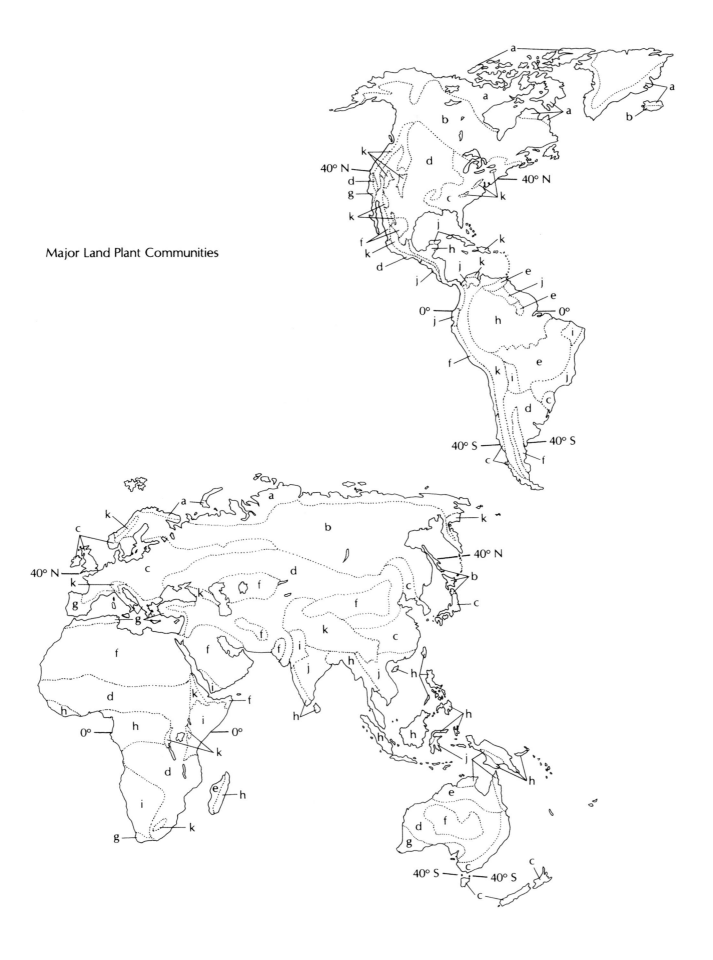

Major Land Plant Communities

Magnolia Family (Magnoliaceae)

This family of plants is considered to be among the most primitive of the flowering plants. It consists of woody shrubs and trees that may have deciduous or evergreen leaves. The flower parts (sepals, petals, stamens, pistils) are spirally arranged and usually without a definite number. These are considered to be primitive characteristics. There are 12 genera and over 200 species.

The alternate leaves are simple and usually have smooth (entire) margins. They are pinnately (feather-like) veined. Petioles are present. Stipules (see 23) enclose young leaf buds and are shed as the leaves expand. Bracts that enclose flower buds are shed as the flower opens. Flowers are often large, showy, and solitary. These features attract pollinators, usually primitive insects, such as beetles.

Flowers are usually bisexual, having both male and female structures. Sepals and petals look alike or there may be 3 sepals and 6 to many petals. Many stamens are spirally arranged around a raised axis with many pistils. Again, these are considered to be primitive characteristics.

Each pistil has an ovary of one carpel containing one to many ovules in parietal placentation (see 28) and one style and stigma. Fruit types in this family include follicles (as in *Magnolia*), samaras (as in *Liriodendron*) and berries.

Of interest... **lumber:** *Liriodendron tulipifera* (tulip tree, yellow poplar); **cabinet work:** *Magnolia* spp.

(magnolias); **ornamentals:** *Liriodendron tulipifera* (tulip tree), *Magnolia acuminata* (cucumber tree), *M. denudata* (Yulan magnolia), *M. grandiflora* (bull-bay), *M. stellata* (star magnolia), *M. tripetala* (umbrella tree), *M. virginiana* (sweet-bay), *M.* × *soulangeana* (saucer magnolia, a hybrid between *M. denudata* and *M. liliflora*, has large, deep pink, sterile flowers), *Michelia fuscata* (banana shrub), *Talauma, Illicium vernum* (Chinese anise, star anise).

Magnolia spp.   Magnolia

The illustrations represent three showy ornamental *Magnolia* species: *M.* × *soulangeana* (saucer magnolia), *M. stellata* (star magnolia), and *M. grandiflora* (bull-bay, southern magnolia).

In the spring, magnolia flowers open before the leaves (a) expand. Bud coverings of leaf stipules (b) and flower bracts (c) are fuzzy with hairs (**pubescence**).

The star magnolia flower has many similar appearing sepals and petals (e) and many stamens (f) with anthers (g) that open longitudinally on both sides. Pistils (h) are arranged spirally on a central axis and occur in a position superior to other flower parts. Each pistil has a style (i) extending from the ovary (j).

In fruit, the pistils form a cone-like aggregate of follicles. A **follicle** (k) splits along one seam. Inside each are one or two, bright red, flesh-covered seeds (l) attached by thin strands (called **funiculi**, m) to the ovary wall (n).

COLOR CODE

green:	leaf (a), pistils (h)
gray-green:	leaf bud stipules (b), flower bracts (c)
brown:	stem (d)
white:	sepals and petals (e)
light yellow:	stamens (f), anthers (g)
pale green:	style (i), ovary (j)
pink-brown:	outer follicle surface (k)
red:	seed (l)
tan:	ovary wall (n)

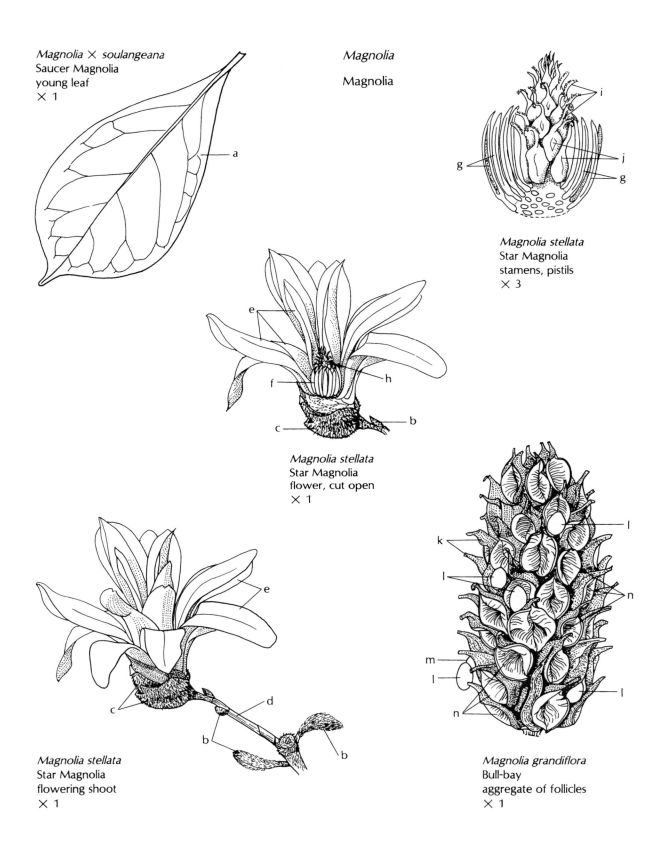

Magnolia × *soulangeana*
Saucer Magnolia
young leaf
× 1

Magnolia

Magnolia

Magnolia stellata
Star Magnolia
stamens, pistils
× 3

Magnolia stellata
Star Magnolia
flower, cut open
× 1

Magnolia stellata
Star Magnolia
flowering shoot
× 1

Magnolia grandiflora
Bull-bay
aggregate of follicles
× 1

Laurel Family (Lauraceae)

Aromatic bark and leaves, and small flowers, with parts in whorls of three are characteristics of this family. A distinguishing feature of the family is the **valvate** anther dehiscence (see below in *Sassafras*). Plants are mostly evergreen, except in the Temperate Zone where they are deciduous. Forms are trees, shrubs, or vines. Usually, the leaves are alternate and have entire margins. There are about 32 genera with 2,500 species with most occurring in the tropics and subtropics of Southeast Asia and Brazil.

The flowers are usually bisexual and develop as clusters in the leaf axils. A single flower has 6 sepal-petal parts and 12 stamens in 3 whorls. But one or two stamen whorls may be sterile "honey leaves" called **staminodes.**

The single pistil usually has an ovary above the other flower parts (a superior position), containing one ovule in parietal placentation, one style and one stigma. For fruit types, there are **drupes** or **berries**, which are enclosed at the base by the persistent sepal tube.

Of interest . . . **aromatic oils:** *Cinnamomum camphorum* (camphor), *C. zeylanicum* (cinnamon bark spice), *Lindera Benzoin* (spice bush), *Laurus nobilus* (bay laurel, bay tree, sweet bay), *Persea* spp. (avocado), *Sassafras albidum* (sassafras); **ornamentals:** *Laurus, Lindera benzoin* (spicebush), *Umbellularia californica* (California laurel); **food:** *Persea americana* (avocado); **lumber:** *Beilschmiedia, Endiandra* (walnut bean, oriental walnut), *Litsea* (pond spice), *Ocotea rodioei* (greenheart, was used to make the original gates of the Panama Canal locks), *O. bullata* (South African stinkwood).

Sassafras albidum   Sassafras

Before the deciduous leaves are shed, this tree is easily recognized by the shape of the alternate leaves (a). They are entire to five-lobed, and all types may be present on one plant. In autumn, the leaves turn yellow.

The bark is thick, aromatic, and dark reddish brown. Bark of the roots was distilled for oil of sassafras, once used to flavor candy, root beer, and soap. Sassafras tea was popular, but now is regarded as possibly carcinogenic. Usually seen as a small tree along forest edges, it may reach a height of 30 meters.

Unlike most members of the laurel family, sassafras has only male flower clusters or only female flower clusters on a plant. This is a **dioecious** condition (of two households). The male flower has 6 sepals (b) and 3 whorls of stamens; an outer whorl of 6 functional stamens (c), an inner whorl of 6 stalked, gland-like staminodes (d), and 3 more functional stamens (e) in the center. The 4 cells (g) of the anther (h) have flap-like valves (i), that flip up to release pollen.

Found on another tree, the female flower has 6 sepals (k), 6 staminodes (l), and a single pistil with an ovary (m) containing an ovule. At maturity, the ovule is the fruit of the purple **drupe** (q). This fruit is enclosed at the base by an expanded cup (r) made up of sepals and pedicel (s).

COLOR CODE

green:	leaf (a), stem (u)
yellow-green:	sepals (b, k), stamens (c, e), pedicel (f, p), cells (g), anther (h), valves (i), filament (j)
yellow:	staminodes (d, l)
pale green:	ovary (m), style (n), stigma (o)
purple:	drupe (q)
red:	cup (r), pedicel (s), peduncle (t)

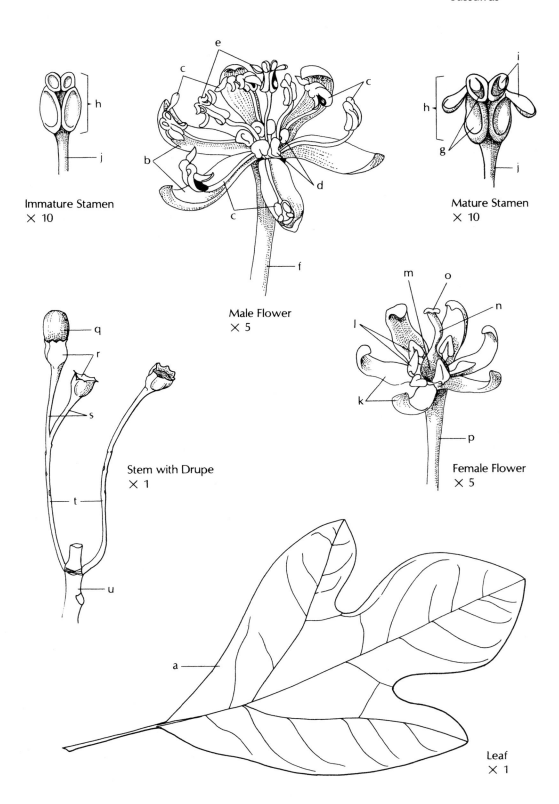

Sassafras albidum

Sassafras

Immature Stamen
× 10

Male Flower
× 5

Mature Stamen
× 10

Stem with Drupe
× 1

Female Flower
× 5

Leaf
× 1

Water Lily Family (Nymphaeaceae)

Water lilies are primitive, fresh water flowering plants. Long petioles are attached inside the blade margin (peltate leaf), or falsely so, permitting the leaves to float on the surface. The flower bud grows upward to the water surface on a long peduncle. Floating or submerged leaves are simple and alternate and often have milky latex ducts. These herbaceous plants are usually perennial. There are 9 genera and over 90 species.

The flower is often showy, single and bisexual. Flower part numbers are indefinite, a primitive characteristic, but mostly in 3's. The sepals are usually green and the ovary is usually superior. Fruits are **follicles**, an aggregate of indehiscent **nutlets** or a **berry**.

Of interest... **ornamentals:** *Brasenia* (water shield), *Cabomba* (fanwort, used as an oxygenator in aquaria), *Nelumbo lutea* (lotus lily, water chinquapin), *N. nucifera* (oriental sacred lotus), *Nuphar* (yellow water lily, spatterdock), *Nymphaea* spp. (water lilies), *Victoria amazonica* (Amazon water lily, giant water lily); **food:** rhizomes of *Nelumbo pentapetala* were used a starchy food source by native Americans.

Victoria amazonica, a native of the Amazon River in South America, was named after Queen Victoria of England when it was first discovered in 1801. Leaves may reach 7 feet (214 cm) in diameter. Leaf margins have upturned edges with notches that allow water to drain off the leaves. They live along river margins in shallow lakes.

White flower buds open at sunset, heating to aid in spreading their strong sweet scent. This attracts the main pollinator, a one-inch scarab beetle, *Cyclocephala hardyi.* As the flower temperature falls, the scent disappears; the flower closes with beetles trapped inside. The next day, the now dark pink flower reopens and releases the pollen-covered beetles, ready to fly off to another new white flower.

Nymphaea odorata   Water Lily

This freshwater perennial plant has a stout underground stem (**rhizome**) with roots that anchor the plant in soil. Long petioles (a) make it possible for the buoyant leaf blades (b) to float on the water surface. The petiole is attached at a cleft in the blade where veins radiate from it. Large air passages run the length of both leaf petioles and flower peduncles (c) making them buoyant. Specialized parenchyma tissue with intercellular air spaces also aids water plants to float.

The flower has 4 sepals (d), many petals (e), and many stamens (f). The outer whorls of stamens have smaller anther sacs and tend to be more petal-like than stamens of the inner whorls. The stamen's filament (g) extends above the anther sacs (h). The inner, female part of the flower has a compound ovary (i) with many outwardly radiating carpels. A stigmatic surface (j, k) covers the upper portion of each carpel (l), which has numerous ovules (m) inside.

After fertilization, the flower grows downward below the water surface to the mud substratum, where a many-seeded berry (n) matures. Sepals and petals disintegrate, leaving scars (o) on the outer surface.

COLOR CODE

tan:	petiole (a), peduncle (c)
green:	blade (b)
yellow-green:	sepals (d), berry (n)
light lavender:	petals (e)
dark lavender:	petal tips (shaded), stamen tips (shaded)
yellow:	stamens (f), filament (g), anther sacs (h), ovary (i), stigmas (j), stigmatic surface (k), carpel (l)
white:	ovules (m)

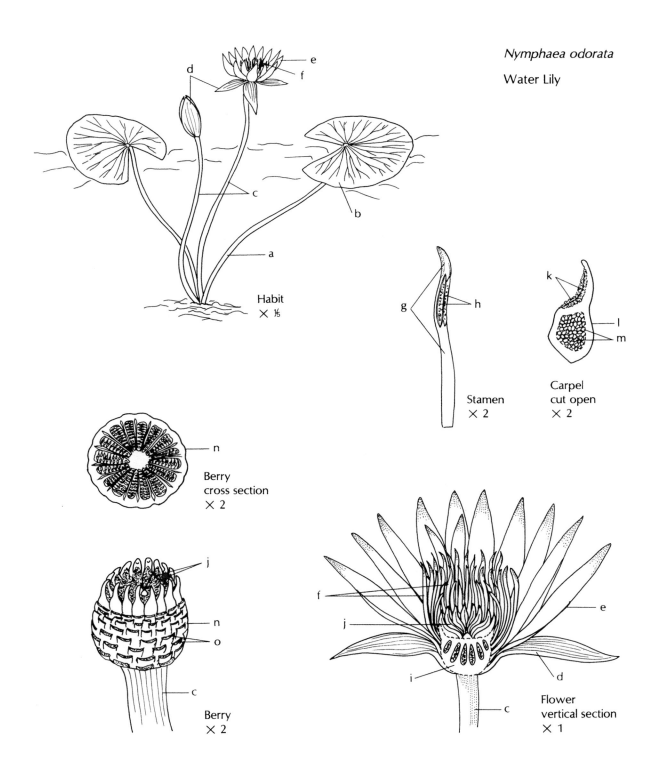

Nymphaea odorata

Water Lily

Habit
× ⅕

Stamen
× 2

Carpel
cut open
× 2

Berry
cross section
× 2

Berry
× 2

Flower
vertical section
× 1

Buttercup Family (Ranunculaceae)

Plants in this primitive family are mostly annual or perennial herbs, but a few are woody vines. They prefer moist habitats. The leaves are usually alternate, compound, and have palmate venation. Typically, the flower is bisexual with spirally arranged parts. There are over 50 genera and over 1800 species found in temperate and cold areas of the Northern Hemisphere.

There are usually 5 sepals (but there may also be 3 to many) and 5 petals or varying numbers or none. The petals sometimes serve as nectaries. There are many spirally arranged stamens.

Pistils are 3 to many. The ovary is superior with one to many carpels and one to many ovules in parietal placentation. The fruits are follicles, achenes, berries, or capsules.

Of interest... **ornamentals:** *Aconitum* (monkshood, extremely poisonous), *Actaea* (baneberry), *Anemone* (windflower), *Anemonella* (rue anemone), *Aquilegia* (columbine), *Caltha* (marsh marigold, cowslip), *Clematis* (clematis), *Cimicifuga* (bugbane), *Coptis* (golden thread), *Delphinium* (delphinium, larkspur), *Helleborus* (Christmas rose), *Hepatica* (hepatica), *Hydrastis* (golden seal), *Isopyrum* (false rue anemone), *Nigella* (love-in-a- mist), *Paeonia* (peony), *Ranunculus* (buttercup), *Thalictrum* (meadow rue), *Trollius* (globe flower).

Ranunculus pensylvanicus   Buttercup

This wild plant is an erect, branched perennial growing 30 to 70 centimeters high. Its hairy stem (a) has alternate leaves (b, c), which are deeply lobed forming three leaflets (d) that have toothed margins and palmate venation.

The flower (f) has reflexed sepals (g) that are longer than the petals (h). Numerous stamens (i) surround the base of a central aggregate of numerous pistils (j). The carpel of each pistil has one ovule (k). At maturity, the fruit consists of an aggregate of achenes (m).

COLOR CODE

light green:	stem (a), pistils (j), ovules (k)
green:	petioles (b), leaflets (d), sepals (g), peduncle (l)
tan:	roots (e), achenes (m)
yellow:	flower (f), petals (h), stamens (i)

Ranunculus pensylvanicus

Buttercup

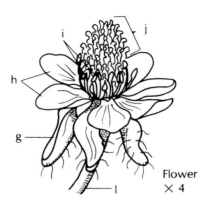

Flower
× 4

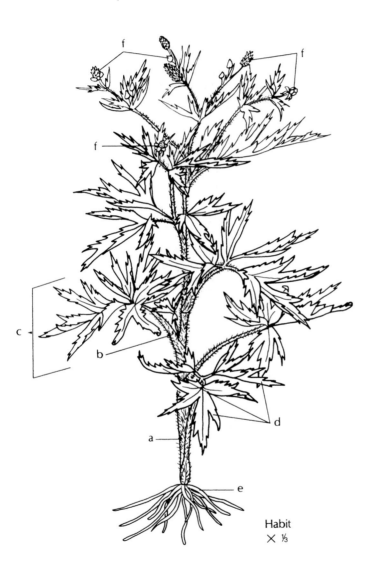

Habit
× ⅓

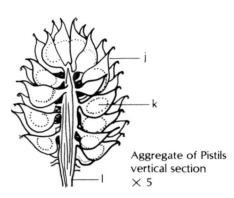

Aggregate of Pistils
vertical section
× 5

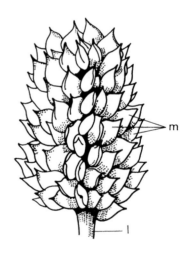

Aggregate of Achenes
× 4

Witch Hazel Family (Hamamelidaceae)

Trees and shrubs are represented in this family. There are 23 genera and about 100 species, mostly found in Asia. The alternate leaves are simple, deciduous or evergreen, and have deciduous, reduced leaves (**stipules**) at the petiole base. Often, star-like (**stellate**) or tree-like (**dendrite**) hairs occur on leaf surfaces.

Flowers are bisexual or unisexual on the same plant (**monoecious**) or on separate plants (**dioecious**). Flower parts include 4-5 sepals, 4-5 petals or none, 2-8 stamens, and a single pistil.

Significant flower characteristics are a pistil with 2 styles that curve away from each other and an ovary consisting of 2 carpels and chambers that develop into a woody capsule.

Of interest... **extract:** *Hamamelis virginiana* (witch hazel); **lumber:** *Altingia excelsa* (rasamala), *Bucklandia*, *Liquidambar styraciflua* (sweet gum); **ornamentals:** *Corylopsis* (winter hazel), *Disanthus*, *Distylium*, *Fothergilla* (fothergilla, witch alder), *Hamamelis virginiana* (common witch hazel), *H. vernalis* (vernal witch hazel), *Liquidambar styraciflua* (sweet gum, with multicolored autumn foliage), *Loropetalum*, *Parrotia*, *Sinowilsonia*.

Hamamelis virginiana   Common Witch Hazel

Forked branches of this shrub or small tree were used in the past as divining rods for water-well sites. An astringent compound is extracted from leaves, stems, and bark for use in ointments and lotions. Leaves (a) have an uneven blade base (b) and microscopic dentrite hairs on the surfaces.

In this autumn-flowering species, two-loculed capsules (d) mature from the previous year's flowers at the same time as flower buds (e) are opening. As the capsule opens, two shiny seeds (f) are forcibly ejected.

Hamamelis vernalis   Vernal Witch Hazel

Unlike the common witch hazel, this shrub's flowers open in the spring. Stemless (sessile), bisexual flowers form in small clusters. Below each flower are hairy bracts (g). The flower parts include 4 triangular sepals (h), 4 strap-shaped petals (i), 4 stamens (j), which open (**dehisce**) by upward-turning anther valves (k), and a single pistil (m).

The pistil has 2 styles (n) and stigmas (o) and a half-inferior ovary consisting of 2 carpels (p). In each carpel is an ovule (q) with axile placentation (see 28). After fertilization, the ovule will form the seed that is later ejected. Surrounding the ovary are 4 nectar flaps (r), which attract insect pollinators.

COLOR CODE

green:	leaves (a), pistil (m)
tan:	stem (c), capsule (d), flower buds (e), bracts (g), peduncle (s)
dark brown:	seeds (f)
dark red:	sepals (h)
yellow:	petals (i), stamens (j), anther valves (k), filament (l), nectary (r)
light green:	style (n), stigma (o), carpel (p), ovule (q)

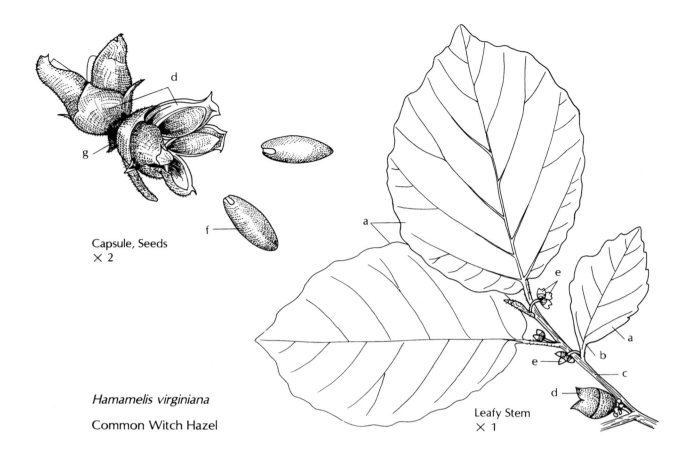

Capsule, Seeds
× 2

Hamamelis virginiana

Common Witch Hazel

Leafy Stem
× 1

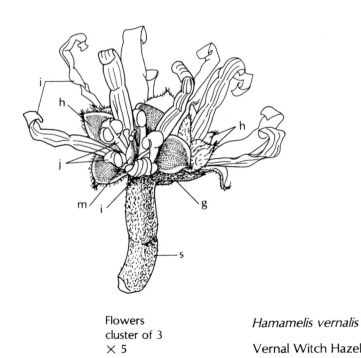

Flowers
cluster of 3
× 5

Hamamelis vernalis

Vernal Witch Hazel

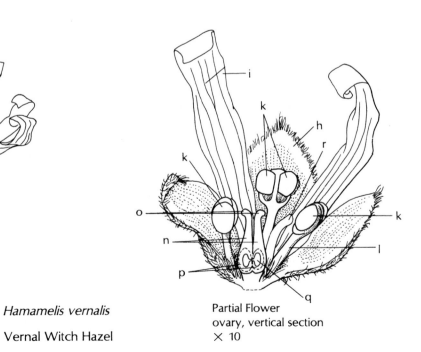

Partial Flower
ovary, vertical section
× 10

Elm Family (Ulmaceae)

There are only about 16 genera in this family of trees and shrubs. Most of the genera have unisexual flowers. The alternate leaves usually have toothed (**dentate)** margins, uneven blade bases, and deciduous stipules.

Parts of the flower include fused sepals of 4-8 lobes, stamens the same number as sepal lobes, and one pistil. The flower has no nectaries and no petals, but does have abundant pollen, that is wind-carried to the pistil's plumose stigmas. A single ovule develops in a superior ovary consisting of two fused carpels. Fruit types are a winged samara, drupe, or nutlet.

Of interest... **lumber:** *Ulmus* spp. (elms), *Planera abelica* (false sandalwood); **ornamentals:** *Ulmus glabra* (Scotch elm), *U. parvifolia* (Chinese elm); **weed tree:** *Ulmus pumila (Siberian elm).*

Ulmus americana   American Elm

This once common American tree has been greatly reduced in numbers by the Dutch elm disease caused by an ascomycete fungus, *Ceratocystis ulmi* (see 48).

The leaf has a short petiole (a), an uneven blade base (b) and doubly serrated margins (c). Its upper surface feels rough to the touch, while the lower surface is smooth. At the end of each growing season, the terminal bud falls off. The first lateral bud then forms a false "terminal" bud (d). This accounts for the zigzag stem (f) in elms.

In the spring, clusters of tiny flowers (g) dangle out from the bud scales (h) before the leaves expand. Flowers are grouped in clusters (**fascicles**, i) of 3 or 4. The bisexual flower has 8 lobes (j) on joined sepals (k) and 8 stamens (m) with the anthers (o) extended on long filaments (p).

The pistil produces one ovule in the ovary (q) and has many stigmatic hairs along 2 styles (r). The fruit is a winged (s) samara with a single seed (t). It is wind-dispersed in the spring.

Celtis tenuifolia   Dwarf Hackberry

This small tree has variably margined (entire and toothed) leaves (v) with lobed veins (w) and a drupe (x) type fruit which matures in the autumn.

COLOR CODE

dark green:	petiole (a), blade (b)
tan:	bud (d), lateral bud (e), stem (f), sepals (u)
orange:	sepal lobes (j)
red:	flowers (g), joined sepals (k)
yellow-green:	sepal base (l)
brown:	bud scales (h), seed (t)
yellow:	stamens (m), pollen (n), anther (o)
white:	filament (p), style (r)
light green:	ovary (q)
yellow-tan:	samara (s)
green:	leaf (v)
orange-red:	drupe (x)

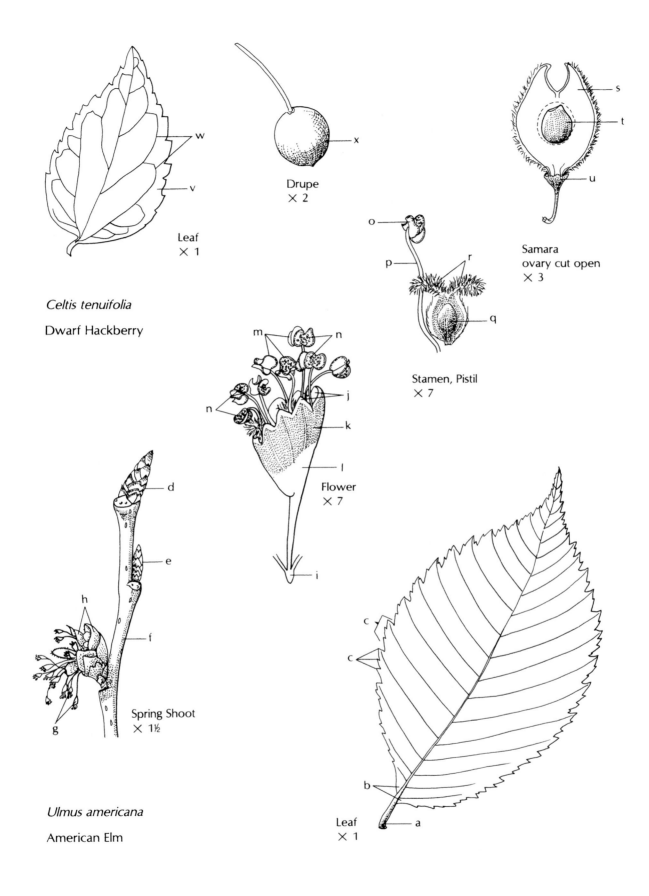

Leaf
× 1

Celtis tenuifolia

Dwarf Hackberry

Drupe
× 2

Samara
ovary cut open
× 3

Stamen, Pistil
× 7

Flower
× 7

Spring Shoot
× 1½

Ulmus americana

American Elm

Leaf
× 1

Beech Family (Fagaceae)

A nut fruit partially enclosed in basal bracts (acorn, beechnut, chestnut) is the most familiar characteristic of this family of eight genera. Genera are *Fagus* (beech), *Castanea* (chestnut), *Quercus* (oak), *Lithocarpus, Nothofagus, Castanopsis, Trigonobalanus,* and *Chrysolepsis*. Fossils indicate an origin during the middle of the Cretaceous.

Alternate, simple leaves have lobed, toothed, or entire margins and deciduous stipules. The flowers are usually unisexual and wind-pollinated.

The male flower occurs in **catkins** or singly and has 4–7 sepals, 4–40 stamens, and a vestigial pistil. The female flower can be single in a cup of bracts or in a cluster of 2–3 flowers, which may have bracts. Fused sepals of the female flower have 4–6 lobes. The pistil has an inferior ovary of 3–6 carpels, chambers (**locules**), and styles. Two ovules occur in axil placentation. They develop in each chamber, in which one develops and the other aborts.

Of interest... **hardwood lumber:** *Castanea spp.* (chestnuts), *Fagus* spp. (beeches), *Quercus* spp. (oaks; in the United States, oaks are the second most important source of lumber after conifers); **food:** nuts of chestnut, beech and oaks. Native Americans used acorns to make bread. Acorns are also an important source of food for wild animals and birds. Commercially, *Castanea sativa* (sweet chestnut), of southern Europe, is a source of nuts for purees, stuffings and stews; **cork:** bark of *Quercus suber* (cork oak) is stripped and dried to produce cork; **ornamentals:** *Castanea* spp. (chestnuts), *Fagus* spp. (beeches), *Lithocarpus* (tanoak), *Quercus* spp. (oaks); **plant diseases:** *Castanea dentata* (American chestnut) has been devastated to near extinction by chestnut blight, caused by an ascomycete fungus, *Endothia parasitica* (see 48); sudden oak death is a disease caused by *Phytophthora ramorum*, an oomycete (see 46).

Quercus spp.   Oaks

Oak forms vary from small shrubs to large trees. They have deep roots and occur in dry sites. There are over 400 species and many hybrids between species. Oaks are easily identified, even in winter, by the cluster of buds (a) at the end of the stem (c). The leaves (e, f, g) are deciduous or persistent. Young oaks tend to retain their leaves over winter and shed them in spring.

Oaks are divided into two groups: white oaks and red oaks. White oaks have leaves with rounded lobes and acorns that mature in the autumn of their first year. Red oak leaves have bristle tips and acorns that mature in the autumn of the second year.

Oaks are **monoecious** with separate male and female flowers on one tree. Male flowers form in pendant clusters, an indication of wind pollination. The male flowers (h) open before or with expanding leaves (k). The male flower has four to seven sepals (m) and six to twelve stamens (n).

The female flowers form in the axils of new spring leaves. This flower has joined sepals of 6 lobes (o) and a pistil consisting of 3 styles (p) and 3 stigmas (q). The ovary is enclosed in a whorl of bracts (**involucre**, r) which matures into a scaly cup (s). The fruit of cup and nut (t), commonly known as an **acorn**, may mature in one or two seasons.

COLOR CODE

tan:	buds (a, b), leaf scar (d), leaf (g), cup (s)
brown:	stem (c), peduncle (l)
green:	leaf (e)
red:	leaf (f)
yellow:	male flowers (h)
light tan:	stem (i), sepals (m), nut (t)
tan-orange:	bud scales (j)
grey-green:	young leaves (k)
yellow-green:	stamens (n)
light green:	sepals (o), styles (p), stigmas (q), involucre (r)

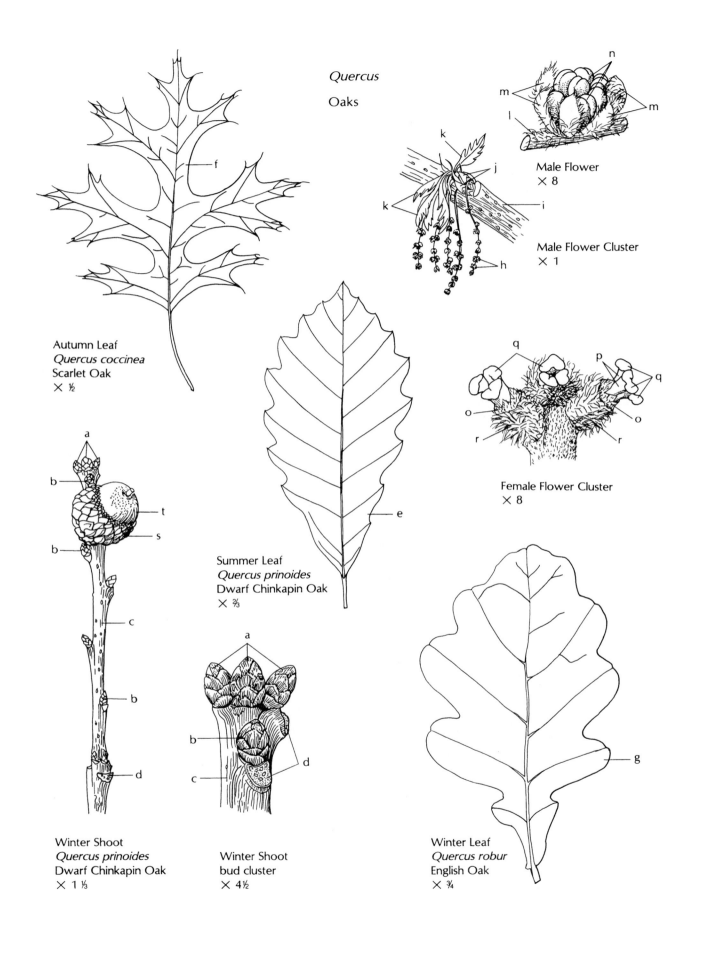

Quercus

Oaks

Male Flower
× 8

Male Flower Cluster
× 1

Autumn Leaf
Quercus coccinea
Scarlet Oak
× ½

Female Flower Cluster
× 8

Summer Leaf
Quercus prinoides
Dwarf Chinkapin Oak
× ⅔

Winter Shoot
Quercus prinoides
Dwarf Chinkapin Oak
× 1 ⅓

Winter Shoot
bud cluster
× 4½

Winter Leaf
Quercus robur
English Oak
× ¾

Birch Family (Betulaceae)

Trees and shrubs that comprise this family have deciduous leaves. Leaves are alternate and simple and have saw-tooth (serrate) margins. Stipules are present at petiole bases. The bark is prominently marked by **lenticels** (openings for gas exchange) and in some species it peels off in thin papery layers. There are six genera with about 170 species. Genera are *Alnus* (alder), *Betula* (birch), *Carpinus* (hornbeam), *Corylus* (hazelnut, filbert), *Ostrya* (hop-hornbeam), and *Ostryopsis.*

The flowers in this family are borne in catkin-like clusters of **cymules**. Many cymules make up a **cyme** (cluster of clusters). Each cymule has two to three flowers and several bracts. Clusters of male flowers and clusters of female flowers occur on the same plant (**monoecious**).

The male flower may have 4 **tepals** (sepals and petals that look alike) with 2–20 stamens per cyme. The female flower has a single pistil with an inferior ovary, 2 styles, and 2 stigmas. These spring flowers utilize wind for pollination and produce a nutlet or a winged **samara** fruit.

Of interest... **lumber:** *Betula* (birch), *Alnus oregona* (red alder), *Ostrya* (hop-hornbeam); **charcoal:** *Alnus* (alder), *Betula;* **food:** *Corylus* (hazelnut, filbert); **ornamentals:** *Betula papyrifera* (paper birch, white birch), *B. pendula* (European white birch), *B. nigra* (river birch), *Corylus, Ostrya.*

Betula pendula   European white Birch

This tree, often used as an ornamental, has several trunks typically occurring together in a clump. The white paper bark has dark horizontal lenticular markings and peels off in layers. Trees have graceful spreading or pendant branches.

The flowering shoot (a) shows the leaves (b) with serrated (toothed) margins and stipules (c). Male flower clusters (**catkins,** d) are pendant, while the female cluster (e) is smaller and upright. The male cyme shown has 7 stamens enclosed by bracts (f). At the end of the stamen's filament (g), 2 anther sacs (h) separate and open in longitudinal slits. Three pistils and 3 bracts (i) make up the female **cymule**. The pistil has an inferior ovary (j) and 2 styles (k).

At fruiting time, the female cluster disintegrates, freeing the samaras (l) from the woody bracts (m).

COLOR CODE

tan:	stem (a), male cluster (d), bract (f), seed (n)
green:	leaves (b), stipules (c), female cluster (e)
yellow:	filaments (g), anther sacs (h), samara (l)
dark green:	bract (i)
white:	ovary (j), styles (k)
brown:	bracts (m)

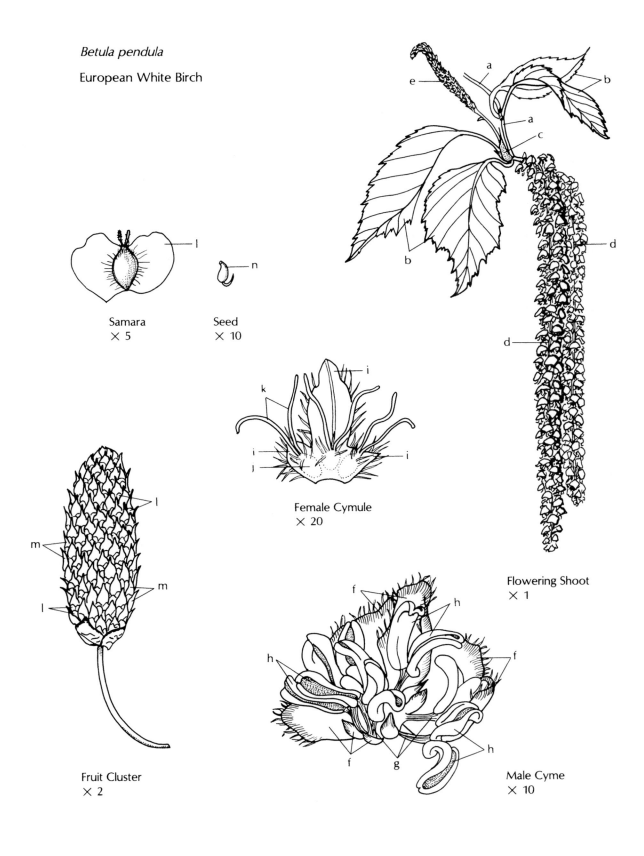

Betula pendula

European White Birch

Samara
× 5

Seed
× 10

Female Cymule
× 20

Fruit Cluster
× 2

Male Cyme
× 10

Flowering Shoot
× 1

Cactus Family (Cactaceae)

Native only to the Western Hemisphere, cacti have fleshy stems with spines or barbed hairs (**glochids**) arranged in cluster sites (**areoles**). Spines may occur as hairs, hooks, bristles, and be short, straight, curved, long. They shade the plant and reduce drying effects of wind. The usually solitary flower has numerous sepals and petals that look alike (**tepals**) and the fruit type is a berry.

As the stems usually function in photosynthesis, the leaves are mostly small, scale-like, or absent. Root systems are usually extensive and close to the soil surface, adapted to absorb rainwater quickly.

To prevent water loss in dry climate conditions, there is an outer surface **cuticle** that is thick and waxy. Water absorbed by the roots is stored in tissue as mucilaginous sap.

Of interest...cacti number about 80 genera with at least 2,000 species; a few of the **ornamentals** are: *Echinocereus, Echinopsis, Ferocactus sp.* (barrel cactus), *Lemaireocereus thurberi* (organpipe cactus), *Mammillaria* (pincushion cactus), *Neoporteria, Opuntia versicolor* (staghorn cactus), *Peniocereus greggii* (night blooming cereus), *Pereskia* (rose cactus), *Rubutia* (red crown cactus), *Schlumbergera bridgesii* (Christmas cactus); **food:** *Opuntia* (prickly-pear) has edible berries, that can be made into jam; young stems, "pads," are eaten as food, and can be propagated by drying overnight, then placed in sand to root; considered a weed pest on rangeland; many are **endangered:** *Carnegiea gigantea* (saguaro, see 36).

There are many forms of cacti: single, clumps, low to the ground, upright, even tall as trees such as *Carnegiea gigantea* (saguaro). At an age of 200 years saguaro can reach 50 feet (152 meters) high and weigh 8 tons (45 tonnes). "Skeletons" of dead saguaro stem look like cylinders of woody rods. People of the desert use them as fence-building material.

Birds nest in cacti. Even though cactus spines deter contact, squirrels and other rodents, rabbits, and peccaries eat the fleshy stems.

The cactus craze began in the 1800's. Hobbyists collect different types of cacti. Cacti are easy to grow, needing sparse water, sandy soil, and produce beautiful flowers in many colors. And, they are perennials.

Cacti are transported to other desert areas of the world. In some places they are serious invaders, displacing native species.

Opuntia decumbens   Prickly-pear

The fleshy stem (a) is flat. A stem node in cacti is called an **areole** (b). From it emerges a leaf (c), if present, spines (d) and **glochids** (reduced short shoots), hairs (e), and the flower.

Schlumbergera truncata   Thanksgiving Cactus

On this plant, the stems (f) are flat and form a branching pattern as new stems emerge from areoles at ends of the stem pieces. With short daylength periods, flowers are produced (see 26). The flower has numerous tepals (g), numerous stamens (h), and a single pistil. The stigma (i) of the pistil extends beyond the stamens.

Pereskia grandifolia   Rose Cactus

This cactus plant has round fleshy stems (j) with spines (k) and a large thickened simple leaf (l) emerging from the areole.

COLOR CODE

light green:	leaf (c)
green:	stem (a, f, m j), leaf (l)
yellow:	areoles (b), spines (d, k), hairs (e)
red:	tepals (g)

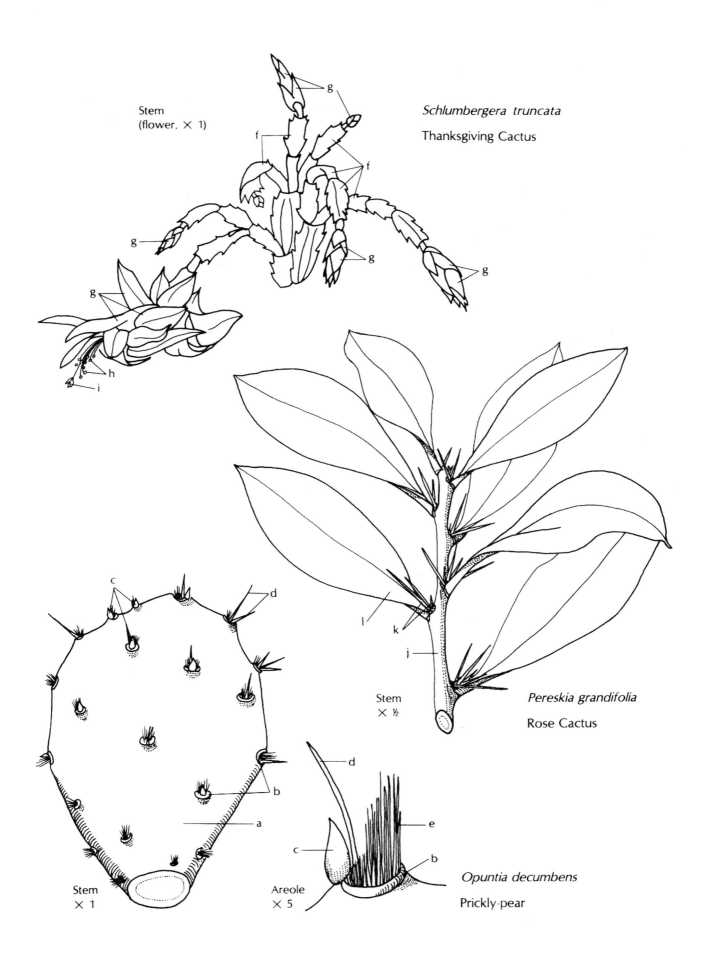

Stem
(flower, × 1)

Schlumbergera truncata

Thanksgiving Cactus

Pereskia grandifolia

Rose Cactus

Stem
× ½

Stem
× 1

Areole
× 5

Opuntia decumbens

Prickly-pear

Cactus Family (continued)

Mammillaria prolifera   Pincushion Cactus

This perennial herb from North American deserts is globe-shaped. Areoles (a), with hairs (b) and glochids (c), occur raised on fleshy tubercles (d).

The typically bisexual cactus flower consists of numerous tepals (e), numerous stamens (f), and one pistil with an inferior ovary (g) consisting of three fused carpels forming one chamber (**locule**) in which numerous ovules (h) develop in parietal placentation. The pistil has one style (i) and a branched stigma (j).

The fruit is a red berry (k) with black seeds (l).

Of interest... **houseplants:** there are many species to choose from with flower colors from white to cream, reds, pinks and yellows. They are easy to grow in porous, sandy/gravely soil. Some include *Mammillaria albescens* (greenish-white flowers), *M. bocasana* 'Inermis' (snowball cactus, yellow flowers), *M. columbiana* (deep pink flowers), *M. compressa* (purplish-red flowers), *M. elegans* (red flowers), *M. geminispina* (carmine flowers), *M. heyderi* (coral cactus, white with red or pink flowers), *M. magnimamma* (cream-white flowers), *M. zeilmanniana* (rose-pincushion, purple flowers).

COLOR CODE

tan:	glochids (c)
pale green:	tubercles (d)
yellow:	tepals (e)
white:	stamens (f), ovules (h), style (i), stigma (j)
red:	berry (k)

Mammillaria prolifera

Pincushion Cactus

Tubercle
× 5

Berry
cut open
× 5

Habit
× 2

Flower
vertical section
× 5

Pink Family (Caryophyllaceae)

Leaves of plants in this family are usually opposite and joined at the base on the stem. The flower's ovary has 2–5 carpels joined to form one chamber (**locule**) with the ovules usually in free-central placentation (see 28 and *Dianthus*). There are more than 2,000 species.

This is one of the families in the subclass Carophyllidae that does not have **betalain** pigments (see 75). Instead, there are **anthocyanins** (see 4). These Temperate Zone plants are mainly annual or perennial herbs, or sometimes shrubs. The fruit is usually a one-chambered capsule with valves.

Of interest... **ornamentals:** *Arenaria* (sandwort), *Dianthus barbatus* (sweet William), *D. caryophyllus* (carnation), *D. gratianopolitanus* (cheddar pink), *D. plumarius* (cottage pink), *Gypsophila* (baby's-breath), *Lychnis chalcedonica* (Maltese Cross), *L. coronaria* (rose campion, dusty miller), *Silene acaulis* (cushion pink); **weeds:** *Cerastrium* (mouse-ear chickweed), *Scleranthus, Stellaria* (chickweed, starwort), *Spergula arvensis* (corn spurry), *Silene pratensis* (white cockle); **prairie plants:** *Dianthus armeria* (Deptford pink), *Silene dioica* (red campion), *S. latifolia* (evening campion), *S. regia* (royal catchfly), *S. stellata* (starry catchfly), *S. virginica* (fire pink); **poisonous:** *Agrostemma githago* (corn cockle), *Crymaria pachyphylla, Saponaria vaccaria* (cow cockle), *S. officinalis* (bouncing Bet).

Dianthus　Pink

Plants in this genus are annual or biennial herbs. The arrangement of leaves (a) on the stem (b) is **decussate**, which means 4 pairs of opposite leaves alternate at right angles. Leaf bases form swollen nodes (c) on the stem.

The bisexual flower has 5 joined sepals (d) and overlapping bracts (e) that prevent nectar-robbing bees from biting into the flower. The illustration of the capsule shows the sturdy construction. The 5 petals (f) are notched or "pinked" at the edges as if cut by pinking shears and consist of **claw** (g), **corona** (h) and **limb** (i).

Ten stamens (j) have 5 joined filaments at the base. The pistil has a superior ovary (k) with carpels united into one locule in which many ovules (l) are attached in free-central placentation. Two styles (m) elongate into feathery (plumose) stigmas (n).

The capsule is enclosed by sepals and bracts and splits into valves (o) adapted to release the seeds.

COLOR CODE

dark green:	stem (b)
green:	leaves (a), nodes (c), sepals (d), bracts (e)
red:	petals (f), claw (g), corona (h), limb (i)
yellow:	stamens (j)
light green:	ovary (k)
white:	ovules (l), style (m), stigmas (n)
tan:	valves (o)

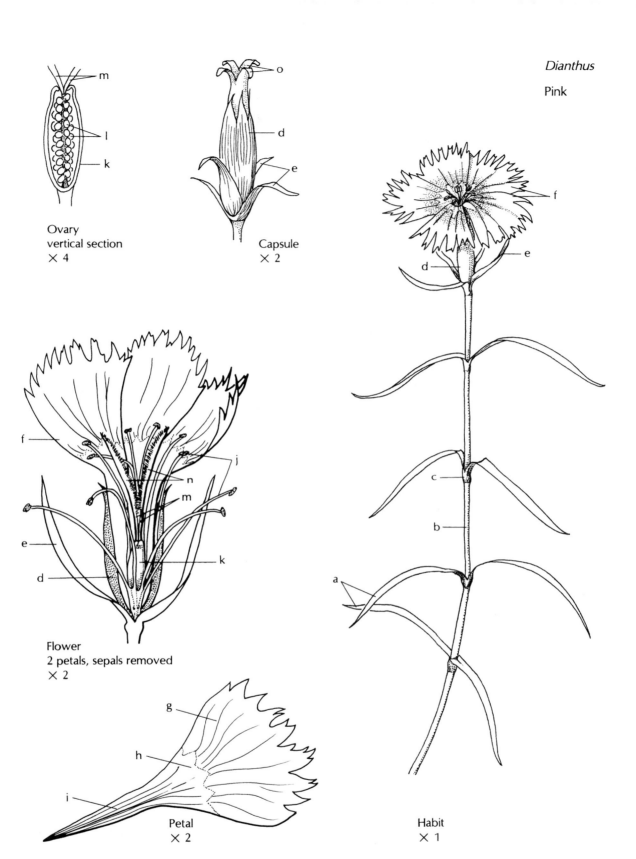

m

l

k

Ovary
vertical section
× 4

o

d

e

Capsule
× 2

Dianthus

Pink

f

d

e

c

b

a

f

n

m

j

e

d

k

Flower
2 petals, sepals removed
× 2

g

h

i

Petal
× 2

Habit
× 1

Goosefoot Family (Chenopodiaceae)

Plants in this family are mainly salt-tolerant annual or perennial herbs with vesicular (glandular) hairs on the leaves that store salt as sodium or potassium chloride. **Betalain** pigments are also present (see 4, 75). This family has about 100 genera and about 1500 species.

Tiny, wind-pollinated flowers form in dense clusters. The usually bisexual flower has 5 joined sepals or none, 5 stamens, and a single pistil. In the superior ovary is one chamber (**locule**) containing one ovule that develops into a nutlet fruit.

Of interest... **food:** *Beta vulgaris* (beet, eaten as a vegetable and also as a sugar beet variety that is a source of sugar), *Beta vulgaris* variety *Cicla* (Swiss chard, rhubarb chard), *Chenopodium quinoa* (quinoa), *Spinacia oleracea* (spinach), *Tetragonia expansa* (New Zealand spinach); **wild food:** *Chenopodium album* (lamb's quarters); **ornamental:** *Chenopodium amaranticolor, Kochia* (cypress spurge).

Chenopodium album   Lamb's Quarters

This annual herb grows to one meter in height and is found in dry, waste places and as a common garden weed. The 5-sided stem (a) has alternate, "goose-foot"-shaped leaves (b) with white spots of salt-containing glands (c) on the lower surfaces (d).

Spikes of sessile green flowers (e) produce black, shiny nutlets. The flower consists of 5, gland-covered sepals (f), 5 stamens (g) and a single pistil with a superior ovary (h), and 2 styles (i).

Chenopodium capitatum   Strawberry-blite

Also an annual, this plant is found along roadsides and in clear-cut and burned-over forest areas. Its alternate "goose-foot" leaves (k) lack salt-containing glands.

Round clusters of flowers (l) are bright red, the color of the flower's fleshy sepals. Dull black nutlets (m) are produced.

COLOR CODE

green:	stems (a, j), leaf upper surface (b), leaves (k)
pale green:	leaf lower surface (d), flowers (e), sepals (f), ovary (h), styles (i)
yellow:	stamens (g)
red:	flowers (l)
black:	nutlet (m)

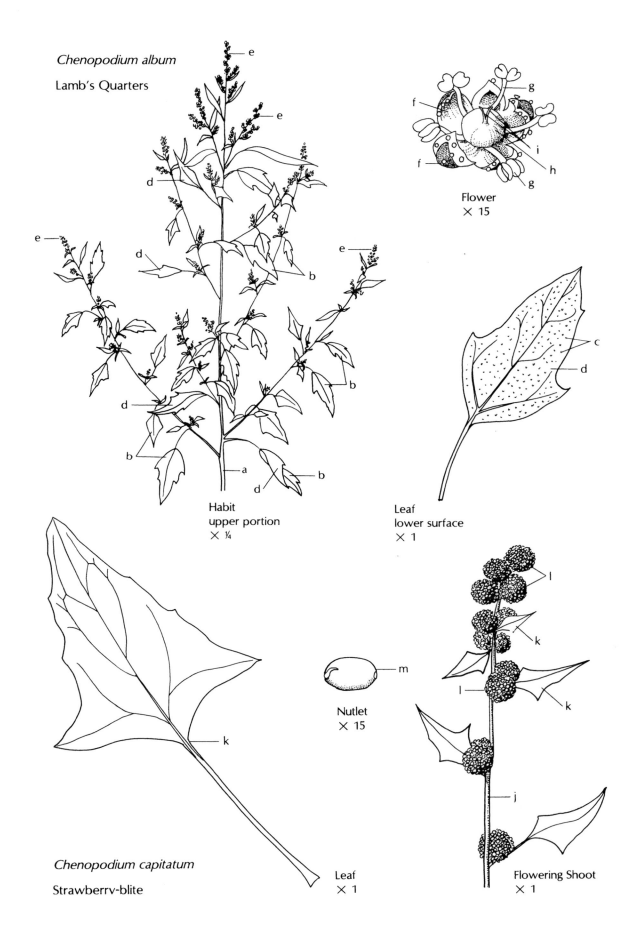

Chenopodium album

Lamb's Quarters

Flower
× 15

Habit
upper portion
× ¼

Leaf
lower surface
× 1

Chenopodium capitatum

Strawberry-blite

Leaf
× 1

Nutlet
× 15

Flowering Shoot
× 1

Buckwheat Family (Polygonaceae)

Most of the family's plants are weedy with small flowers and are adapted to many different soil types. The stem nodes have a sheathing membrane consisting of two fused **stipules**, called an **ocrea**. While mainly herbs, there are also a few shrubs, trees, and vines in this family. There are about 30 genera.

Small, wind-pollinated flowers are borne in clusters or in heads. Usually the flower is bisexual and with a variable number of parts of 3-5 sepals and petals that look alike (called **tepals**), 6-9 stamens, and a single pistil. The superior ovary of fused carpels has one chamber (**locule**) with one ovule which develops into an achene fruit (see dry fruit types, 38)

Of interest... **food:** *Fagopyrum esculentum* (buckwheat), *Rheum rhaponticum* (rhubarb), *Rumex acetosa* (garden sorrel); **ornamentals:** *Antigonon* (mountain-rose vine), *Atraphaxis frutescens, Coccoloba uvifera* (sea grape), *Eriogonum, Muehlenbeckia axillaris, Persicaria bistorta* 'Superba' (knotweed), *Polygonum aubertii* (silver lace vine), *P. orientale* (prince's feather), *P. sachalinense* (sacaline), *Rheum palmatum* (Chinese rhubarb, a medicinal plant), *Rumex hydrolapathum*; **weeds:** *Polygonum* spp. (knotweeds, smartweeds, wild buckwheat), *Polygonum convolvulus* (black bindweed), *Rumex acetosella* (sheep sorrel, red sorrel, an edible plant with tart-tasting, arrowhead-shaped leaves), *R. crispus* (curly dock).

Polygonum persicaria   Lady's Thumb, Redleg

Lady's thumb is known as a smartweed for its sap, which irritates, "smarts," on contact with human eye and nose tissues. This annual's stems (a) have a jointed appearance. *Polygonum* means "many joints" or "knees," the name given for the knobby stems with swollen nodes.

At each node (b) is an ocrea (c) fringed with bristles. Short-petioled, linear leaves (d) have a spot (e) of dark green pigmentation near the center, the so-called lady's thumb-print.

Pink flower clusters (f) have a minute membranous sheath (ocreola, g) at the base of inner clusters. The small flower has 5 tepals (h), 2 enclosing the inner 3; 8 stamens (i); and a single pistil (j) with 2 stigmas (k). The fruit is an achene that may be lentil-shaped (l) or 3-sided (m).

COLOR CODE

green:	stem (a), leaf (d)
light green:	ocrea (c), ocreola (f), pistil (j)
dark green:	spot (e)
pink:	flowers (f), tepals (h)
white:	stamens (i)
black:	achene (l)
tan:	achene (m)

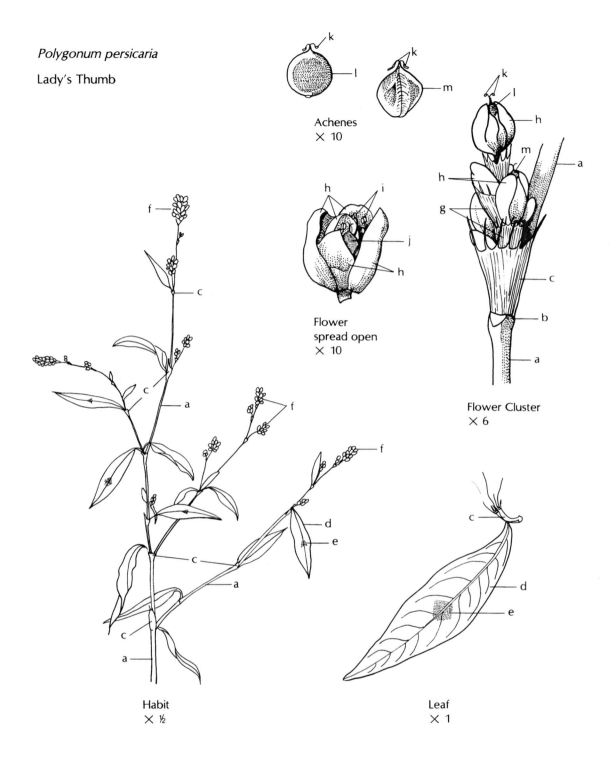

Polygonum persicaria

Lady's Thumb

Achenes
× 10

Flower
spread open
× 10

Flower Cluster
× 6

Habit
× ½

Leaf
× 1

Mallow Family (Malvaceae)

The flower's column of fused staminal filaments is a common feature in this family (see 29). The leaves are alternate, simple, entire or variously lobed with palmate venation. Sap is often mucilaginous. There are about 80 genera of herbs, shrubs and trees.

Usually the flower is bisexual with 5 valvate sepals, 5 separate petals, numerous stamens, and a single pistil with 2-many carpels. Fruit types include a **loculicidal capsule**, often with hairy (**comose**) seeds as in cotton, a **schizocarp**, and a **berry** (see 38, 39).

Of interest... **crops:** *Gossypium hirsutum* (cotton, cottonseed oil), *Hibiscus esculentus* (okra); **ornamentals:** *Abutilon* (flowering maple), *Althea rosea* (hollyhock), *Callirhoe* (poppy mallow), *Hibiscus moscheutos* (common rose mallow, swamp mallow, mallow rose), *M. rosa-sinensis* (rose-of-China, see 29), *Malope, Malva alcea* (hollyhock mallow), *M. moschata* (musk mallow), *Sidalcea malviflora* (prairie mallow, miniature hollyhock); **weeds:** *Abutilon theophrasti* (velvet-leaf), *Malva neglecta* (common mallow, cheese weed).

Abutilon × *hybridum*   Flowering Maple

This plant is a tropical woody shrub, hybridized in cultivation for its speckled (a) leaves and showy flowers. Resembling maple leaves (maple family, see 105), the blades (b) are palmately lobed and veined; the petioles (c) are long.

Showy orange flowers have 5 joined sepals (d), which separate into valves and 5 petals (e). Anthers (f) extend from the center of the flower.

In the vertical section of the flower, the column (g) of fused staminal filaments (**monadelphous** condition) is shown. Anthers (h) are separate at the top of the column. Within the column is the pistil's style (i) and, emerging above the anthers, there are 5 stigmas (j). The superior ovary (k) is composed of 5 fused carpels, and has numerous ovules (l).

Althea rosea   Hollyhock

The numerous carpels of a compound ovary produce a **schizocarp** fruit (see 38) of seeded **mericarps** (m) attached to a central axis and enclosed by persistent sepals (n). When the sepals wither, the mericarps separate from the axis.

Gossypium hirsutum   Cotton

A cotton "boll" is a loculicidal capsule (o) of 4 carpels that split to release hair-covered (p) seeds (q).

COLOR CODE

yellow:	speckles (a)
green:	leaf blades (b), petioles (c)
pale green:	sepals (d, n), ovary (k), style (i)
dark red:	stigmas (j)
orange:	petals (e)
pale orange:	column (g), anthers (f, h)
white:	hairs (p), ovules (l)
tan:	mericarps (m), seed (q)
brown:	capsule (o)

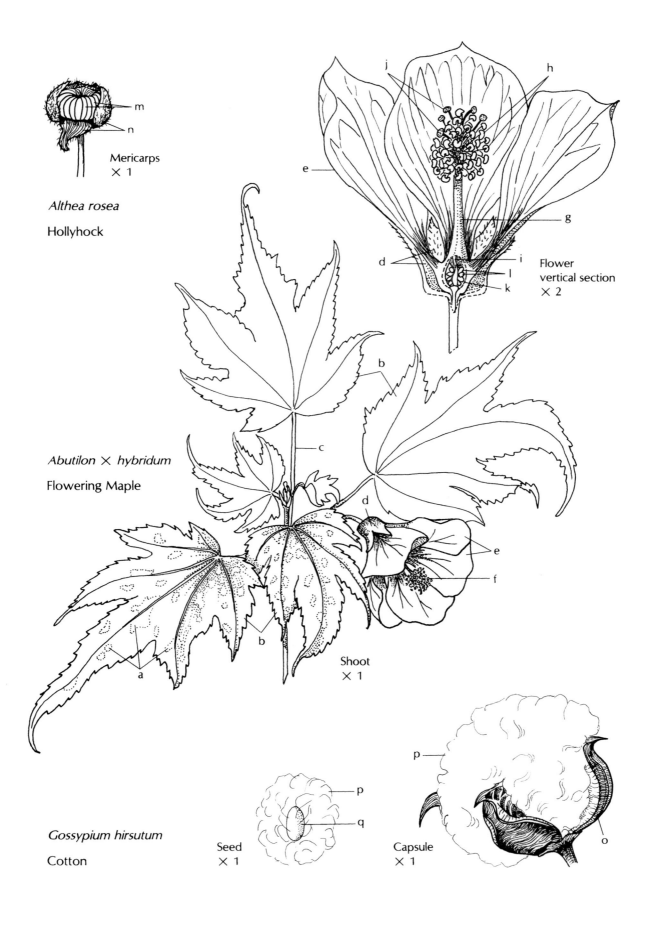

Mericarps
× 1

Althea rosea

Hollyhock

Flower
vertical section
× 2

Abutilon × hybridum

Flowering Maple

Shoot
× 1

Gossypium hirsutum

Cotton

Seed
× 1

Capsule
× 1

Pitcher-plant Family (Sarraceniaceae)

Although there are only three genera in this family, these perennial herbs of bogs are of interest for their unusual insect-trap leaves. Forming in a basal rosette, the fluid in the modified leaves digest insects for nutritional nitrogen.

The bisexual flower consists of 4-5 sepals, 5 petals or none, numerous stamens, and a single pistil. The style is lobed or expands into an umbrella-shape. A **loculicidal capsule** is formed from 3-5 carpels.

Of interest . . . as **novelty ornamentals,** pitcher-plant genera are: *Darlingtonia* (native to California and Oregon), *Heliamphora* (native to northern South American), and *Sarracenia* (native to eastern North America).

Sarracenia purpurea   Pitcher-plant

Early in leaf development, the blade forms a tube with the upper leaf surface on the outside and the hairy lower surface on the inside. Within the leaf tube (a), with downward-pointed hairs (b), is a solution (c) of rainwater and enzymes in which insects drown and are digested. The leaf consists of petiole (d) and blade tube (e), wing (f), and lip (g). It is pale green with red veins (h).

Appearing as a hanging umbrella covered with flaps, the flower towers above the leaves on a long peduncle (i). The flaps are 5 sepals (j). In the habit drawing, the 5 petals (k) and numerous stamens (l) have been shed.

The flower structure drawing shows their location. The pistil consists of a superior ovary (m) composed of 5 carpels (n) with ovules (o) in **axile placentation**, and an expanded, umbrella-shape style (p, q) with 5 stigmatic tips (r).

COLOR CODE

red:	hairs (b), veins (h), sepals (j), petal (k)
light green:	petiole (d), tube (e), wing (f), lip (g), peduncle (i), style (p, q), stigma (r),
blue:	liquid (c)
white:	ovules (o)
yellow:	stamens (l)

Sarracenia purpurea

Pitcher-plant

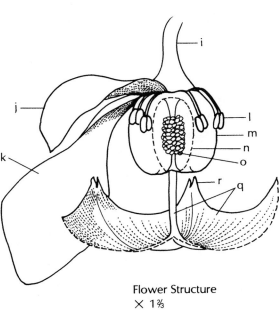

Flower Structure
× 1⅔

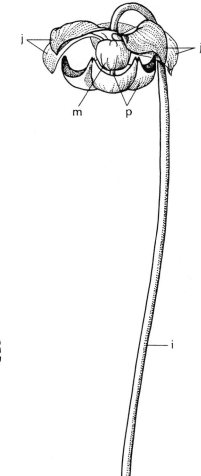

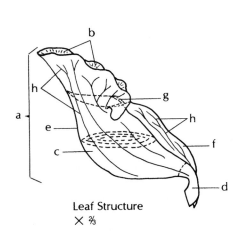

Leaf Structure
× ⅔

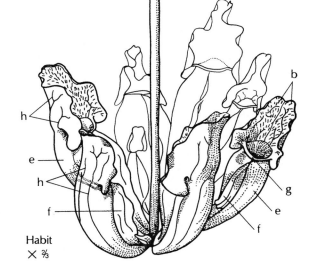

Habit
× ⅔

Violet Family (Violaceae)

Members of this family are herbs and shrubs and usually perennial. Leaf-like appendages (**stipules**) occur at the base of simple, usually alternate leaves.

Violet flowers are bisexual with flower parts in 5's. The lowermost petal is often spurred. The flower's pistil has a superior ovary of 3–5 fused carpels. The stamens, close around the pistil, are connected at the base and open inwardly. The fruit is a one-chambered capsule or a berry.

Of interest...**ornamental:** *Viola* spp. (pansies, violets); **essential oils:** *Viola oderata,* grown for oil in the manufacture of perfumes, flavorings, and liqueur; **food:** *Viola* (leaves and flowers are edible, containing copious amounts of vitamins A and C) and crystallized flowers are used for decoration; **wild:** *Hybanthus* (green violet), *Rinorea, Viola* spp. (violets).

Viola papilionacea Common Blue Violet

This perennial herb has a horizontal **rhizome** (a) and produces flowers in the spring and fertile **cleistogamous** flowers in the summer and fall. Cleistogamous flowers are small, lack petals, do not open, are self-pollinated and occur near the base of the plant in soil substratum.

The leaf blades (b) have heart-shaped bases with scalloped margins and are attached to long petioles (c) with **stipules** (d) at the bases.

Flowers are borne on long peduncles (e) that arise from the base of the plant. The 5 sepals (f) have small projecting lobes (**auricles**, g). There are 5 unequal petals (h) with the 2 lateral petals bearded with hairs (i) the base. The lower petal has a spur (j). (See bud drawing.) Purple nectar guides converge toward the base of the white-throated petals.

Five basally-fused stamens (k) enclose the ovary (l) (shown in a cross-section) and the style (m). The end of the style is enlarged and on the lower side is a beak, the stigma (n). Two of the stamens have nectar horns (o), which project into the lower petal spur (j). Pollen from the stamens sheds into the inner cone.

As an insect probes for nectar, it moves the stigma aside and pollen in the cone sticks on its tongue. Cross-pollination results when the next flower is visited.

The pistil's ovary (l) has 3 fused carpels with many ovules (p) in **parietal placentation**. At maturity, the fruit is a **loculicidal capsule** (q) with numerous seeds (r).

COLOR CODE

white:	rhizome (a), roots, stipules (d), beards (i) ovules (p)
light green:	petioles (c), peduncles (e), stamens (k), nectar horns (o), style (m), stigma (n)
green:	leaves (b), sepals (f, g)
violet:	petals (h), capsule (q)
violet-brown:	petal spur (j), ovary (l)
dark brown:	seeds (r)

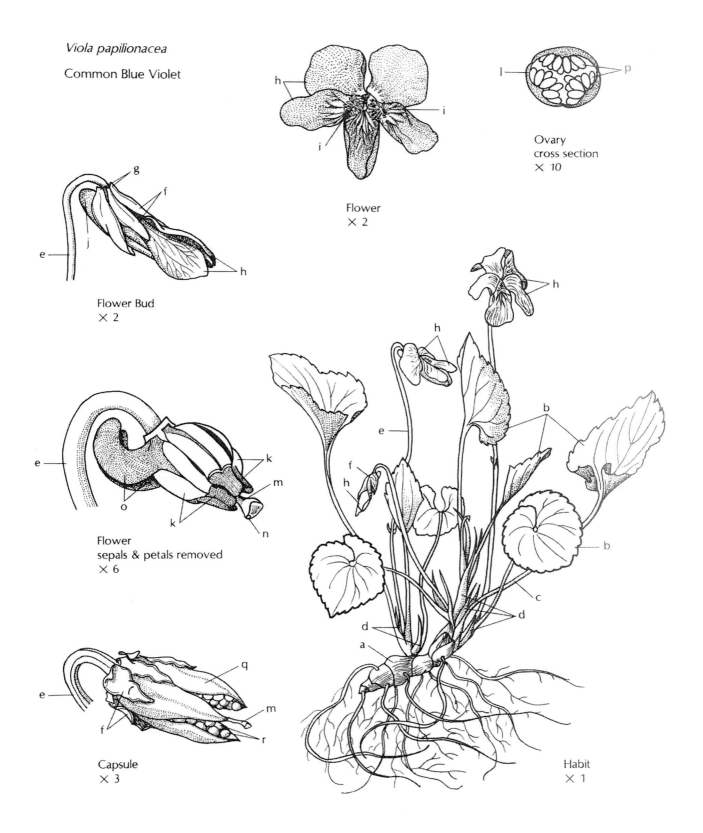

Viola papilionacea

Common Blue Violet

Flower Bud
× 2

Flower
× 2

Ovary
cross section
× 10

Flower
sepals & petals removed
× 6

Capsule
× 3

Habit
× 1

Begonia Family (Begoniaceae)

This family of plants is native to the tropics and subtropics across the continents. It is popularly known for its *Begonia* cultivars grown as house plants. Plants in this family are perennial herbs or small shrubs. Most have succulent stems with alternate leaves and with deciduous **stipules**. Leaves are mostly palmately veined and usually asymmetrical with an oblique base.

Unisexual male and female flowers occur on the same plant (**monoecious**). The male flower has 4 irregular tepals and many stamens in whorls. The pistil of the female flower has an inferior, angled or winged ovary consisting of usually 3 carpels and twisted stigmas. Fruit types are a loculicidal capsule and a berry.

Of interest...there are over 900 species within five genera: *Begonia, Begoniella, Hillebrandia, Semibegoniella* and *Symbegonia;* **ornamental:** *Begonia* (begonia) species are widely hybridized in horticulture for flowering and foliage plants. Flowering plants are tuberous- and fibrous-rooted, such as *B. semperflorens* (wax begonia) and *B. tuberhybrida* (tuberous begonia). Foliage plants have rhizomes, such as *B. rex* (rex begonia) cultivars.

Begonia semperflorens cultivar 'Scarletta'

This small plant flowers continuously (**semperflorens** = always flowering). Its waxy leaves (a, b) are alternate and 2-ranked, which means the leaves arise in one plane on opposite sides of the succulent stem (c). At the petiole (d) base is a pair of stipules (e). Veins (f) of the leaf blade radiate out from a common point (**palmate venation**). Winged capsules are also shown on the stem portion. The stalk (**peduncle**, g) supporting a cluster of flowers or capsules has bracts (h) where the flower stems (**pedicels**, i) arise.

A male flower has 2 petal-like sepals (j) and 2 smaller petals (k), together called **tepals**, and about 20-30 stamens (l). The stamen's anther sacs (m) open (dehisce) longitudinally and have an enlarged **connective** (n) between them. There is a short filament.

The four remaining drawings illustrate parts of the female flower. At the base of the flower are 3 bracts (p, x). Extending from the pistil's inferior ovary are 3 unequal wings (q, v–w, z) partially covered by 5 tepals (r, y). The pistil's 3 styles each have 2 twisted stigmas (s, t).

A cross-section of an ovary shows the 3 carpels (u) with numerous ovules. As the flower matures, pigments fade in the ovary wings (v, w) and bracts (x), and the tepals (y) wither. The capsules (z) have a papery texture and contain many seeds.

COLOR CODE

dark green:	upper blade surface (a)
green:	lower blade surface (b), stem (c)
light green:	petiole (d), peduncle (g), pedicel (i), inner ovary wing (v)
tan:	stipules (e), bracts (h, x), tepals (y), capsule (z)
red:	tepals (j, k, r), bracts (p), ovary wings (q)
yellow:	stamens (l), anther sacs (m), connective (n), filament (o), stigma (s, t)
white:	carpels (u)
red-brown:	outer ovary wing (w)

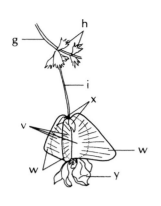

Female Flower
× 1

Style, Stigmas
× 5

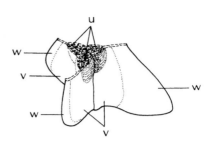

Ovary
cross section
× 2

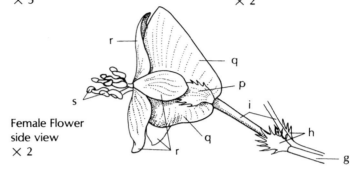

Female Flower
side view
× 2

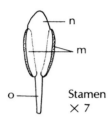

Stamen
× 7

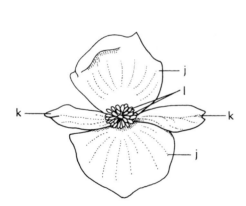

Male Flower
× 2

Begonia semperflorens

Common Wax Begonia

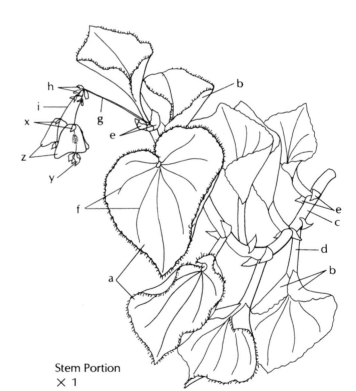

Stem Portion
× 1

Gourd Family (Cucurbitaceae)

Plants in this family are mostly perennial herbaceous vines with spirally coiled **tendrils**. Tendrils are modified stipules (see 23) that may be simple, forked or branched. Leaves are alternate, with usually **palmate** or **pinnate venation**. Plants lie flat (prostrate) and creep or climb (**scandent**). There are about 100 genera with over 700 species.

Separate male and female (unisexual) flowers are most common in this family. This is a monoecious condition. Male flowers have unusual stamens. Where there may be 1-5, there are usually 3 stamens with one stamen having a one-chambered anther of 2 pollen sacs and 2 stamens with 2-chambered anthers of 4 pollen sacs each.

The female flower's pistil has an inferior ovary consisting of 3-5 carpels, usually united into a single chamber (**locule**) of ovules in parietal placentation.

The fruit is a berry with a soft fleshy **pericarp** or with a hardened pericarp and classified as a **pepo**. Before Linnaeus' classification system, pepo was the name for pumpkin. Pepos that are dried for ornamental use or hollow vessels are commonly called **gourds**.

Of interest . . . economically important as **food:** *Citrullus lanatus* (watermelon, citron), *Cucumis anguiria* (West Indian gerkin), *C. melo* (cantaloupe, honey-dew melon, melon), *C. sativus* (cucumber), *Cucurbita* spp. (pumpkins, squashes, vegetable marrows, vegetable spaghetti); **ornamentals:** gourds: *Benincasa* (Chinese watermelon), *Coccinea* (ivy gourd), *Lagenaria siceraria* (calabash gourd, one of the earliest cultivated plants, dried to use as containers), *Luffa cylindrica* (loofah, dishcloth gourd, used as a dried sponge), *Momordica* (balsam apple), *Sechium* (chayote), *Sicana* (cassabanana), *Sicyos* (bur cucumber), *Trichosanthes* (snake gourd); *T. kirilowii*, the tree-of-joy from China, is an important anti-cancer plant containing the drug, trichosanthin.

Cucumis sativus   Cucumber

The hairy stems (a) of this plant climb by means of tendrils (b) that arise at the petiole (c) bases. Five-lobed leaf blades (d) have palmate venation. Both male and female, unisexual flowers occur on the same plant (**monoecious**). A flower bud and a young female flower are on the shoot that is illustrated.

Male flowers have 5 sepals, 5 petals (f), and 3 stamens (g). The stamens' stout filaments (h) arise from the petals with the anthers joined together in the center of the flower. The stamen illustrated has a 2-chambered (bilocular) anther (i) with 4 pollen sacs (j).

The female flower has 5 sepal lobes (k), 5 petals (l), and a single pistil. The pistil's inferior ovary (m) consists of 3 united carpels (n) with ovules (o) in **parietal placentation**. Glandular hairs (p) dot the ovary surface. Inside the fused sepal base (k) is a nectar disc (q). The style (r) emerges above and terminates in a 3-lobed stigma (s). After fertilization, the ovary develops into a pepo, the familiar cucumber fruit (see 39).

COLOR CODE

green:	stem (a), petiole (c), leaf blade (d), peduncle (e), sepals (k), ovary (m)
light green:	tendril (b), filament (h), anther (i, j), glandular hairs (p), style (r), stigma (s)
yellow:	petals (f, l), stamens (g), disc (q)
white:	carpels (n), ovules (o)

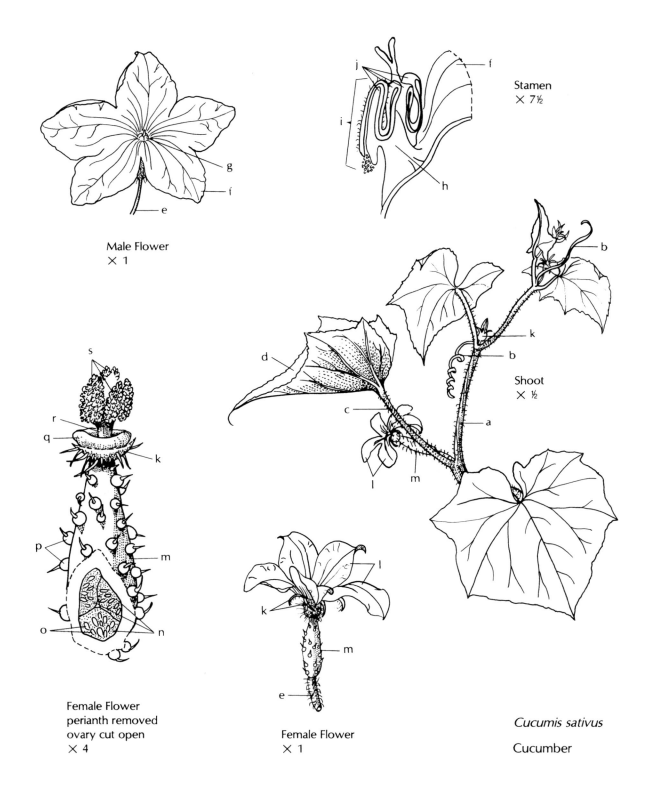

Male Flower
× 1

Stamen
× 7½

Shoot
× ½

Female Flower
perianth removed
ovary cut open
× 4

Female Flower
× 1

Cucumis sativus

Cucumber

Willow Family (Salicaceae)

Plants in this family are woody trees or shrubs that are **dioecious**, producing male and female flowers on separate plants. Specialized and reduced, unisexual flowers are aggregated into **catkins (aments)**. Leaves are alternate, simple, and deciduous with stipules and petioles.

Familiar plants include the aspens, poplars and willows. There are four genera. In the North Temperate Zone are *Populus* with 300 species and *Salix.* In Northeast Asia, *Chosenia* and *Toisusu* occur.

Populus flowers are wind-pollinated, while *Salix* flowers, with nectar glands, are insect-pollinated. The fruit is a 2-4-valved capsule with numerous, hairy (**comose**) seeds.

Of interest... **ornamentals:** *Salix* spp. (willows) with familiar species, *S. babylonica* (weeping willow), *S. discolor* (pussy-willow); *Populus* spp. (poplars, cottonwoods, aspens), *P. grandidentata* (large-toothed aspen), *P. deltoides* (cottonwood), *P. tremuloides* (quaking aspen); **pulpwood:** *Populus tremuloides* (quaking aspen), *P. balsamifers* (balsam poplar); **wickerware and basket-making:** *Salix* (willow); **aspirin:** acetylsalicylic acid was originally derived from the bark of *Salix alba* (white willow).

Salix Willow

The leaves (a) are alternate on the stem (b) and have toothed **(dentate)** margins. Flowers (c) of this male **catkin** have 3 stamens (d) with a hairy bract (e) and a nectar gland (f) below.

The flowers (g) of the female catkin each have a single pistil with numerous ovules (h) in the ovary (i), a single style (j) and 2 stigmas (k). At the base of the flower pedicel (l) is a bract (m) and a nectar gland (n). With the evolutionary reduction of parts, these specialized flowers consist of only stamens or a pistil and are without sepals and petals.

During spring, the female catkin becomes a fuzzy mass of hairy (comose) seeds (o) released from 2-valved (p) capsules to be dispersed by wind.

COLOR CODE

green:	leaf (a), female catkin flowers (g), pedicel (l)
brown:	stem (b), capsule valves (p)
yellow:	male flowers (c), stamens (d)
dark brown:	bracts (e, m)
light green:	nectaries (f, n), ovules (h), ovary (i), style (j), stigmas (k)
tan:	seed (o)

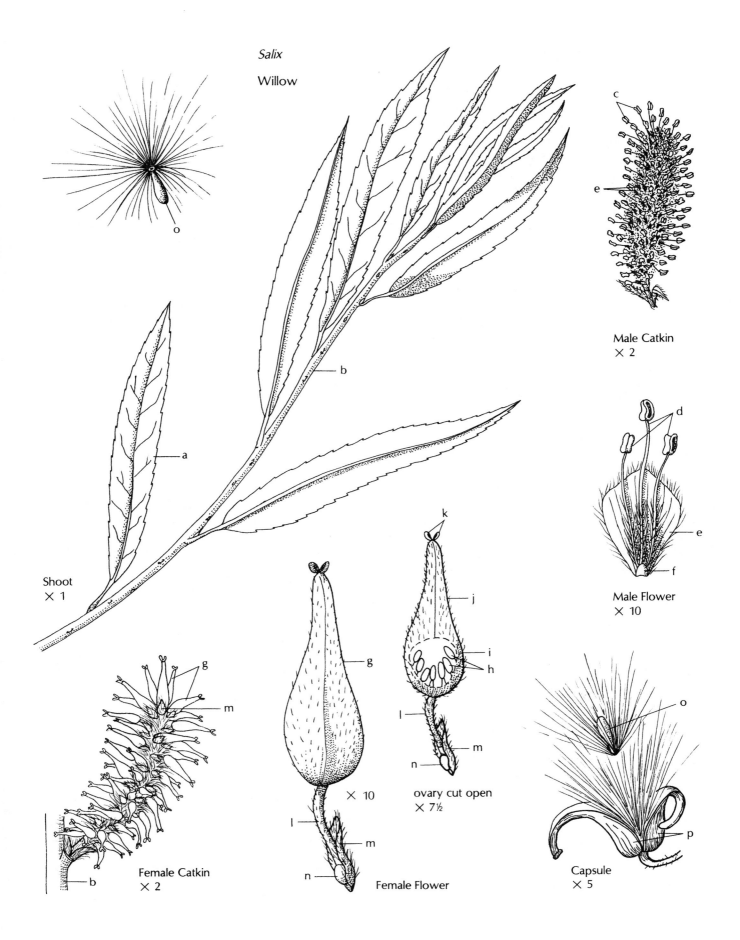

Salix

Willow

o

Male Catkin
× 2

c

e

Male Flower
× 10

d

e

f

Shoot
× 1

a

b

Female Catkin
× 2

g

m

b

Female Flower

g

l

m

n

ovary cut open
× 7½

k

j

i

h

l

m

n

× 10

Capsule
× 5

o

p

Mustard Family (Brassicaceae)

The old family name, Cruciferae, refers to the cross form (cruciform) of the 4 diagonally opposed petals. Flowers also have 4 sepals. Stamen number and lengths and the single pistil's ovary with a false partition are notable characteristics found in this family (the illustrated plant shows these features). The specialized type of fruit produced from the ovary is a narrow **silique** or a round **silicle**, in which seeds are separated into 2 chambers by a partition and covered on each side by a valve (see dry fruit types, silique, 38).

This family consists of annual, biennial, or perennial herbs with pungent oils in the sap. There are about 380 genera and about 3,000 species. The leaves are alternate and simple and usually have forked or star-like, one-celled hairs.

Of interest . . . **food:** *Armoracia rusticana* (horseradish); *Brassica nigra* (mustard); *Brassica napus* (rape, rapeseed oil, canola oil); *Brassica oleracea* (wild cabbage, colewort) selected varieties produced such cole crops as *B. capitate* (head of leaves—cabbage), *B. acephala* (without a head—kale), *B. botrytis* (cluster of white flower buds—cauliflower), *B. gemmifera* (axillary buds—Brussels sprouts), *B. italica* (green flower buds—broccoli), *B. caulorapa* (swollen stem—kohlrabi), *B. napobrassica* (rutabaga); *Eruca sativa* (arugula); *Nasturtium officinale* (water cress); *Raphanus sativus* (radish); **weeds:** *Alliaria petiolata* (garlic mustard); *Brassica* spp. (wild mustards), *Capsella bursapastoris* (shepherd's purse), *Lepidium densiflorum* (pepper-grass), *Draba* (hoary cress), *Thlaspi arvense* (penny cress); **ornamentals:** *Arabis caucasica* (rock cress), *Aubrieta* (false rockcress), *Erysimum* (wallflower), *Hesperis matronalis* (rocket), *Iberis sempervirens* (candytuft), *Lobularia maritima* (sweet alyssum), *Lunaria annua* (honesty, money plant), *Matthiola* (stock); **dye:** *Isatis tinctoria* (dyer's woad); **endangered:** *Cardamine diphylla* (two-leaved toothwort), *C. dissecta* (divided toothwort).

Alliaria petiolata (garlic mustard) is the scourge of the forest, replacing native wildflowers by shading them out of existence. As a biennial, it produces a rosette of leaves the first year, and flowers and seeds the second year, then dies. One plant can produce several thousand seeds that are viable for seven years or more.

Garlic mustard eliminates native food plants for native animals. In state parks overrun with garlic mustard, it is not unusual to see deer begging for edible handouts from tourists.

Lobularia maritima   Sweet Alyssum

A native of the Mediterranean area, this plant is an annual or perennial herb. The leaves (a) are linear and alternate to almost opposite on the stem (b). The habit drawing shows the pedicelled flowers (c), borne on a stalk (**raceme**) and the **silicle** (d) fruit.

The pedicel (e) supports a flower with 4 hairy sepals (f) that alternate with the 4, diagonally opposed petals (g). The 6 stamens are **tetradynamous**, which means there are 4 long stamens (h) and 2 outer short stamens (i). At the base of the filament are nectaries (j), seen when the sepals and petals are removed.

The single pistil has a superior ovary (k) and a capitate stigma (l). When the silicle fruit develops. the ovary's false partition (**replum**, m) can be seen separating the seeds (n). Two outer valves (o) are shed to release the seeds.

COLOR CODE

green:	leaves (a), stem (b), pedicel (e), sepals (f)
white or lavender:	flowers (c), petals(g)
yellow:	silicle (d), stamens (h, i), replum (m), valves (o)
pale green:	nectaries (j), ovary (k), stigma (l)
tan:	seeds (n)

Lobularia maritima

Sweet Alyssum

Habit
× 1

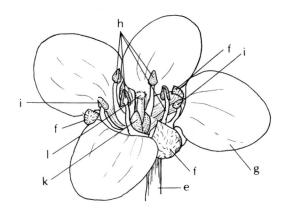

Flower
× 10

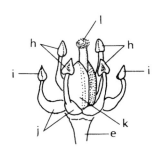

Flower
perianth removed
× 10

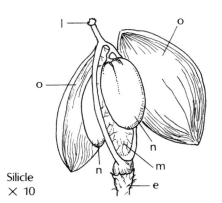

Silicle
× 10

Heath Family (Ericaceae)

Shrubs are the most common in this family, but some plants are perennial herbs, trees, or vines. Most prefer acidic substrate and many have fungus-root (**mycorrhizal**) associations (see 12). Leaves are simple, alternate, sometimes opposite or whorled, and often leathery and evergreen. There are about 100 genera and about 3,000 species.

Flowers are bisexual, borne, singly or in clusters. The 4-5 sepals are distinct or joined at the base. There are usually 4-5 petals joined in a funnel or urn-shape, but in some genera, the petals are separate. Stamens are the same or twice the number of petals. The single pistil has a superior ovary of 4-10 carpels with ovules usually in **axile placement**, and a single style and stigma. Fruit types (see 38, 39) include capsules, berries, or drupes.

Of interest... **food:** Gaultheria procumbens (oil-of-wintergreen), Gaylussacia (huckleberry), Vaccinium spp. (bilberry, blueberry, cranberry, lingonberry); **ornamentals**: Calluna (heather), Enkianthus campanulatus (redvein enkianthus), Kalmia latifolia (mountain laurel), Leucothoe catesbii (leatherleaf), Oxydendrum arboreum (sourwood), Pieris japonica (andromeda), Rhododendron spp. (rhododendron, azalea); **wild:** Arctostaphylos (bearberry), Epigaea repens (trailing arbutus); **poisonous:** Andromeda glaucophylla (bog rosemary), Kalmia angustifolia (lamb-kill), K. latifolia (mountain laurel), K. polifolia (swamp laurel), Leucothoe spp., Lyonia mariana (stagger-bush), Menziesia ferruginea (fool's huckleberry), Rhododendron spp. It is advised not to roast food on sticks of any of the plants listed.

Here is how to tell a rhododendron from an azalea. Although they share the same genus (Rhododendron),

rhododendron flowers have 10 stamens while azalea flowers have 5 stamens. Rhododendrons are mostly evergreen plants, while azaleas are mostly deciduous.

Arctostaphylos uva-ursi   Bearberry, Kinnikinik

This plant forms mats of prostrate shrubs, cloned by shoots arising from the root system. The alternate leathery leaves (a) are evergreen. Young stems (b) and leaf blade margins and petioles are covered with fine hairs.

Urn-shaped flowers are borne in small clusters at the end of the shoot. The flower has 5 basally fused, pink sepals (c), a fused petal tube (d) with 5 lobes (e), 10 stamens, and a single pistil with a superior ovary. The stamen's anther (f) has 2 cells each with a pore (g) for pollen dispersal and a bristly awn (h) at the base of each. The stamen's filament (i) is dilated at the base.

The pistil's ovary (j) has 5-8 chambers (**locules**) with one ovule (k) in each, in **axile placement**. The pistil has a single columnar style (l) and stigma (m). The ovary is elevated on a 2-lobed nectar disc (n). Hairs (o) line the inner surface of the petal tube and cover the stamen's dilated filament base.

As the drupe (r) type fruit matures, the skin color changes from yellow to red, then to purple as it decays. Inside is a fleshy **mesocarp** (s) surrounding the fused nutlets (t). Yellow veins (u) run as a band along the back of each and as seams between the nutlets.

COLOR CODE

green:	leaves (a), disc (n)
brown:	stem (b)
dark pink:	sepals (c), petal lobes (e)
pale yellow:	petal tube (d), mesocarp (s), vein (u)
purple:	anther (f)
pink:	awn (h)
red-brown:	pedicel (p), bract (q)
red:	drupe (r)
tan:	nutlet (t)
white:	filament (i), ovary (j), ovule (k), style (l), stigma (m)

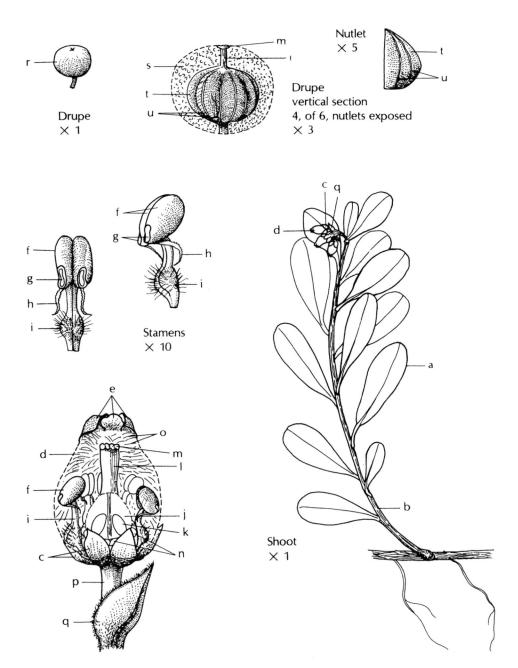

Drupe
× 1

Drupe
vertical section
4, of 6, nutlets exposed
× 3

Nutlet
× 5

Stamens
× 10

Flower
vertical section
× 7

Shoot
× 1

Arctostaphylos uva-ursi

Bearberry

Saxifrage Family (Saxifragaceae)

Distinctive characteristics to identify this family are lacking. The flower parts tend to be in 4's or 5's with stamens one or two times the number of sepals and with a pistil consisting of 2-5 united carpels having numerous ovules.

The leaves are mostly without stipules and usually alternate on these commonly perennial herbs and deciduous shrubs. The fruit type is a capsule or berry.

Of interest... **food:** *Ribes* spp. (gooseberry, currant); **ornamentals:** *Astilbe* (false goat's-beard), *Bergenia cordifolia* (pigsqueak), *Darmera, Deutzia,* × *Heucherella (Heuchera* × *Tiarella), Heuchera* (coral bells), *Hydrangea, Mukdenia, Philadelphus* (mock orange), *Rodgersia* (Roger's flower), *Saxifraga* spp. (saxifrages), *S. stolonifera* (see 16, strawberry begonia, Aaron's-beard, mother-of-thousands); **wild:** *Mitella* (bishop's-cap), *Tiarella* (foamflower, false miterwort).

Ribes americanum   Wild Black Currant

This shrub has alternate leaves (a) with yellow resin dots (b) on the lower surface of the blades.

Small flowers (d) have 5 petal-like sepals (e), 5 petals (f), and 5 stamens (g). The single pistil has an inferior ovary (h) of 2 united carpels with ovules (i) in **parietal placentation**, and a style (j) that divides into 2 stigmas (k).

Heuchera sanguinea   Coral Bells

This perennial herb has basal leaves (m) and develops flowering shoots on long peduncles (n). The peduncle is green at the base and fades into dark pink at the top where the flowers occur. Like *Ribes*, the 5 sepals (p) are united at the base and petal-like. With magnification, delicate, pink, glandular hairs can be seen on the sepals. Inside the sepal tube are 5 tiny petals (q), 5 stamens (r) and a pistil with 2 stigmas.

Philadelphus   Mock Orange

This deciduous shrub has opposite leaves and flowers that have the fragrance of orange blossoms and produce **loculicidal capsules** (s).

COLOR CODE

yellow:	resin dots (b), stamens (g, r)
green:	leaves (a, c, m) pedicel (l), peduncle (n)
pale yellow:	flowers (d), sepals (e), petals (f)
white:	ovary (h), ovules (i), style (j), stigmas (k)
pale pink:	petals (q), upper portion of peduncle (n)
dark pink:	pedicel (o), sepals (p)
tan:	capsule (s), stem (t)

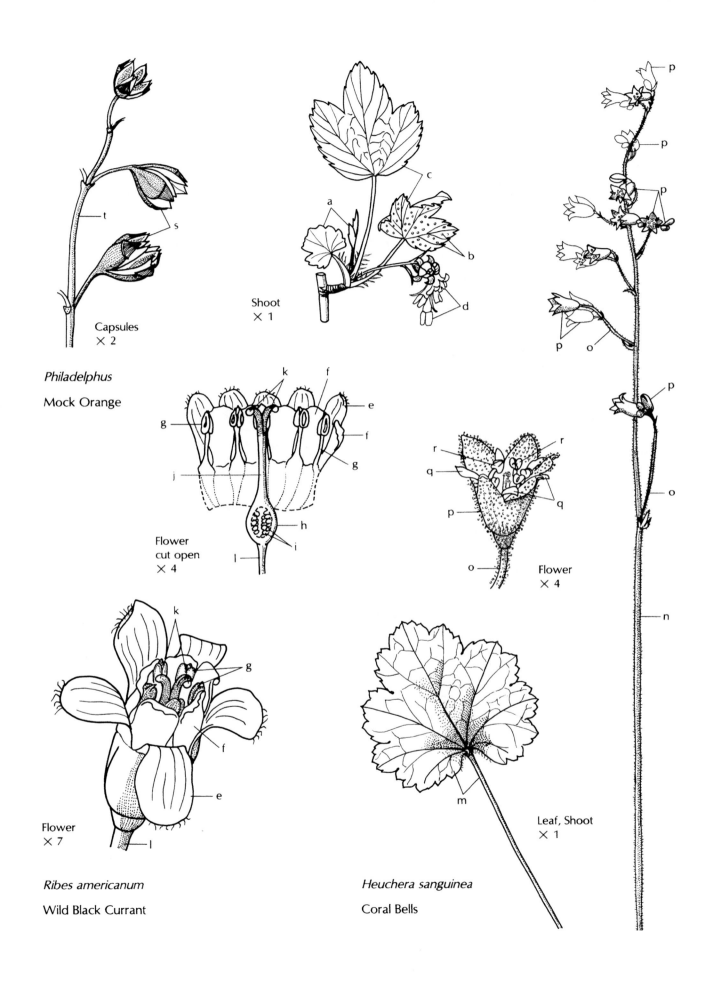

Capsules
× 2

Philadelphus

Mock Orange

Shoot
× 1

Flower
cut open
× 4

Flower
× 4

Flower
× 7

Ribes americanum

Wild Black Currant

Leaf, Shoot
× 1

Heuchera sanguinea

Coral Bells

Rose Family (Rosaceae)

Common features in this family include the following: leaf stipules are usually present, flower parts are in 5's and the flower has a cup-like receptacle or floral tube, called a **hypanthium**, which develops around the ovary.

Plants in the rose family range from herbs and shrubs to trees. Many have spines and a few are climbers. Usually, the leaves are alternate, simple or compound, with toothed (**dentate**) margins. The flower is usually bisexual and has an enlarged nectar cup that inflates the floral tube, has stamens in whorls of five, and possesses a single compound pistil or many pistils together.

There are over 3,000 species in the Rose Family.

Of interest...**fruit types:** achene, drupe, follicle and pome. Rose family plants have English common names for the **edible fruits.** In combination with the fruit type, some examples follow. Strawberry (Fragaria) is an aggregate of achenes. Pear (Pryus), apple (Malus), and quince (Cydonia) are pomes. Cherry (Prunus avium), peach (Prunus persica), plum (Prunus spp.) and apricot (Prunus armeniaca) are drupes. Blackberry (Rubus ulmifolius), raspberry (Prunus idaeus), loganberry (Prunus loganobaccus), dewberry (Rubus caesius) and cloudberry (Rubus chamaemorus) are aggregates of drupelets. The fleshy fruit of the almond (Prunus dulcis) is not eaten; the common name represents the seed inside the drupe's pit, while a similar appearing seed in the peach drupe contains a poison, cyanide, as do the pome seeds of apple (Malus).

Ornamental shrubs whose fruits are not usually eaten are named for other characteristics. Named for thorns, hawthorn (Crataegus) and firethorn (Pyracantha) have pome fruits. Bridal wreath (Spiraea) is cultivated for its sprays of white flowers and has a follicle fruit, as does ninebark (Physocarpus). Mountain ash (Sorbus), with a pome fruit, has compound leaves similar to those of ash (Fraxinus) in the olive family. Roses (Rosa spp.) are cultivated for their showy flowers, which are edible (candied rose buds) and develop into "hips" composed of many carpels, which form an aggregate of achenes high in vitamin C. Other ornamental shrubs are Amelanchier (with many common names: Juneberry, serviceberry, shadblow, shadbush), Aronia (chokeberry), and Kerria japonica from China. Some ornamental herbaceous perennials include lady's mantle (Alchemilla), goatsbeard (Aruncus), meadowsweet (Filipendula), avens (Geum), cinquefoil (Potentilla), and burnet (Sanguisorba).

Malus pumila   Apple

This is a tree of spreading branches. The stems (a) have characteristic short, spur shoots (b) that bear leaves, flowers, and fruits. Small stipules (d) occur at the petiole base.

The flower has 5 basally joined sepals (e) and 5 petals, which are white above and pink below (f). In bud (g), the flowers appear pink. When the flower opens, the petals appear white and are blushed pink from the recurved tips. There are many stamens (h) in whorls of 5.

The vertical section shows the stamens (j) attached to the floral tube (k) outside the nectar cup (l). The pistil has an inferior ovary (m) with 5 carpels (n) and a style (o) that projects through the nectar cup and divides into 5 branches topped with stigmas (p).

The cross-section of an ovary shows the floral tube tissue (k), carpellary tissue (q) and ovules (r) in the 5 carpels. Other features are vascular bundles of the carpels (s), petals (t), and sepals (u). A hairy epidermis (v) covers the floral tube.

The cross and vertical sections of a pome show how the structures of a small flower develop into a fruit after fertilization. The ovules develop into seeds (w) in the carpels (n). Carpellary tissue (q) enlarges, the floral tube tissue (k) expands, and the epidermis (x) becomes pigmented (yellow, red, or green).

COLOR CODE

brown:	stem (a), short shoot (b), seeds (w)
green:	leaves (c), stipules (d), sepals (e), peduncle (i), sepal bundles (u), epidermis (v)
pink:	lower surface of petals (f), bud (g)
yellow:	stamens (h, j)
white:	floral tube tissue (k), nectar cup (l), carpels (n)
light green:	stigmas (p), petal bundles (t)
red:	epidermis (x)
pale yellow:	carpellary tissue (q), ovules (r), carpel bundles (s)

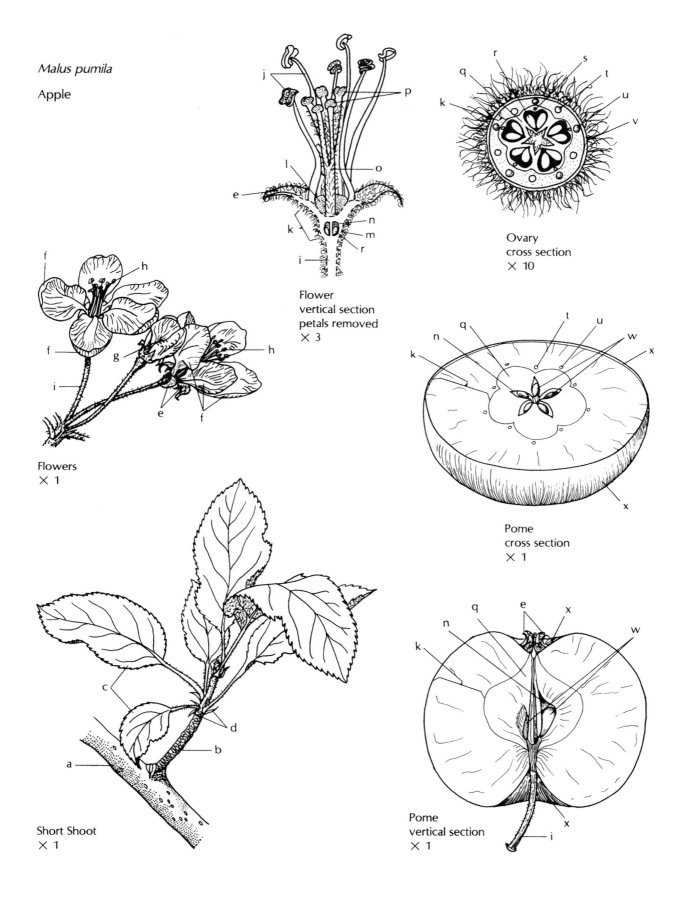

Malus pumila

Apple

Flower
vertical section
petals removed
× 3

Ovary
cross section
× 10

Flowers
× 1

Pome
cross section
× 1

Short Shoot
× 1

Pome
vertical section
× 1

Pea Family (Fabaceae)

The old family name was Leguminosae, a reference to the legume fruit (see 38). Members of this family are easily recognized by the usually alternate, compound leaves divided into leaflets, the typical pea flower, and the pea pod (**legume**) fruit. The form of plants ranges from small herbs to shrubs and trees. There are 700 genera and over 17,000 species.

Looking closely at the leaves, the base of each has a pair of leaf-like **stipules**. The leaf petiole has a swollen base (**pulvinus**) that has the ability to change the position of the leaf and leaflets when stimulated by light or gravity. When touched, *Mimosa*, the sensitive plant, demonstrates this ability, dramatically (movement occurs rapidly).

A pea flower has 5 fused sepals, with 5 separate petals or petals in 3 groups of one **banner** petal, two **wing** petals, and a **keel** of 2 fused petals that enclose the stamens after fertilization. The ovules are attached to the side of the carpel wall in two alternating rows.

Some species have only two seeds per legume pod; others may have many. Pods may be straight such as in garden beans and peas, or straight and sectioned into compartments called a **loment**, or coiled in a spiral (see 38). When mature, the pod splits into halves.

Pea family plants such as clover or alfalfa are often used in crop rotation because they increase soil nitrogen, a plant nutrient, by the presence of bacterial nitrogen-fixing nodules on the roots (see 12).

Of interest... economically important for **food:** *Arachis* (peanut), *Cicer* (chick pea), *Glycine max* (soybean), *Lens* (lentil), swollen roots of *Pachyrhizus erosus* (jicama), *Phaseolus* spp. (beans), *Pisum* (pea), *Tamarindus* (tamarind), *Vicia faba* (broad bean, fava bean), *Vigna angularis* (adzuki bean); **native American foods:** tubers of *Apios americana* (ground nut), tubers of *Psoralea esculenta* (prairie potato); **fodder:** *Lupinus* (lupine), *Medicago* (alfalfa), *Melilotus* (sweet clover), *Prosopsis juliflora* (Mexican mesquite tree pods and beans), *Trifolium* (clover, see 33), *Vicia* (vetch); **ornamentals:** *Acacia* (wattle), *Albizzia*, *Baptisia* (false indigo), *Bauhinia* (orchid tree), *Caesalpinia pulcherrima* (pride of Barbados), *Cassis* (senna), *Cercis* (redbud), *Cytisus scoparius* (Scotch broom), *Delonix* (poinciana), *Genista*, *Gleditsia (honey locust)*, *Laburnum* (golden-chain tree), *Lathyrus* (sweet pea), *Lupinus* (lupine),

Mimosa (sensitive plant), *Wisteria* (wisteria); **dye plants:** *Genista tinctoria* (yellow dye), *Indigofera* (indigo); **poisonous:** *Abrus precatorius* (precatory bean), *Astragalus* (locoweed), *Gymnocladus* (Kentucky coffee tree), *Lupinus argenteus* (silvery lupine), *Robinia* (locust); **invasive weeds:** introduced to the United States for erosion control, *Coronilla varia* (crown vetch) is now an invasive plant overrunning natural prairies and savannas; *Pueraria montana* (kudzu vine, "the vine that ate the south.")

Lathyrus latifolius   Sweet Pea

A perennial herb, the sweet pea plant climbs by means of tendrils (a) on its broadly winged stems (b). Double-lobed stipules (c) emerge from nodes on the stem. The compound leaf has a winged petiole (d) and two leaflets (e).

The raceme inflorescence has a long peduncle (f) that supports many flowers on short pedicels (g). In addition to a sepal tube (h) with 5 lobes, the flower has a corolla divided into a banner petal (i), 2 wing petals (j), and a keel (k) composed of 2 fused petals. Petal color varies from white to pink to dark pink-red.

With the petals removed from a flower bud, 10 2-chambered stamens (l) enclose the pistil in 2 bundles of one above plus nine basally fused ones. A flower, with sepals (h) also removed, and ovary (m), cut open, shows the ovules (n) attached to the inner carpel wall. Extending from the ovary are the style (o) and stigma (p).

When the sweet pea legume (q) matures, the carpel wall splits into two halves and twists, thus expelling the seeds(r).

COLOR CODE

green:	tendril (a), stem (b), stipules (c), petiole (d), leaflets (e), peduncle (f), pedicel (g), sepals (h)
white, pink, or pink-red:	petals (i, j, k)
yellow:	stamens (l)
light green:	ovary (m), ovules (n), style (o), stigma (p)
tan:	legume (q)
dark brown:	seeds (r)

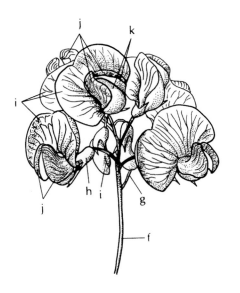

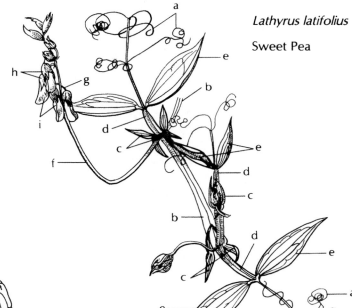

Lathyrus latifolius

Sweet Pea

Inflorescence
× 1

Flowering Shoot
× ½

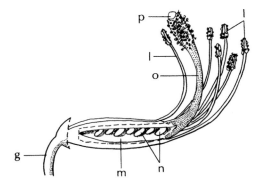

Flower Bud
petals removed
× 3

Flower
perianth & 4 stamens removed
ovary cut open
× 3

Legume
× ⅔

Dogwood Family (Cornaceae)

Plants in this family are usually woody trees and shrubs with simple, usually opposite leaves. The small flowers are arranged in clusters that are sometimes subtended by bracts that may be quite showy. The flower is bisexual or has unisexual male and female flowers on one plant (**monoecious**) or on separate plants (**dioecious**). There are about 13 genera and over 100 species.

Attached to the ovary are 4–5 sepals and 4–5 petals or none. The number of stamens is the same as the number of petals. There is a single pistil with an inferior ovary consisting of 2–4 carpels. A glandular disc surrounds the one or more styles. Fruits are drupes or berries.

Of interest... **ornamentals:** *Acuba* (Japanese laurel), *Cornus canadensis* (bunchberry), *C. florida* (flowering dogwood), *C. sericea* (red-osier dogwood), *C. mas* (Cornelian cherry), *C. kousa* (Japanese dogwood), *C. racemosa* (grey dogwood), *Griselinia* spp. (kapuka), *Helwingia*.

Cornus florida variety *rubra*
Flowering Dogwood

This small tree has widely spreading branches and is valued as an ornamental specimen tree in landscaping. In nature, this understory tree of the forest has white flowers. Mutant pink and red varieties have been selected from natural populations of flowering dogwood and propagated for horticultural use.

The opposite leaves are simple with entire, wavy margins. In autumn they turn a deep red color.

What appear to be pink flower petals are actually 4 large bracts (b) around a cluster of 20-30 bisexual flowers (c) with yellow petals. Each flower has 4 joined sepals (d), 4 small petals (e, f), 4 stamens (g), and a single pistil (h). The stamens arise from the base of a nectar disc (k). Parts of the pistil include a 2-carpelled ovary (l) with an ovule (m) in each locule, a style (n), and stigma (o).

Birds are very fond of the bright red drupe fruits (p).

COLOR CODE

dark pink:	bract veins (a)
pink:	bracts (b)
yellow-green:	flowers (c), bud petals (e)
light green-pink:	sepals (d)
white:	filament (j), ovary (l), ovules (m)
yellow:	petals (f), stamens (g), anther (i), nectar disc (k)
light green:	pistil (h), style (n), stigma (o)
red:	drupes (p)
gray:	peduncle (q)
optional:	leaf (green in summer; or dark red with yellow veins in autumn)

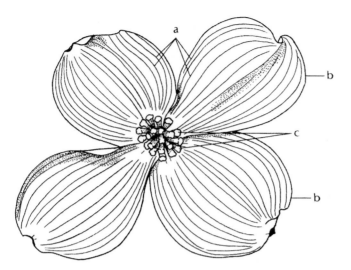

False Flower
× 1

Cornus florida var. *rubra*

Flowering Dogwood

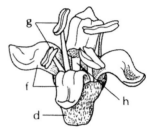

Flower
× 5

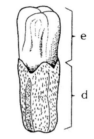

Flower Bud
× 5

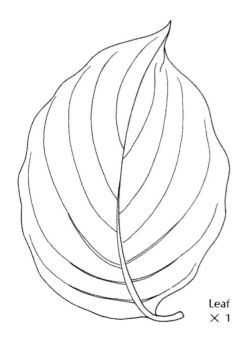

Leaf
× 1

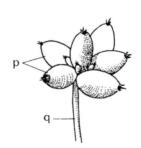

Drupes
× 1

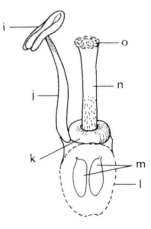

Pistil, Stamen
ovary vertical section
× 10

Staff-tree Family (Celastraceae)

The most obvious characteristic this family is the pulpy orange or red **aril** usually covering the seed. Small clusters of green flowers usually have flower parts in 4's or 5's. The flowers are bisexual or sometimes **polygamodioecious** (see *Celastrus,* below). There is one pistil with a superior ovary surrounded by a nectar disc.

Plants in this family are woody shrubs, trees, and twining vines. Leaves are simple, alternate or opposite. There are 55 genera and 850 species.

Of interest... **ornamentals:** *Catha* (Cafta), *Celastrus* (bittersweet), *Elaeodendron, Euonymus alatus* (burning bush), *Gymnosporia, Maytenus, Pachystima, Tripterygium;* **furniture making:** *Euonymus europaeus* (European spindle tree); **food:** *Catha edulis* (Khat tree), leaves for tea.

Euonymus alatus   Burning Bush, Winged Wahoo

This shrub's stems (a) have a winged appearance because of the presence of corky ridges (b) of **periderm.** Its opposite leaves (c) have finely toothed (**dentate)** margins and hair-like stipules (d) at the base of the petioles. In autumn, the leaves are a brilliant red.

The flowers are borne in a cluster, with the central flower opening first (**cyme**). Parts of the bisexual flower consist of 4 sepals (e), 4 petals (f), and 4 stamens (g) surrounding the single pistil (h) with 4 carpels (i).

Stamens and pistil are situated on a disc (j), which is a fleshy **receptacle** with the sepals and petals arising around it.

After the leaves are shed in autumn, the **loculicidal capsule** (m) is more obvious. A red outer covering (aril, n) encloses the seeds.

This vigorous growing native of the Temperate Zone in East Asia can crowd out plants that are native to the American Temperate Zone because of its prodigious seed production and high seedling survival rate.

Celastrus scandens   Bittersweet

Bittersweet is a twining vine with alternate leaves (o). **Polygamodioecious** is the condition describing the plants' flowers. This means that unisexual male and female flowers are mainly on separate plants, but bisexual flowers are also present on both plants.

The male flower consists of 5 sepals (r), 5 petals (s), 5 stamens (t), and a vestigial pistil (u). The female flower (not shown) has a pistil composed of 3 carpels, a single style, and a 3-lobed stigma with vestigial stamens present.

The fruit type is a **loculicidal capsule.** Three orange valves (v) open to expose a red aril (w) covering the seeds. Autumn branches of orange fruits are used for decoration.

COLOR CODE

brown:	bark ridges (b)
green:	stem (a, p), leaves (c, o), stamens (g), pistil (h), carpels (i), pedicel (k, q), disc (j), pedicel (k, q),
light green:	sepals (e, r), petals (s)
yellow-green:	petals (f)
pink:	pedicel (l)
lavender:	capsules' valves (m)
orange:	capsules' valves (v)
red:	arils (n, w)
yellow:	pedicel (x), stamens (t)
white:	pistil (u)

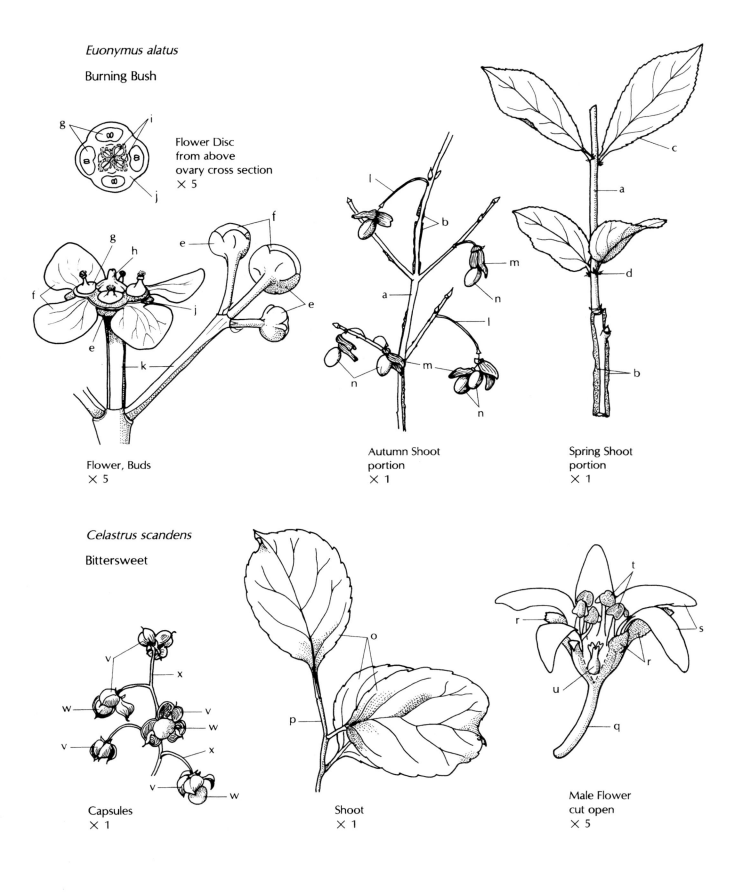

Euonymus alatus

Burning Bush

Flower Disc
from above
ovary cross section
× 5

Flower, Buds
× 5

Autumn Shoot
portion
× 1

Spring Shoot
portion
× 1

Celastrus scandens

Bittersweet

Capsules
× 1

Shoot
× 1

Male Flower
cut open
× 5

Spurge Family (Euphorbiaceae)

Euphorbs, a large family of 283 genera, often have a poisonous, milky, latex sap. Very diverse forms are represented. There are weedy herbs, commonly called **spurges**, showy ornamental herbs, trees, and cactus-like succulents. Most are tropical. Stems usually have stipules that take the form of hairs, glands, or spines.

False flowers, often with petal-like appendages, have unisexual, male and female flowers aggregated in a **cyanthium** (a cup-like structure). Male flowers, of one stamen each, may number 1 to many in a cyanthium. The female flower is a single pistil composed of a superior ovary with usually 3 fused carpels, 3 styles, and 3-6 stigmas. The fruit is usually a **schizocarp** in which the three carpels, bearing seeds, break apart.

Of interest . . . **economic:** *Aleurites fordii* (tung oil, used in paints and quick-drying varnishes), *Hevea brasiliensis* (para rubber, the most important source of natural rubber), *Manihot esculenta* (tapioca, bitter cassava, manioc, sweet-potato tree; the starchy roots yield many food products), *M. glaziovii* (ceara rubber), *Ricinus communis* (castor oil); **ornamentals:** *Acalpha hispida* (chenille plant), *Codiaeum* (croton), *Euphorbia* with 1550 species (poinsettia, crown-of-thorns, spurges), *Pedilanthus tithymaloides* (redbird flower, slipper flower, redbird cactus, Japanese poinsettia), *Phyllanthus* (Otaheite gooseberry, gooseberry tree), *Ricinus communis* (castor-bean).

poisonous: *Aleurites fordii* (tung oil), *Euphorbia* spp., The milky sap of *Euphorbia* species causes dermatitis with severe blistering on contact with human skin,

E. marginata (snow-on-the-mountain, ghostweed), a garden ornamental, is especially corrosive to the skin; *Ricinus communis* (castor-bean, castor-oil-plant) contains **ricin**, a blood poison. All parts of the plant are poisonous, but particularly the "bean" (seeds). If ingested severe poisoning, ending in convulsions and death, can occur.

Euphorbia milii   Crown-of-thorns

This native of Madagascar is a spiny, succulent shrub. The stem (a) has alternate, simple leaves (b), spiny stipules (c), and latex sap (d). Two flowering shoots (e) are shown, one with the petal-like appendages (f) in bud and the other with the appendages expanded (g).

The drawing shows an enlarged false flower, called a **pseudanthium**, has five nectar glands (h) and 5 bracts (i) that surround the flowers. At the center of each of the female flowers are 3 styles (j), which divide into 2 stigmas (k). With the petaloid appendages removed, the cyanthium can be seen. Male flowers (l), three are complete in the illustration, develop before the female flower.

The female flower, in an enlarged drawing, shows the ovary's 3 fused carpels (o), which contain ovules in **axile placentation**, the styles, and stigmas. The ovary develops into a **schizocarp** (p) fruit, which has green petaloid appendages (q) in this related species.

COLOR CODE

green:	stem (a), leaves (b), peduncle (e), nectar glands (h), pedicel (m, n), petal-like appendages (q)
gray:	stipules (c)
white:	latex (d)
red:	petal-like appendages (f, g)
yellow:	bracts (i), male flowers (l)
light green:	style (j), stigma (k), ovary (o)
tan:	schizocarp (p)

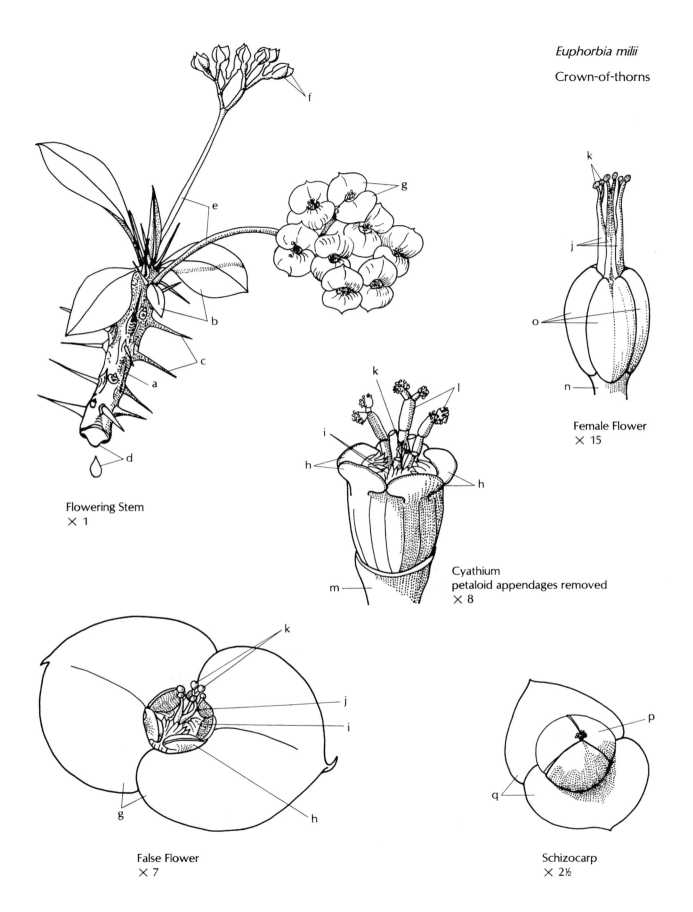

Euphorbia milii

Crown-of-thorns

Flowering Stem
× 1

Female Flower
× 15

Cyathium
petaloid appendages removed
× 8

False Flower
× 7

Schizocarp
× 2½

Grape Family (Vitaceae)

This family consists mostly of climbing vines with tendrils that develop opposite the leaves. The tendrils may be twining, as in grape *(Vitis),* or end in attaching discs (see 16), as in Virginia creeper *(Parthenocissus).* Flower clusters also develop at the node opposite a leaf. The leaves are alternate, simple or **pinnately** or **palmately compound**. There are twelve genera and about 700 species.

The minute flowers are bisexual or unisexual on the same plant (**monoecious**). Usually there are 4–7 sepals, 4–7 petals, 4–7 stamens, and a single pistil, which develops into a berry fruit.

Of interest... **food:** *Vitis* spp. (grape berries are used for making wine and jellies; dried gapes for raisins and currants); **ornamentals:** *Cissus* spp. (grape ivy), *Parthenocissus quinquefolia* (Virginia creeper), *P. tricuspidata* (Boston ivy).

Vitis riparia   River-bank Grape

Lobed leaves (a) with **palmate venation** are alternate to each other and opposite tendrils (b) on the stem (c). Tiny, unisexual, male and female flowers occur on this plant.

Above the flower pedicel (d) joined sepals (e) appear as a wavy-margined cup in bud. The flower is covered with petals (f) joined at the top. As the petals separate into 5-6 lobes (g), the stamens' filaments (h) and the nectar disc (i) are exposed. When the petal cap is shed, the 5-6 anthers (j) can be seen.

The female flower, with the petal cap shed, consists of sepal cup (k), nectar disc (l), and pistil composed of a superior ovary (m) and a short style with a **capitate** (formed like a head) stigma (n).

Clusters of female flowers develop into the familiar bunch of grapes, botanically called berries (o, p).

COLOR CODE

green:	leaves (a), stem (c), pedicels (d), sepal cup (e), peduncle (q), underdeveloped berrry (p)
red:	tendril (b)
light green:	petals (f, g), ovary (m), stigma (n)
dark green:	nectar disc (i, l)
white:	filaments (h)
yellow:	anthers (j)
purple:	sepal cup (k)
dark blue:	berries (o)

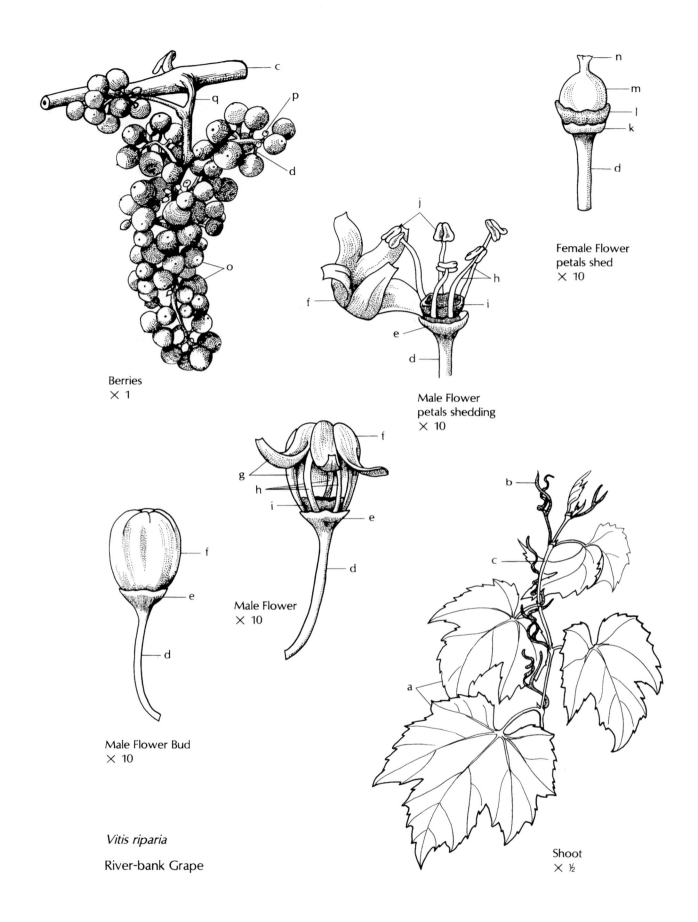

Berries
× 1

Female Flower
petals shed
× 10

Male Flower
petals shedding
× 10

Male Flower Bud
× 10

Male Flower
× 10

Vitis riparia

River-bank Grape

Shoot
× ½

Maple Family (Aceraceae)

Maples have opposite branches and leaves. The leaves may be simple, with palmate venation, or compound as in box elder (*Acer negundo*, see 35). Species differ in leaf shape with margins entire or variously toothed or variously lobed. The two genera (*Diperonia* from China and *Acer*) in this family are small to large deciduous trees found mainly in the North Temperate Zone.

The flower in some species is bisexual, but usually it is unisexual with both male and female flowers on the same tree (**monoecious**) or separated on different trees (**dioecious**). Unisexual flowers may have vestigial organs of the opposite sex. There are usually 5 sepals, 5 petals or none, 4–12 stamens, and a single pistil with a superior ovary. The 2 fused carpels of the ovary develop into a double **samara** fruit.

Of interest... **lumber:** *Acer saccharum* (sugar maple), *A. macrophyllum* (big leaf maple); **food:** *A. saccharum* (sugar maple, all maple species yield sap with a high sugar content); **ornamentals:** *A. saccharum* (sugar maple), *A. platanoides* (Norway maple), *A. pseudoplatanus* (sycamore maple), *A. ginnala* (amur maple), *A. palmatum* (Japanese maple) with many cultivars for variously cut leaves and colors.

Acer saccharinum   Silver Maple

This maple tree grows to 37 meters high. The main trunk divides near the ground into several upright stems. It is a fast-growing tree with narrow-angled brittle branches and a spreading root system. Therefore, it should not be used for landscaping, even though many people use it for this purpose by mistake.

Opposite leaves (a) are 5-lobed with **palmate venation** (b) and toothed (**dentate**) margins. The leaf is green on the upper surface and silvery on the lower surface (c). In one location with two silver maple trees growing next to each other, the tree bearing male flowers in the spring produces yellow autumn leaves, while the other tree bearing female flowers in the spring has red autumn leaves.

This species is **dioecious** and wind-pollinated, with flowers opening before the leaves (e) expand on the stem (f). Hair-fringed bud scales (g) enclose clusters of male flowers (h). The male flower has joined sepals (i), a variable number of stamens (j), and a vestigial pistil (k).

Clusters of female flowers (l) are produced on another tree. The female flower consists of a sepal cup (m) and vestigial stamens (n), and the single pistil has an ovary consisting of 2 fused carpels (o) and 2 **plumose** (feathery) stigmas (p). Two ovules occur in each carpel, but only one develops, so that the fruit is a double samara (q) with a seed (r) on each side. In suitable soil, there is almost 100% germination of the seeds, another good reason not to use this plant in the landscape.

COLOR CODE

yellow:	autumn leaves (a), male flowers (h), stamens (j, n)
green:	veins (b), leaf bud (e), stem (f), female flowers (l), stigmas (p)
white:	leaf lower surface (c), pistil (k), carpels (o)
brown:	stem (d), samara (q, r)
red:	bud scales (g)
pale yellow:	sepals (i, m)

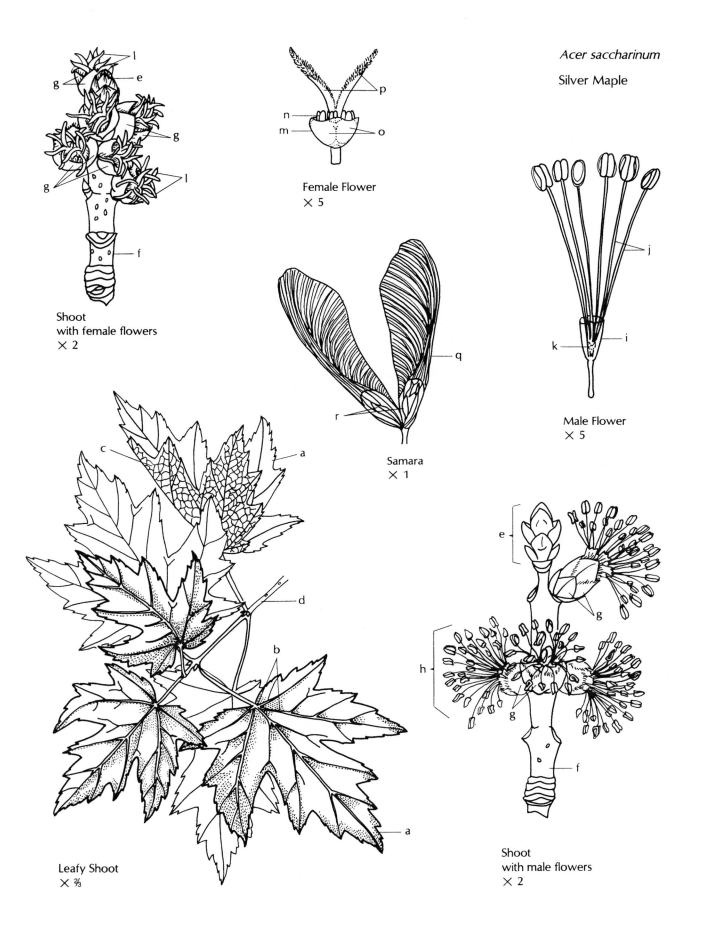

Acer saccharinum

Silver Maple

Shoot
with female flowers
× 2

Female Flower
× 5

Samara
× 1

Male Flower
× 5

Leafy Shoot
× ⅔

Shoot
with male flowers
× 2

Cashew Family (Anacardiaceae)

Plants in this family usually have **resin ducts** in the tissues. The resin in some of these plants causes contact dermatitis in humans. Poison ivy, poison oak, and poison sumac are familiar examples.

Trees, shrubs, and vines, with usually alternate and compound leaves, are present in this family. The small flowers form in branched clusters (**panicles**) and are bisexual or unisexual. Flower features are a prominent nectar disc and a pistil of fused carpels that form one seed cavity in which one ovule develops. Fruits are usually **drupes**. There are about 80 genera and about 600 species.

Of interest... **food:** *Anacardium occidentale* (cashew), *Mangifera indica* (mango), *Pistacia vera* (pistachio); **resins:** *Pistacia lentiscus* (mastic tree), *Toxicodendron vernicifera* (varnish tree); **tannic acid:** *Rhus typhina* (staghorn sumac), *Schinopsis;* **ornamentals:** *Cotinus coggygria* (smoke tree), *Rhus aromatica* (fragrant sumac), *Rhus copallina* (shining sumac), *R. typhina* 'Laciniata' (lacy-leafed staghorn sumac); **poisonous:** *Metopium toxiferum* (poison-wood, doctor gum coral-sumac), *Toxicodendron verniciflua* (Japanese lacquer-tree), *T. radicans* (poison-ivy, poison-oak), *T. diversiloba* (Pacific poison-oak), *T. vernix* (poison-sumac).

Rhus typhina   Staghorn Sumac

An extensive **clone** of shrubs may be formed from one plant by new shoots arising from the root system. The common name refers to the very hairy stem that has the appearance of new antlers. The leaves are alternate and compound, with opposite pairs of leaflets (a) on the petiole (b). In autumn, leaves become a brilliant red.

In the spring, clusters of flowers form at the ends of shoots. An individual plant has either male or female flowers with vestigial organs of the opposite sex. The male flower has 5 sepals (c), 5 petals (d), 5 stamens (e), and a vestigial pistil (f). Staminal filaments (g) arise beneath the nectar disc (h).

The female flower has 5 sepals (j), 5 petals (k), 5 vestigial stamens (l) and a single pistil with 3 stigmas (m). Arising from the nectar disc (n), the superior ovary (o) consists of three fused carpels containing one ovule.

In autumn, plants with female flowers produce conspicuous upright clusters of **drupes** (p) on the ends of shoots. The drupe's epidermis is covered with red hairs (q). When partially removed, a single seed (r) is exposed.

Cotinus coggygria   Smoke Tree

In late summer, this low structured tree is covered with what appears to be clouds of pink smoke. As the drupes (s) form, the surrounding hairy, filamentous stems (t) change from green to pink.

COLOR CODE

red:	petiole (b)
green:	leaflets (a), sepals (c, j), pedicel (i), leaf (u)
yellow:	anthers (e)
white:	pistil (f), filaments (g)
light green:	petals (d, k), stigmas (m), ovary (o), anthers (l)
orange:	nectar disc (h, n)
dark red:	drupes (p), hairs (q)
tan:	seed (r)
brown:	drupes (s)
pale pink:	stems (t)

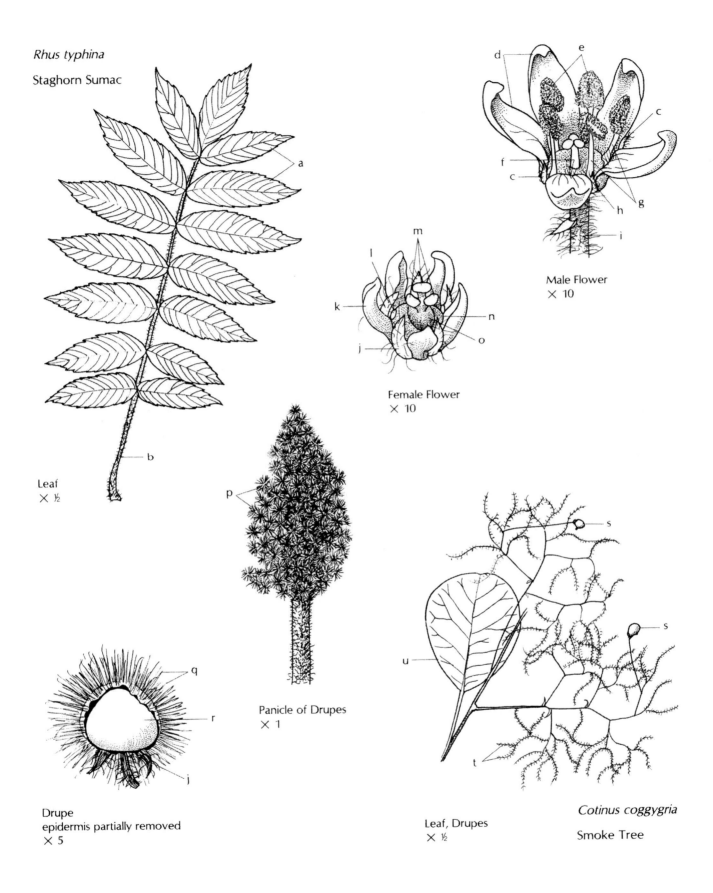

Rhus typhina

Staghorn Sumac

Leaf
× ½

Female Flower
× 10

Male Flower
× 10

Panicle of Drupes
× 1

Drupe
epidermis partially removed
× 5

Leaf, Drupes
× ½

Cotinus coggygria

Smoke Tree

Rue Family (Rutaceae)

Aromatic **essential oils** are produced in glands that appear as clear "dots" on the leaves. Another characteristic feature of this family is that within the usually bisexual flowers, the pistil has a superior, usually lobed ovary situated on a nectar disc.

Rue family plants include trees, shrubs, and herbs, with many different fruit types represented. They include capsules, leather-rind berries (**hesperidia**), **drupes**, and **samaras**. There are 150 genera and about 900 species.

Of interest...**food:** *Citrus* 60 species, with many hybrids between species, *Citrus sinensis* (sweet orange), *C. limon* (lemon), *C. paradisi* (grapefruit), *C. aurantifolia* (lime), *C. aurantium* (Seville orange), *C. reticulata* (tangerine, manderin, satsuma), *C. medica* (citron), *Fortunella* (kumquat), *Pummelo*; **ornamentals:** *Calodendrum* (Cape chestnut), *Choisya ternata*, *Dictamnus* (gas plant, fraxinella, dittany), *Diosma, Murraya* (orange jessamine), *Phellodendron* spp. ("cork" trees), *Poncirus* (trifoliate orange), *Ptelea* (hop tree, wafer ash), *Ruta graveolens* (common rue), *Skimmia, Zanthoxylum americanum* (prickly ash); **lumber:** *Zanthoxylum flavum* (West Indian silkwood); **poisonous:** *Dictamnus, Ruta graveolens, Ptelea.*

Citus limon   Lemon

The small lemon tree has alternate, simple leaves (a) with translucent glands containing aromatic oil. *Citrus* species' leaves have various sizes of petiole wings (b). A grapefruit *(C. paradisi)* leaf (c) is shown for comparison of the winged petiole (d).

Small, bisexual flowers, produced in clusters, are very fragrant. The 5 joined sepals (f) form a cup. Inside the cup are 5 petals (g) and many stamens (h) of unequal lengths. Some of the filaments (i) are joined together, and arise at the base of the nectar disc (j).

The pistil is situated in the center of the disc surrounded by the stamens, with the stigma (k) at anther (l, m) level. The pistil has a superior ovary (o), a single style (p) composed of fused canals leading to the carpels, and a single stigma (q). As the fruit (r) develops, the petals, stamens, and the style are shed.

The fruit is a berry with a leathery rind (**hesperidium**). The rind or **exocarp**(s) has oil glands. Inside the rind is the spongy **mesocarp** tissue (t). Inside the mesocarp is the **endocarp**, which produces the juice sacs (u) of enlarged cells.

The seed (v) contains a green chlorophyll-containing embryo, which may germinate within the fruit. This is called **vivipary**. The grapefruit leaf shown here is from a plant that was grown from a seed that germinated within a fruit.

COLOR CODE

green:	leaf blade (a, c) petiole wings (b, d), peduncle (n), nectar disc (j), ovary (o), fruit (r)
light green:	pedicel (e), sepal cup (f), stigma (k)
white:	petals (g), stamens (h), filaments (i), style (p), mesocarp (t) seeds (v)
yellow:	anthers (l, m), exocarp (s), juice sacs (u)
tan:	stigma (q)

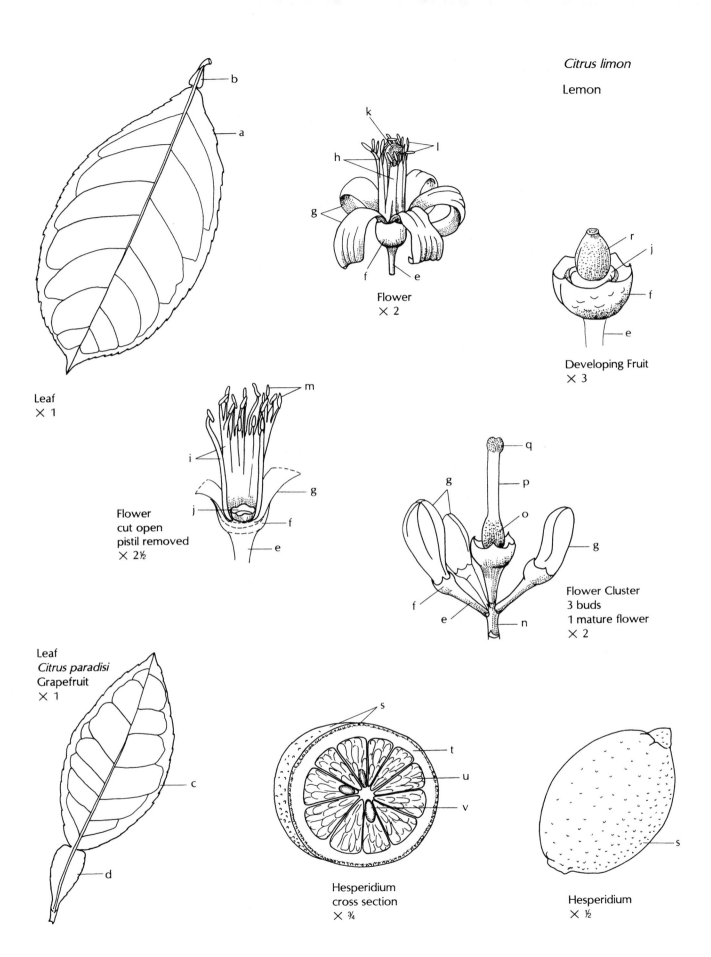

Citrus limon

Lemon

Leaf
× 1

Flower
× 2

Developing Fruit
× 3

Flower
cut open
pistil removed
× 2½

Flower Cluster
3 buds
1 mature flower
× 2

Leaf
Citrus paradisi
Grapefruit
× 1

Hesperidium
cross section
× ¾

Hesperidium
× ½

Geranium Family (Geraniaceae)

Mostly herbs and some small shrubs are represented in this family. Scented leaves in some species, a flower with parts in 5's, and a beaked or lobed fruit are some of the characteristics. There are eleven genera and about 750 species.

Leaves are alternate or opposite, and compound or simple with lobes or divisions. They usually have **palmate venation** and **stipules**. Usually in clusters, the bisexual flower has 5–15 stamens in whorls of 5 and a single pistil with 3–5 carpels in a superior ovary. Adhering to the ovary axis, the 3–5 styles aid in an unusual seed dispersal mechanism. The fruit is a **septicidal** or **loculicidal capsule**.

Of interest ... **ornamentals:** *Erodium* spp. (stork's-bill, heron's-bill), *Geranium* spp. (hardy geraniums, 400 species). Some popular garden geraniums are *G. cantabrigiense* 'Biokovo', *G. cinereum* 'Ballerina,' *G. himalayense* × *G. pratense* 'Johnson's Blue,' *G.* 'Rozanne,' *G. sanguineum, G. macrorrhizum, G.* × *magnificum. Pelargonium zonale* is the geranium of commerce and native to South Africa. Some with scented leaves include *P. crispus* (lemon-scented), *P. graveolens* (rose-scented), *P. fulgidum* (spice-scented), *P. tomentosum* (peppermint-scented).

Geranium maculatum Wild Geranium, Crane's-bill

This spring-flowering herb produces annual shoots from a thick perennial rhizome. Leaves (a) at the base of the plant have 5–7 lobes and long petioles (b), while upper leaves (c) have 5 lobes and shortened petioles (d). The leaves have **stipules** (f) and **palmate venation**.

Flowers form in a flat-topped cluster with the central flower opening first (**cyme**). The flower has 5 persistent sepals (j) with spine-like tips (k), which are clearly seen when the petals are removed. Five lavender petals (l) have green veins (m).

The 10 stamens (n) are in two whorls of 5 each around the base of the pistil (o). Staminal filaments (p) are attached at the center of the anther (q) and flare outward at the bases. In the illustration, the anthers of the inner whorl of stamens are open, exposing pollen grains, while the anthers of the outer whorl are not yet mature.

The single pistil has a superior ovary (r) of 5 fused carpels (s), 5 styles (t), which form a column around the ovary axis (u), and 5 stigmas (v). Between the petal bases are 5 nectar glands (w).

Crane's-bill, one of the common names, applies to the fruit, a beaked, **septicidal capsule**. At the time of seed dispersal, the styles (t, x) attached to the carpels (s, y) coil elastically, which propels the seeds (z) out of the carpel chambers.

COLOR CODE

green:	blade (a, c), petiole (b, d), stem (e), peduncle (g), pedicel (h), nectar glands (w)
light green:	stipules (f), bracts (i), sepals (j), petal veins (m), filament (p), pistil (o), carpels (s), styles (t), stigmas (v)
light lavender:	petals (l)
red:	sepal tips (k)
yellow:	stamens (n), anther (q)
tan:	ovary axis (u), styles (x), carpels (y), seed (z)

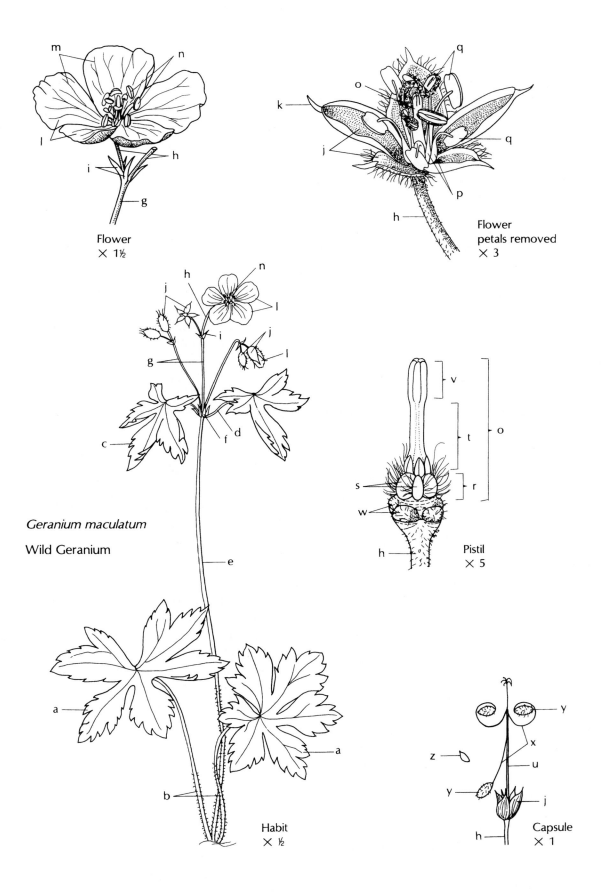

Flower
× 1½

Flower
petals removed
× 3

Geranium maculatum

Wild Geranium

Pistil
× 5

Habit
× ½

Capsule
× 1

Carrot Family (Apiaceae)

Members of this family form flowers in a dense, flat-topped cluster (**umbel**, the reason for the old family name, Umbelliferae). Below the umbel, there is often a whorl of bracts. The number of sepals, petals and stamens of an individual flower in a cluster are each 5. The single pistil has an inferior ovary made up of 2 carpels, an ovule in each. After fertilization, the ovary develops into a **schizocarp fruit**, which splits apart into 2 **mericarps** that function as **achenes** (see 38).

Also characteristic of plants in this family are the usually stem-sheathing leaf **petioles**. Compund leaf blades are usually several times divided into leaflets. Aromatic oils are present in leaves, stems, flowers, and fruits of these mostly biennial or perennial herbs. There are about 300 genera and over 3,000 species.

Of interest... **food** and **flavoring**: *Anethum graveolens* (dill), *Angelica, Anthriscus cerefolium* (chervil), *Apium graveolens* (celery), *Carum carvi* (caraway), *Coriandrum sativum* (coriander), *Cuminum cyminium* (cumin), *Daucus carota* cultivars (carrot), *Foeniculum vulgare* (fennel), *Levisticum officinale* (lovage), *Myrrhis odorata* (sweet cicely), *Pastinaca sativa* (parsnip), *Petroselinum crispum* (parsley), *Pimpinella anisum* (anise); **ornamentals**: *Aegopodium podograria* (goutweed, bishop's weed), *Ammi majus* (lacy false bishop's weed), *Astrantia major* (masterwort), *Heracleum lanatum* (cow parsnip), *Trachymeme* (blue lace-flower); **poisonous**: *Cicuta maculata* (water hemlock), *Conium maculatum* (hemlock), *Heracleum mantegazzianum* (giant hogweed). Cases of poisoning of humans and animals by plants are mainly caused by ingestion or skin contact. Some plants are poisonous only at certain stages of growth or certain times of the year. *Cicuta maculata* (water hemlock) grows mainly in wet areas and poisons cattle foraging in wet pastures in the spring. A small piece the size of a walnut is enough to kill a cow. **Cicutoxin** is concentrated in the roots, but is present in all parts of the plant. In humans if mistaken as roots of parsnips, Jerusalem artichokes or other roots, when eaten, the results are fatal. *Conium maculatum* (poison hemlock) was the plant that caused the death of Socrates. The plant, native to Eurasia, has become naturalized in the United States. It causes paralysis of the lungs. *Heracleum mantegazzianum* (giant hogweed) is not only an invasive plant but causes painful blisters on contact with human skin. This is called "contact dermatitis."

Daucus carota   **Wild Carrot, Queen Anne's Lace**

Wild carrot plants are biennial herbs, producing a basal rosette (a) of leaves and a tap root (b) the first year, with stem-borne leaves, flowers, and fruits the second year. The alternate, feathery-compound leaves (c) have petiole bases (d) that sheath the stem (e). This plant is not considered a "wild food," as it greatly resembles other extremely poisonous plants in this family, which can be confused with it. Moreover, the fibrous white tap-roots are not edible like the orange, β-carotene-containing tap roots of cultivated carrot.

The umbel of flowers is supported by a peduncle (f) and has a whorl of compound bracts (g) below. The stems that branch out above the bracts are primary rays (h). Secondary **rays** or **pedicels** (i) support flowers in small clusters (umbellets, j) subtended by bractlets (k). This arrangement is termed a **compound umbel**. Outer flowers of the umbel have larger petals to attract pollinators, and the one center flower (l) has deep red petals.

Lacking sepals, a carrot flower consists of 5 petals (m), which are occasionally pink instead of white, 5 stamens (n), and a single pistil. In the flower drawing, because of the pistil's inferior (below other flower parts) ovary, only the 2 styles (o) on stylopodia (p) can be seen. The **stylopodium** is a nectar-secreting organ that attracts a variety of pollinating insects. Cross-pollination with other plants is not necessary, because the flowers are self-fertile.

As the fruit develops, the parts of the pistil become more obvious. A stylopodium (p) with style (o) and stigma (q) is atop each carpel, the **mericarp** (r) of the **schizocarp** fruit. Hooked spines (s), which catch in animal fur, facilitate fruit dispersal. Fruit is also wind-dispersed. In autumn, the umbel of schizocarps (t) close inward to form a "bird's-nest" in appearance. It closes in wet weather and opens when dry, further aiding fruit dispersal.

COLOR CODE

green:	leaf (c), petiole (d), stem (e), peduncle (f), bracts (g), ray (h), pedicel (i)
white:	root (b), flowers (j), petals (m)
red:	center flower (l)
yellow:	stamens (n)
light green:	styles (o), stylopodium (p), stigma (q)
tan:	mericarp (r), spines (s), schizocarps (t)

Daucus carota

Wild Carrot

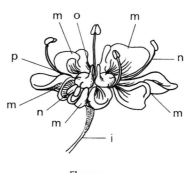

Flower
× 10

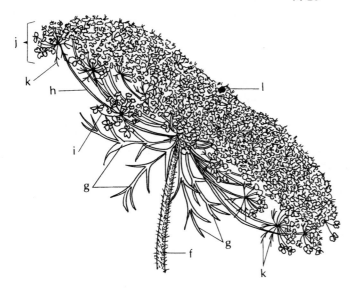

Habit
first year
× ⅕

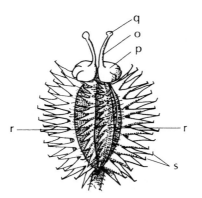

Schizocarp
× 20

Umbel
× 1

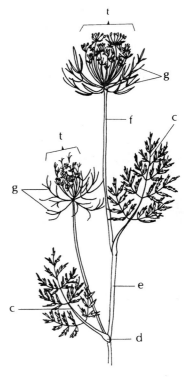

Fruiting Shoot
second year
× ⅕

Milkweed Family (Asclepiadaceae)

The common characteristic in this family is an unusual flower (shown in the illustrated example). Otherwise, the appearance of these plants varies enormously. This family includes sturdy perennial herbs, shrubs, dainty vines, small trees, and succulent plants that resemble cacti. Usually, the stems have a milky, latex-containing sap. Simple leaves are in an opposite or whorled arrangement, but are sometimes alternate. There are about 250 genera and 2,000 species. Flowers occur singly or in various types of clusters. The fruit is a **follicle** that splits along one seam.

Of interest . . . **ornamentals:** *Araujia sericofera, Asclepias tuberosa* (butterfly weed), *A. curassavica* (bloodflower), *Ceropegia woodii* (string-of-hearts), *Cryptostegia grandiflora* (purple allamanda), *Hoya carnosa* (wax plant), *Huernia, Oxypetalum caeruleum* (blue milkweed), *Stapelia* (carrion flower), *Stephanotis floribunda* (Madagascar jasmine); **weed:** *Asclepias Cornuti* (common milkweed, silkweed), *Asclepias incarnata* (swamp milkweed). *Asclepias Cornuti* (common milkweed) produces a milky acrid sap. Monarch butterfly larvae feed on the milkweed leaves which renders adults bad tasting to birds. This is advertised in their distinctive orange and black coloration. The somewhat smaller Viceroy butterfly mimics the coloration of the Monarch and avoids being attacked, even though as a willow-feeder, it is edible to birds. The common name silkweed refers to the use in former times of the silky seed as insulation in clothing and down in pillows.

Cryptostegia grandiflora (India rubber vine, purple allamanda) from Madagascar was used to produce a commercial source of poor quality rubber latex. Now, this twining vine is used as an ornamental because of its beautiful flowers. Caution must be used when pruning, as the poisonous sap is very irritating to the skin. *Matelea* has sap once used as arrow poison. The old name *Vincetoxicum* is Greek for "subduing poison."

Asclepias tuberosa Butterfly-weed

This perennial herb is an attractive, orange-flowered plant that does not have milky sap and can be cut for floral arrangements. The hairy stems (a) have alternate, lanceolate leaves (b) and terminate in flat-topped flower clusters (**umbels**). An umbel is a type of cluster in which the peduncle (c) has many pedicels (d) arising at a common point, each supporting a flower (e). Individually, this specialized flower has 5 sepals (f) and 5 reflexed petals (g) that cover the sepals. Above the petals is a structure composed of a **gynostegium** (h) surrounded by a crown of 5 **nectar horns**. Each nectar horn consists of an upright hood (i) and a protruding crest (j). With nectar horns removed, the central portion of the gynostegium can be seen. In the area between the nectar horn attachments are 2 flaps (veils, k) with an opening between. A pin is shown inserted in this side opening. When the pin is raised through the slit opening, it dislodges another structure, a pair of pollinia, tucked into top-opening slits in the gynostegium. A pair of **pollinia** is a male part of the flower; there are five pairs. Each pair consists of 2 anther sacs (l) with arms (m), connecting them to a gland (n).

The female part of the gynostegium cannot be seen from the outside. By opening two side flaps (k), the sticky stigmatic tissue (o) is exposed. There are 5 sets of flaps with stigmatic tissue between each, connecting internally to 2 styles (p). Below each style is a carpel (q), where numerous ovules (r) are attached in **parietal placentation**. So, all together, this flower has a single pistil composed of 2 ovaries (s) of a carpel each, 2 styles, and enlarged 5-lobed stigma.

How does a bee react to this flower? Bees are attracted to the nectar in the nectar horns. To position itself, the bee lands on the gynostegium. When a leg or an antenna slips between the side flaps, it is raised up (as the pin shown) and catches a pair of pollinia. Apparently, this does not bother the bee. With repeated slipping and sliding of many bees on many flowers, pollinia are removed from the top-opening slits and inserted in the side-opening flaps, achieving a high percentage of pollination. The fruits that result are **follicles** (t) of hairy seeds (u).

COLOR CODE

green:	stem (a), leaf (b), peduncle (c), pedicel (d), sepal (f)
orange:	flower (e), petal (g), hood (i), crest (j)
pale green:	gynostegium (h, k), stigma (o), styles (p), carpel (q), ovary (s)
yellow:	anther sacs (l)
tan:	anther arms (m)
brown:	anther gland (n)
white:	ovules (r)

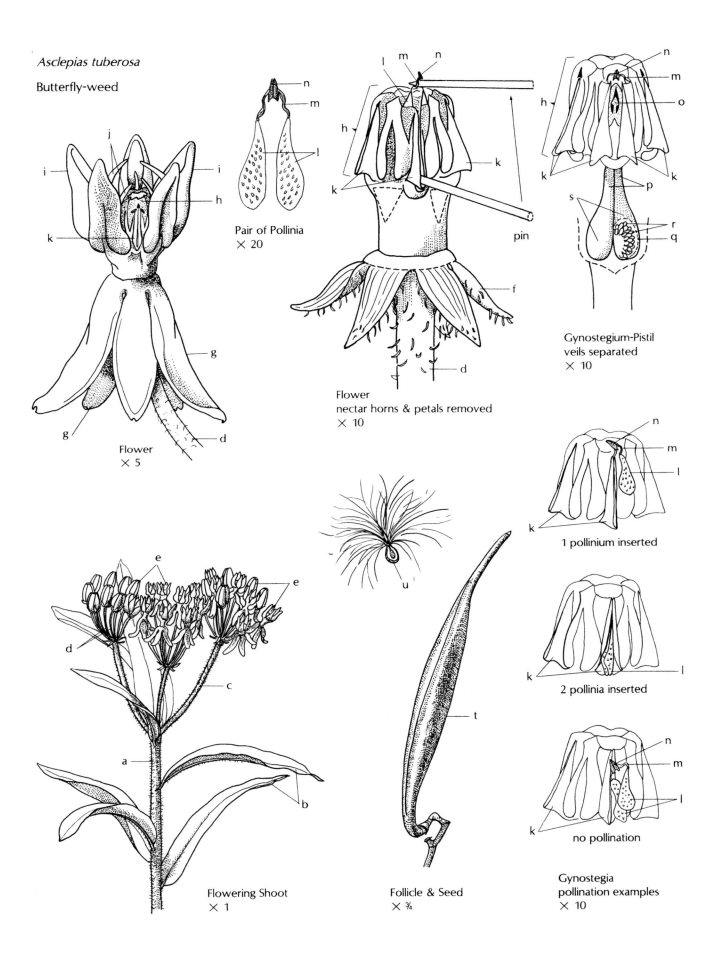

Asclepias tuberosa

Butterfly-weed

Pair of Pollinia
× 20

Flower
× 5

Flower
nectar horns & petals removed
× 10

pin

Gynostegium-Pistil
veils separated
× 10

1 pollinium inserted

2 pollinia inserted

no pollination

Gynostegia
pollination examples
× 10

Flowering Shoot
× 1

Follicle & Seed
× ¾

Nightshade Family (Solanaceae)

There is no single characteristic common only to this family. A number of features are present. Flower petal lobes tend to overlap or have creases in a petal tube that in bud is folded fan-like. And, usually, the stamens are attached to the petals. Numerous ovules develop into numerous seeds in the berry or capsule fruit.

The leaves are mostly alternate and simple, but size and shape vary greatly throughout the family. Usually the leaves are hairy and have a characteristic odor. Plant form ranges from herb to shrub to tree to vine. There are 90 genera and about 3,000 species.

Of interest... **ornamentals:** *Browallia, Datura arborea* (angel's trumpet), *Nicotiana* (tobacco, see 34), *Nierembergia, Petunia, Physalis alkelengi* (Chinese lantern), *Schizanthus, Solandra* (cup of gold); **poisonous:** *Atropa Belladonna* (deadly nightshade, belladonna, atropine), *Datura stramonium* (jimson weed), *Nicotiana* (tobacco, nicotine insecticide), *Physalis heterophylla* (ground cherry), *Solanum Dulcamara* (bitter nightshade), *S. nigrum* (black nightshade), *S. pseudocapsicum* (Jerusalem cherry); **medicinal:** *Hyoscyamus niger* (henbane, a source of alkloidal drugs), *Atropa belladonna* (source of alkaloids, **atropine** and **hyoscyamine**), *Datura Stramonium* (source of alkaloid drug hyoscyamine); **food:** *Capsicum* spp. (bell pepper, chilies, cayenne pepper, paprika), *Lycoperiscon esculentum* (tomato), *Physalis ixocarpa* (tomatillo, jamberry), *P. peruviana* (strawberry tomato, cape gooseberry, ground cherry), *P. pruinosa* (strawberry tomato), *Solanum Melongena* variety *esculentum* (eggplant, aubergine), *S. tuberosum* (potato, Irish potato).

Potatoes are native to the central Andes of Peru. Indians collected wild potatoes before 6,000 B.C.E. There are eight species and as many as 5,000 varieties in many colors. Potatoes can grow almost anywhere except in high humidity areas. Potatoes are 99.9% fat-free and have a higher percentage of protein than soybeans. One potato supplies $1/2$ of the daily vitamin C requirement for an adult. As a crop, potatoes are worth billions of dollars. In the mid-1800s, due to the potato late blight (caused by the fungus *Phytophthora infestans,* see 46) millions of people died of famine in Ireland with the loss of potato crops. Half of the world's crop is fed to livestock. It is also used to make vodka and aquavit, and can be used as starch, paste, and dye. As fuel for cars, one acre (0.40 hectares) of potatoes yields 1,200 gallons (4536 liters) of ethyl alcohol in a year.

Solanum dulcamara   Bitter Nightshade

Leaves on this wild, perennial vine are alternate and divided into 3 lobes (a). Flowers in the **cymose** cluster have jointed pedicels (c, d). (A cyme is a flat-topped cluster with the central flower opening first.) There are 5 joined sepals (e) and 5 reflexed, purple petals (f) with green spots (g) at the base. The 5 stamens with short purple filaments (h) join anthers (i) around the pistil's style (j).

With the flower cut open, the stamens (h, i, k) surround the pistil's superior ovary (l) and style (j, m), with the stigma (n) emerging at the top. Pollen is released from the stamens through pores (o) at the ends of the anthers. Nightshade fruits are small, red, poisonous berries (p).

Lycopersicon esculentum   Tomato

Tomato flowers consist of joined sepals (q) with 6 lobes, joined petals (r) with 6 lobes, 6 stamens, and a single pistil. The stamens have short filaments and elongate, narrow-tipped anthers (s) that converge around the pistil's style (t). Pollen is shed through longitudinal slits on the anthers' inner surfaces.

Physalis alkekengi   Chinese Lantern

A fruit that looks like a Chinese lantern is composed of an enlarged sepal tube (v), which encloses an orange-red berry.

Capsicum grossum   Bell Pepper

Pepper's fruit type is a berry whose rind (w) changes from green to red at maturity. Seeds (x) develop on a central placenta (y).

COLOR CODE

green:	leaf (a), petiole (b), pedicel (c, u), sepals (e, q), spots (g), style (j, t)
purple:	pedicel (d), petals (f), filament (h)
yellow:	anthers (i, s), stamens (k), petals (r)
light green:	ovary (l), style (m), stigma (n)
red:	berries (p), rind (w)
orange:	sepal tube (v)
white:	seeds (x), placenta (y)

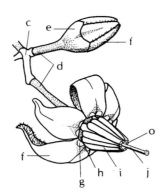

Flower, Bud
× 3

Solanum dulcamara

Bitter Nightshade

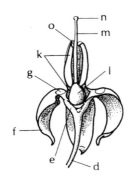

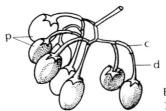

Flower
cut open
× 3

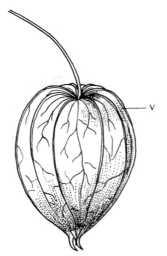

Berries
× 1

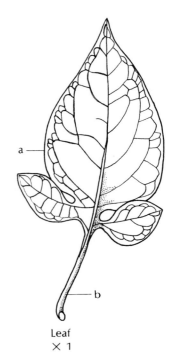

Leaf
× 1

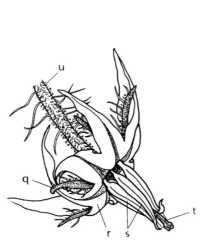

Flower
× 2

Lycopersicon esculentum

Tomato

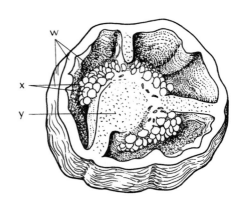

Fruit
sepal tube
encloses berry
× 1

Physalis alkekengi

Chinese Lantern

Berry
cross section
× 1

Capsicum grossum

Bell Pepper

Morning Glory Family (Convolvulaceae)

Flowers in this family often have fused petals that are twisted when in bud and open in a funnel-shape or in a tube shape with a flat top. A pair of bracts usually subtends the usually bisexual flower. Sepals, petals, and stamens are each 5, with the stamens arising from the petals. There is a single pistil with a superior ovary composed of usually 2 carpels, which, when mature, form a **loculicidal capsule**. The plant tissues contain jointed latex-containing cells that produce a milky sap.

Represented in this family are twining herbs, shrubs, and small trees. Heart-shaped, alternate, and simple leaves are most common. There are about 50 genera and about 1,800 species.

Of interest . . . **food:** *Ipomoea batatas* (sweet potato); **ornamentals:** *Calonyction aculeatum* (moonflower), *Convolvulus althaeoides*, *C. cneorum* (silverbush), *C. mauritanicus* (ground morning glory), *Dichondra* (a ground cover), *Ipomea purpurea* (morning glory), *I. tuberosa* (wood rose), *Porana paniculata* (Christmas vine), *Quamoclit pennata* (cypress vine); **weeds:** *Convolvulus* spp. (bindweeds, which are drought-resistant due to their deep and extensive root systems.), *Cuscuta* (dodder, a parasitic plant with orange, thread-like stems that grow en masse in association with many different kinds of flowering plant hosts).

Convolvulus sepium   Hedge Bindweed

This plant is a perennial herb with a twining stem (a) and alternate, triangular leaves (b). Single flowers on long peduncles (c) arise in the leaf axils. At the base of the flower are 2 large bracts (d). Inside the bracts, 5 sepals (e) surround the funnelform tube of 5, pink, fused petals (f). The center of each petal lobe has a white stripe (g).

With the petal tube cut open, the stamens' petal attachment can be seen. Five filaments (h) arise from the petal tube and the anthers (i) meet and enclose the pistil's stigma. Pollen is shed inwardly.

With sepals, petals, and stamens removed, the parts of the pistil can be seen. There are 2 stigmas (j), a single style (k), and an ovary (l) composed of 2 fused carpels, forming one chamber. Within the chamber are 4 ovules (m). An orange disc (n), surrounding the base of the pistil, arises from the receptacle (o).

Ipomoea batatas   Sweet Potato

Cultivated for its edible, underground, tuberous roots (p), the sweet potato plant is a tropical vine, but can be grown in southern temperate regions during summer months.

COLOR CODE

red-green:	stem (a)
green:	leaf (b, q), peduncle (c), bracts (d)
light green:	sepals (e)
white:	stripe (g), filaments (h), anthers (i), stigma (j), style (k), ovary (l), ovules (m), receptacle (o)
pink:	petals (f)
orange:	disc (n), tuber (p)

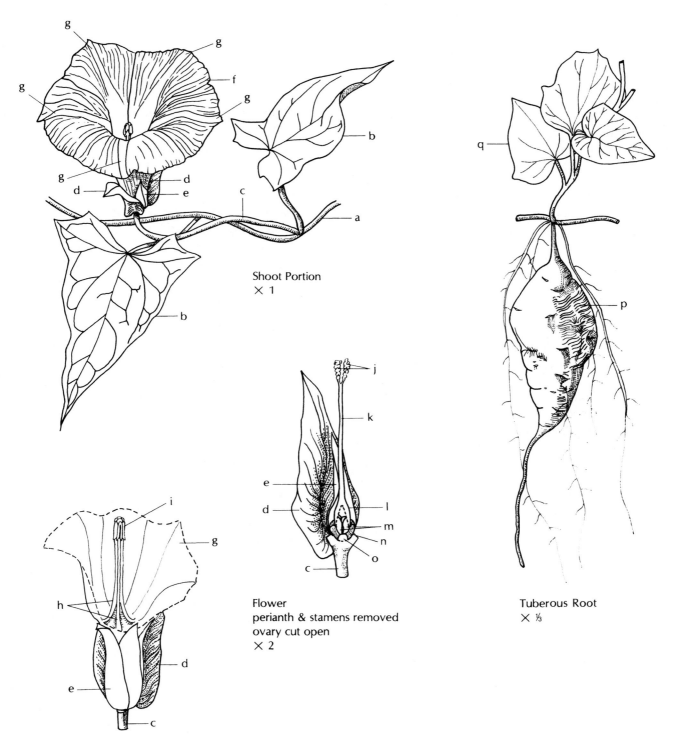

Shoot Portion
× 1

Flower
perianth & stamens removed
ovary cut open
× 2

Tuberous Root
× ⅓

Flower
petal tube cut open
× 2

Convolvulus sepium

Hedge Bindweed

Ipomea batatas

Sweet Potato

Mint Family (Lamiaceae)

Mint plants have square stems and mostly opposite leaves. Aromatic **essential oils** are usually present and account for the distinct and characteristic odors. Flowers have 5 fused petals that diverge into 2 lips (**bilabiate**). (The old family name, Labiatae, means 2-lipped, where the corolla or calyx is divided into two differently shaped parts forming an upper and lower lip.)

The single pistil has a superior, 4-lobed ovary, and a style that arises between the ovary lobes from the base of the flower (**gynobasic style**). The fruits produced are 4 nutlets.

Usually, plants in this family are annual or perennial herbs, although shrubs, trees, and vines are also represented. There are about 200 genera and about 5,600 species.

Of interest... **edible herbs:** *Majorana* (marjoram), *Mentha* spp. (mints), *Nepeta* (catnip, catmint), Ocimum (basil), *Origanum* (oregano), *Rosmarinus officinalis* (rosemary), *Salvia officinalis* (sage), *Satureja* (savory), *Thymus* (thyme); **ornamentals:** *Agastache barberi* (giant hyssop), *Ajuga* (bugloss), *Hyssopus* (hyssop), *Lamium* (dead nettle), *Lavandula* (lavender), *Nepeta* (catmint), *Perovskia* (Russian sage), *Phlomis alpina*, *Physostegia* (false dragonhead, obedient plant), *Salvia splendens* (scarlet sage), *S. viridis* (painted sage), *Scutelaria costaricaca*, *Solenostemon scutellanoides* (coleus), *Stachys macrantha* (big betony), *S. byzantina* (lamb's ears).

Salvia officinalis   Sage

This spring-flowering, perennial herb has square stems (a) with opposite, simple leaves (b). Two small flower clusters arise on opposite sides of the stem. Conspicuous ribs join the flower's 5-lobed sepal tube (c). The 5 fused petals (d) separate into an upper lip composed of 2 petal lobes and a lower lip consisting of 3 petal lobes.

Cut open, the flower reveals several interesting features. Two stamens arise from the lower petal lip. For clarity, one stamen is shaded in the drawing. The filament (e) attaches to an enlarged **connective** (f) that allows for rotation of the anther sac (g), hence facilitating pollen dispersal by insects.

As a visiting bee alights on the lower petal lip, pollen from the rotating anther sacs brushes the insect. After depletion of pollen, the stigmas (h) are lowered by the elongated style (i). Thus, pollen from other flowers is brushed on the stigmas by visiting bees, effecting **cross-pollination.**

Nectar, the source of attraction to insects, is pooled below a ring of hairs (j) in the petal tube, secreted by a disc (k) at the base. The pistil's **gynobasic style** (l) arises between the 4 lobes of the ovary (l).

At fruiting time, when the persistent sepal tube is cut open, 4 nutlets (n) are revealed.

COLOR CODE

green:	stem (a), pedicel (m)
gray-green:	leaves (b)
purple-green:	sepals (c)
purple:	petals (d), nectar disc (k), stigmas (h), style (i)
white:	filaments (e), connectives (f), hairs (j)
tan:	nutlets (n)
yellow:	anther sacs (g)
light green:	ovary (l)

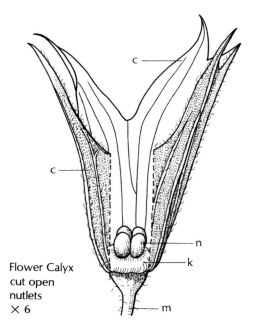

Flower Calyx
cut open
nutlets
× 6

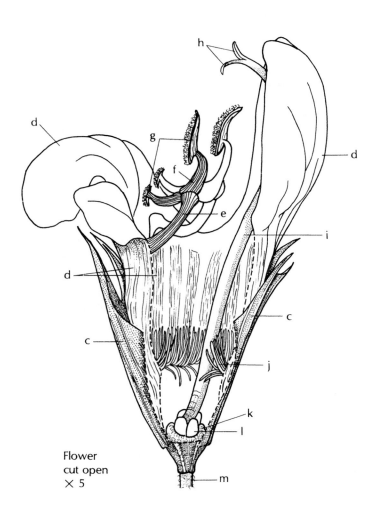

Flower
cut open
× 5

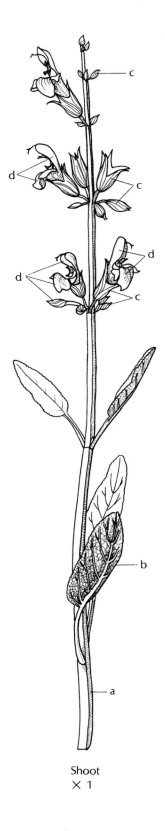

Shoot
× 1

Salvia officinalis

Sage

Olive Family (Oleaceae)

Characteristics in this family include flower parts in 2's with the anthers of the stamens usually touching. The single pistil has a superior ovary of 2 fused carpels with, usually, 2 ovules in each carpel.

Plants in this family are shrubs, trees, and vines with opposite leaves, which may be simple or compound. Some have evergreen, persistent leaves. Usually, the flowers are bisexual and have 4 sepals and 4 petals. Fruit types are **berries**, **drupes**, **capsules** and **samaras**. There are 29 genera with 600 species.

Of interest... **food:** *Olea europaea* (olive); **lumber:** *Fraxinus* (ash); **ornamentals:** *Abeliophyllum* (white forsythia, Korean abelia-leaf), *Chionanthus* (fringe tree), *Forsythia* (golden bells), *Jasminum* (jasmine), *Ligustrum* (privet), *Osmanthus* (fragrant olive, tea olive, devil-weed), *Syringa* (lilac).

In the countries surrounding the Mediterranean Sea, the olive tree, *Olea europaea,* has been in cultivation for over 4,000 years. The drupe fruit has a high oil content that is used both for oil production and preserved whole. Once trees are no longer productive, the wood is used to carve items such as bowls, jewelry, and roseries.

In primitive naturalistic religions, it was believed that "spirits" inhabited certain trees. The olive tree had special significance as one of the most sacred. In summer, if a person slept under the tree, he of she would have pleasant dreams, as the evil spirits are afraid of the olive tree!

Olive leaves were burned for protection from the "evil eye." Accordingly, it was wise to have an olive tree planted near the dwelling place. On the Acropolis in Athens, Greece an olive tree is always growing near the Parthenon. Purportedly, the "evil eye" is the supposed power of some people to harm others by merely looking at them.

Olive leaves, woven into crown wreaths, were used by the Greeks and Romans to reward winners in athletic games.

An olive branch is traditionally a symbol of peace. On the seal of the United States, the American eagle faces toward an olive branch clutched in its right foot as opposed to the cluster of spears in its left foot.

Syringa vulgaris   Common Lilac

Persia (now Iran) is the original home of the common lilac. The leaves (c) of this shrub are opposite, simple, and have entire margins. Clusters of fragrant, tubular flowers (d) branch from a main flower axis (**compound panicle**). Flowers last only a few weeks but are a favorite for the scent and old-fashioned associations. There are many named cultivars. A **loculicidal capsule** (e) is the fruit type.

Forsythia   Golden Bells

Originally from China, *Forsythia* was brought to Europe in 1844. Of the many species, special cultivars and hybrids have been developed. This shrub's flowers emerge along the stem (f) in early spring. Although the shrub is hardy in the Northern Hemisphere, flower buds are frequently killed by late frosts. In its natural form, arching branches are covered with yellow flowers.

The flower has 4 sepals (h) and 4 petals (i) joined at the base. Arising from the petal tube are 2 stamens (j) with back-to-back anthers. The pistil's ovary (k) has many ovules (l) in axile placentation in each of the two carpels (m). There is a single style (n) and a 2-lobed stigma (o).

Ligustrum vulgare   Common Privet

This plant is commonly used as a clipped hedge. If left to grow in its natural form, it produces fragrant flowers and can grow to 15 feet (5 meters) high. A **drupe** (r) is the fruit type of this shrub.

Fraxinus pennsylvanica   Green Ash

The fruit on ash trees is a samara (t). The green ash tree is highly susceptible to the green ash tree borer insect pest; as a result, scores of green ash trees in Eastern United States are now dying or dead.

COLOR CODE

brown:	stem (a, s), capsule (e)
green:	new shoots (b), leaves (c, g, q), sepals (h), pedicel (p)
lilac:	flowers (d)
tan:	stem (f), samara (t)
yellow:	petals (i), stamens (j)
pale green:	ovary (k), style (n), stigma (o)
white:	ovules (l)
dark blue:	drupes (r)

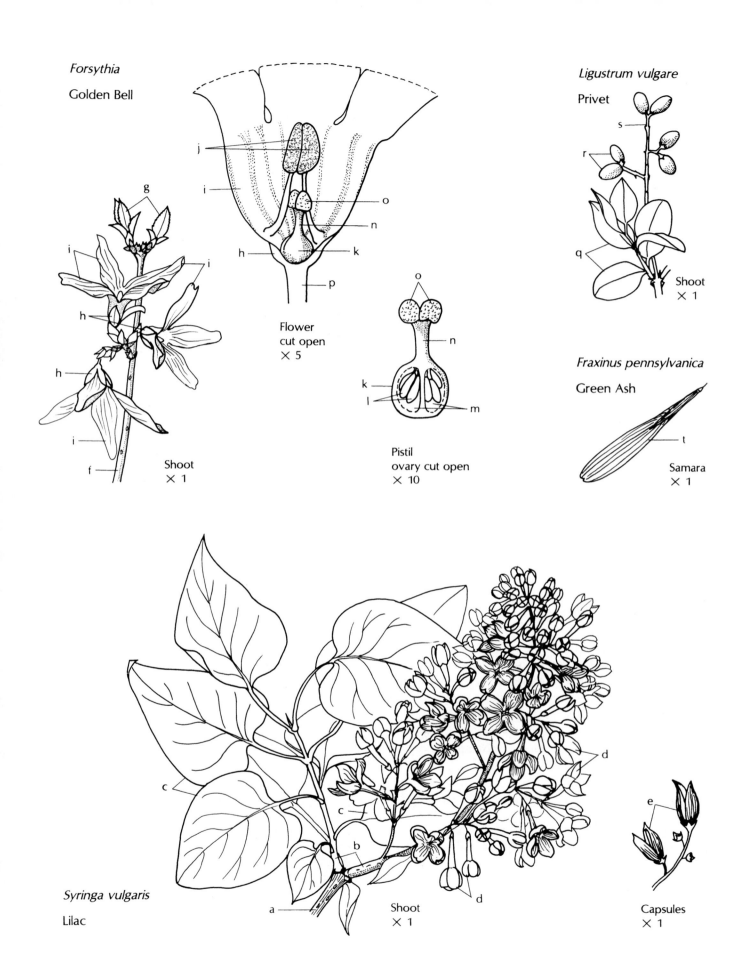

Forsythia
Golden Bell

Flower
cut open
× 5

Shoot
× 1

Pistil
ovary cut open
× 10

Ligustrum vulgare
Privet

Shoot
× 1

Fraxinus pennsylvanica
Green Ash

Samara
× 1

Syringa vulgaris
Lilac

Shoot
× 1

Capsules
× 1

Figwort Family (Scrophulariaceae)

Plants in this large family of 210 genera have characteristics found in many related families. They are mostly herbs or small shrubs that have various methods of obtaining nutrition, which include self-feeding (**autotrophic**), partial dependence on another plant (**hemiparasitic**), total dependence (**holoparasitic**), and dependence on dead or decaying organic matter (**saprophytic**).

Usually, the leaves are alternate on round or square stems. Flower types vary greatly, but all are bisexual. Bracts below flowers are usually present. Flower sepals are usually joined and have 4-5 lobes. Joined petals form a tube that may have 2 lips (**bilabiate**) like a mint flower. Sometimes there are 5 stamens, but usually, there are 4 in 2 pairs of two different lengths (**didynamous**), with, usually, the fifth stamen reduced to a sterile structure or a scale.

The single pistil has a superior ovary, or 2 carpels forming 2 chambers, that usually matures into a **capsule** fruit.

Of interest... **ornamentals:** *Angelonia* (angelonia), *Alonsoa* (mask flower), *Antirrhinum* (snapdragon), *Calceolaria* (slipper-flower), *Cymbalaria* (Kenilworth ivy), *Diascia* (twinspur), *Mazus reptans, Mimulus* (monkey-flower), *Penstamon* (beard-tongue), *Russelia* (coral plant), *Satera* (bacopa), *Torenia* (wishbone flower); **wild:** *Castilleja coccinea* (Indian paint-brush), *Pedicularis* (lousewort), *Scrophularia* (figwort), *Verbascum* (mullein), *Veronica* (speedwell), *Veronicastrum* (Culver's-root); **poisonous:** *Digitalis purpurea* (foxglove). Leaves of this plant have several active and toxic glucosides: **digitoxin, digitalin, digitonin**. The poisons have a cumulative effect when ingested. Digitoxin was used in medicine to treat congestive heart failure and heart rhythm irregularities. It enhances and strengthens the force of the heart's contractions. However, there are serious side effects; so usage is now out of favor in medical practice.

Linaria vulgaris   Butter-and-eggs, Common Toadflax

Numerous, narrow leaves (a) arise on the stem (b) of this wild perennial. Peduncles (c) have bracts (d) below the pedicel (e). The flower's sepal tube (f) has 5 lobes.

The bilabiate petals consist of an upper petal lip (g) with 2 lobes and a lower petal lip (h) with 3 lobes and a conspicuous elevated **palate** (a projection which closes the throat, i). Extending from the lower lip is a nectar-collecting spur (j).

With the 2 petal lips cut apart, the 2 long (k) and 2 short (l) stamens can be seen surrounding the pistil. The pistil has an ovary (m) consisting of 2 carpels containing numerous ovules (n) and a single style (o) and stigma (p). A nectar disc (q) surrounds the ovary base.

A nectar-seeking bee separates the lips, slides its head between the rows of hairs on the palate (i), inadvertently transfers pollen to the stigma, sucks nectar from the spur, and is dusted with pollen from the anthers as it departs.

Castilleja coccinea   Indian Paint-brush

As a saprophyte, annual plants are often found in a circle above buried organic material. The leaves (s) below flowers are 3-parted, bract-like and variable in form. Bright red bracts (u, v) surround the flower (t). This plant is commonly seen in meadows and ravines in mountainous regions.

Verbascum thapsus   Mullein

Mullein is a biennial plant that forms a rosette of leaves the first year, and in the second year, a tall, stout stem (w) with alternate leaves (x), topped by a spike of yellow flowers (y, z).

COLOR CODE

green:	leaves (a, s, x), stem (b, w), peduncle (c), bract (d), pedicel (e), sepals (f, z), disc (q)
pale yellow:	petal lips (g, h), spur (j)
yellow-orange:	palate (i)
yellow:	anthers (k), flowers (y)
white:	filament (l), style (o), ovules (n)
red-purple:	stem (r)
pale green:	ovary (m), stigma (p), flower (t), bract, shaded area (u)
red:	bract, unshaded area (v)

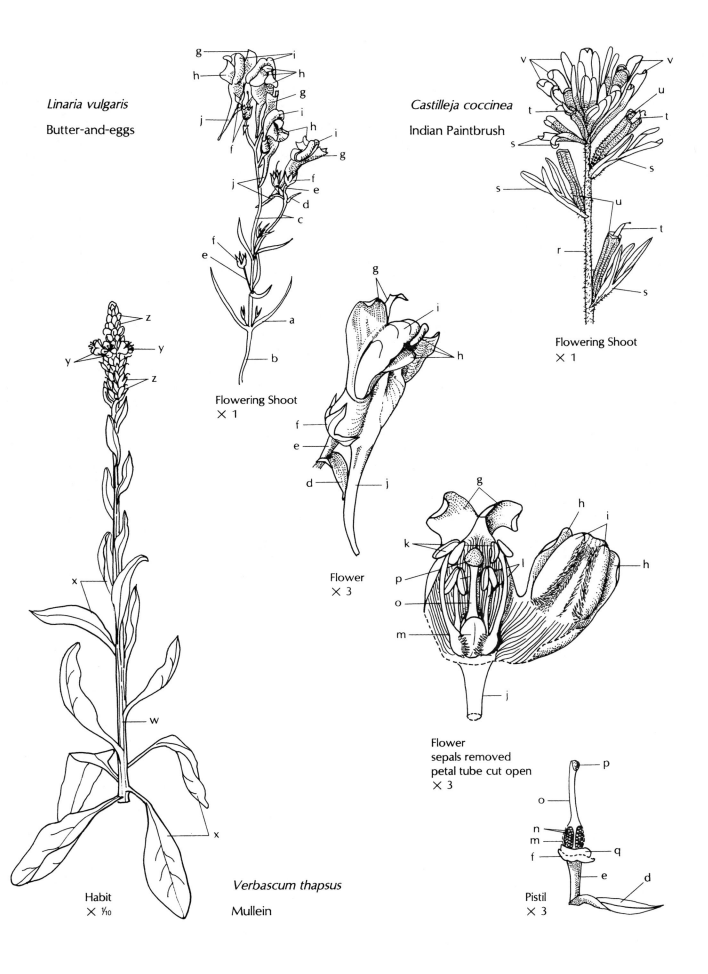

Linaria vulgaris

Butter-and-eggs

Flowering Shoot
× 1

Castilleja coccinea

Indian Paintbrush

Flowering Shoot
× 1

Flower
× 3

Flower
sepals removed
petal tube cut open
× 3

Pistil
× 3

Habit
× ¹⁄₁₀

Verbascum thapsus

Mullein

Gesneria Family (Gesneriaceae)

This family is composed of mainly tropical plants. It includes herbs, shrubs, trees, vines, and epiphytes. In cool climates, the family is known for the ornamental houseplants such as African violets, cape primroses, and gloxinias. There are 125 genera with about 2,000 species.

The generally hairy leaves are usually opposite or occur in basal rosettes. Bisexual flowers arise in clusters or singly and have 5 sepals, 5 petals, 2 to usually 4 stamens, and a single pistil. The ovary has 2 carpels fused into one chamber containing numerous ovules in **parietal placentation**. The position of the ovary may be superior, half-inferior, or interior. A nectar disc is under the ovary or nectar glands surround the ovary. A capsule is the usual fruit type.

Aeschynanthus lobbianus   Lipstick Plant, Blushwort

Because of its trailing stems (a) with bright red flowers, this Indonesian plant can be attractively displayed in a hanging container. The fleshy leaves (b, c) are opposite and have smooth (entire) margins. Borne along the stem in leaf axils, the flower clusters have leaf-like bracts (d) at the base. Five hairy fused sepals (f) form a tube.

The flower developmental series shown reveals why *Aeschynanthus* is commonly called lipstick plant. As the flower develops, fused red petals (g) rise out of the sepal tube. The **bilabiate** (two-lipped) petals open into 2 upper and 3 lower lobes with darkly colored **nectar guide** markings (h), which attract bird pollinators. Protruding from the upper petal lobes are 4 anthers (i).

Four stamen filaments (j) arise from the petals and join to form 2 pairs of anthers (i, k). Inside the petal tube, the pistil has an inferior ovary (l), a single stigma, and a style (m), which is surrounded at the base by nectar glands (n).

Saintpaulia ionantha   African Violet

From Africa, but not a true violet (*Viola*, see 92), this plant forms a basal rosette of hairy leaves (o) with scalloped margins. Clusters of flowers range in color from white to pink to blue to purple, depending on the variety and hybrid cultivar.

Basically, the flower consists of 5 sepals (q) and 5 petals (r), but extra petal-like structures (s), formed from stamens, may be present. The 2 functional stamens each have 2 anther sacs (t, v) and a wide filament (w). The pistil (x) has a superior ovary (y) covered with glandular hairs, and a style (z), which extends at an angle.

COLOR CODE

gray-green:	stem (a), pedicel (e), leaves (o)
green:	leaf upper surface (b)
light green:	leaf underside (c), style (m), ovary (l, y), nectar glands (n), pistil (x, z)
dark purple-red:	bracts (d), sepal tube (f), petal markings (h), anthers (i, k)
red:	petals (g)
white:	filaments (j, w)
brown:	pedicel (p), sepals (q)
purple:	petals (r), staminodes (s), petal and staminode base (u)
yellow:	anthers (t, v)

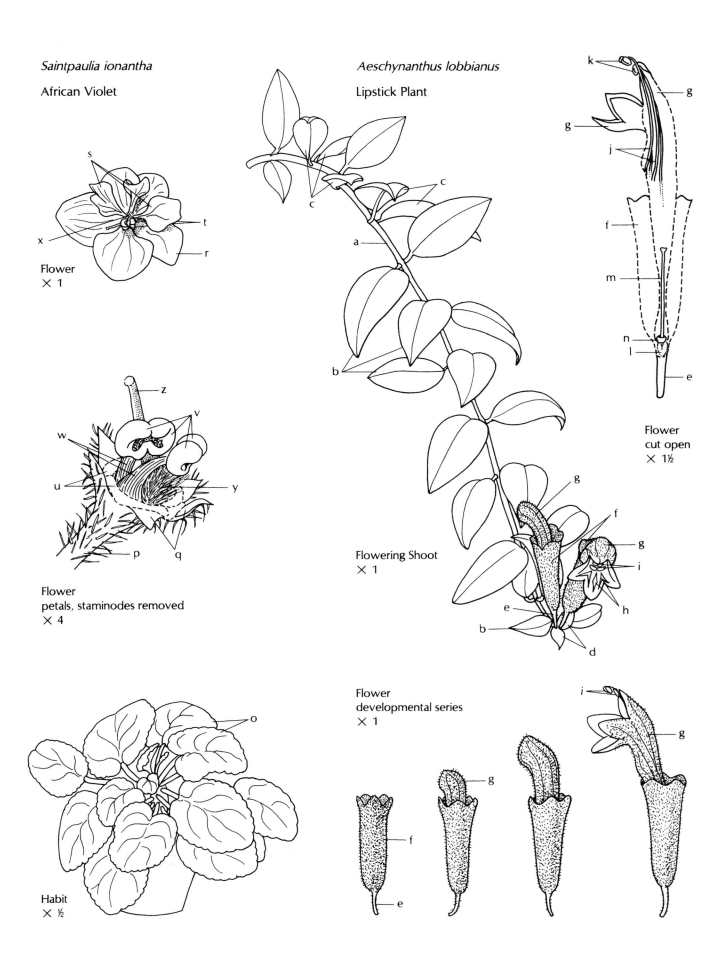

Saintpaulia ionantha

African Violet

Flower
× 1

Flower
petals, staminodes removed
× 4

Habit
× ½

Aeschynanthus lobbianus

Lipstick Plant

Flowering Shoot
× 1

Flower
developmental series
× 1

Flower
cut open
× 1½

Honeysuckle Family (Caprifoliaceae)

Distinguishing features of this family include opposite, simple leaves without **stipules**. Flowers are usually bisexual and often in pairs. The stamens arise from the petals (**epipetalous**) and alternate with the petal lobes.

The pistil's ovary is inferior with 3–5 carpels. There is usually one ovule in each **locule**. The single style has stigmas equal in number to the carpels. Fruit types are berries or drupes and highly desired by birds. Plants in this family are mostly woody shrubs and vines. Many members make wonderful landscaping plants. There are about 18 genera and about 450 species.

Of interest . . . **ornamentals:** *Abelia* × *grandiflora* (glossy abelia), *Linnaea borealis* (twinflower), *Leycesteria*, *Lonicera* × *heckrottii* (goldflame honeysuckle), *Symphoricarpos* (snowberry, coralberry), *Viburnum acerifolium* (mapleleaf viburnum), *V. carlesii* (Korean spice viburnum, and its fragrant hybrids: *V.* × *burkwoodii*, *V.* × *carlcephalum*, *V.* × *juddii*), *V. cassinoides* (witherod viburnum), *V. dentatum* (arrow-wood, so named for its straight stems formerly used by Native Americans for making arrows), *V. lentago* (nannyberry, sheepberry), *V. nudum* (possum haw viburnum), *V. prunifolium* (black haw), *V.* × *rhytidophylloides* (leatherleaf viburnum), *V. opulus* variety *americanum* (formerly *V. trilobum*, American cranberry bush; it has edible, astringent fruit, delicious when made into a jelly), *V. rufidulum* (rusty black haw), *Weigela;* **wild:** *Sambucus canadensis* (elderberry, the fruit is used in jelly, pie, and wine making).

Lonicera japonica   Japanese Honeysuckle

This honeysuckle is a woody, climbing vine. Once cultivated in the U.S., this species has become widely distributed, displacing native plants, and is now regarded as a noxious weed.

The hairy stems (a) have opposite, simple leaves (b) with short petioles (c). A pair of bisexual flowers is borne on a short peduncle (d). At the base of each flower pair is a pair of leafy bracts (e). An individual flower has a tiny sepal cup (f) with even smaller bractlets (g) on each side. The petal tube (h) is divided into a four-lobed lip and a one-lobed lip. Five stamens (i) protrude from the long petal tube.

The cut-open petal tube shows that the stamens arise from the top of the petal tube. The stamen's filament (j) is attached to the center of the anther (k), providing for flexible movement. Extending from the nectar-filled petal tube is the pistil's long, slender style (l) and stigma (m). Pollination is by moths, which are attracted to this highly scented, long-petal tube flower.

A cross-section of the berry (n) fruit shows that the pistil's inferior ovary is composed of 3 fused carpels (o). Paired berries develop from the paired flowers.

COLOR CODE

pale green:	stem (a), style (l), stigma (m)
green:	leaves (b), petioles (c), peduncle (d), bracts (e), sepals (f), bractlets (g)
pale yellow-orange:	petals (h)
white:	stamens (i), filaments (j), anthers (k)
black:	berry (n)

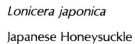

Lonicera japonica

Japanese Honeysuckle

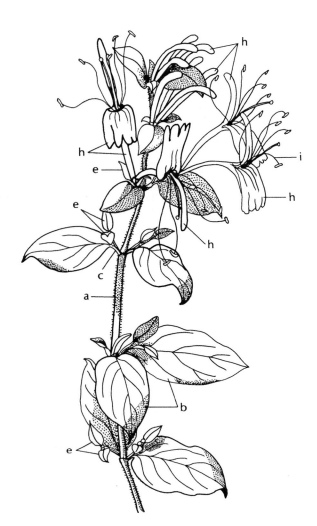

Flowering Shoot
× 1

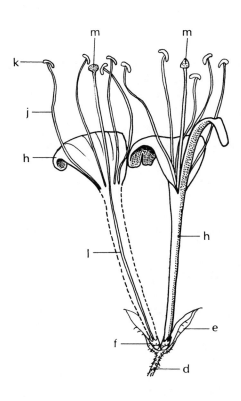

Pair of Flowers
one petal tube cut open
with 1-lobed lip removed
× 2

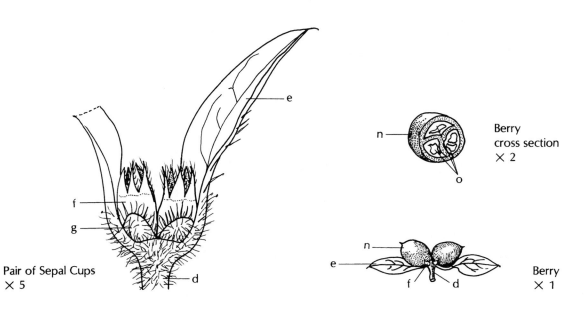

Pair of Sepal Cups
× 5

Berry
cross section
× 2

Berry
× 1

Teasel Family (Dipsacaceae)

There are only 11 general with 350 species in this small family of usually annual, biennial, or perennial herbs. The leaves are opposite.

Bisexual flowers develop in a dense cone-shape head or on a spike. Individual flowers are surrounded at the base by a bract. Small sepals are joined into a cup-shape or divided into 5–10 hairy or bristle-like segments. Petals are united into a tube with 4 or 2 stamens attached. The single pistil's inferior ovary has one locule. The fertilized ovule develops into an achene fruit.

Dipsacus sylvestris   Teasel

Teasel is a tall, biennial herb that grows in low, wet areas along roadsides and in old fields. Its ridged stems (a), covered with spines, have narrow, opposite leaves (b) without petioles (**sessile**). Below the compact flower head are long, upward-curved bracts (c). A narrow, pointed bract (d) at the base of each flower (e) lends a pincushion appearance to the flower head.

An enlarged single flower shows its enclosing bract (d), the silky-haired sepal cup (f), and white petal tube (g), which divides into 4 purple, overlapping lobes (h). Four stamens (i), inserted in the petal tube, emerge free at the top. Pollen (j) is shed inwardly toward the stigma (k). With the petal tube and inferior ovary (l) cut open, the pistil's style (m) and ovule (n) can be seen. The fertilized ovule develops into an achene (o) fruit.

Of interest . . . **ornamental:** The dry cone-like heads (inflorescence) at maturity are collected for use in dried flower arrangements; **utility:** The dry cone-like heads with their bristle-like segments can be used to "card' wool (like a comb) to remove extraneous matter from the wool.

COLOR CODE

light green:	stem (a)
green:	leaves (b), bracts (c), sepal cup (f)
white:	petal tube (g), pollen (j), stigma (k), ovary (l), style (m), ovule (n)
lavender:	bract (d), flowers (e), petal lobes (h), stamens (i)
tan:	achene (o)

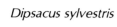

Dipsacus sylvestris

Teasel

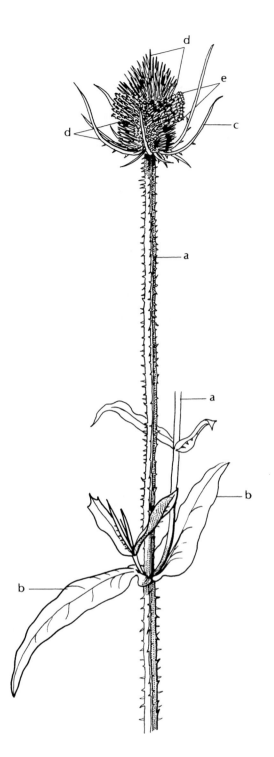

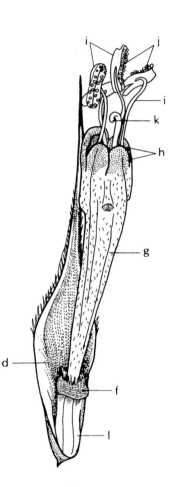

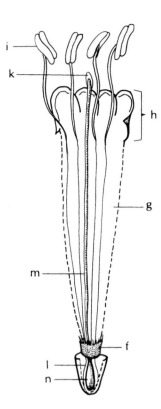

Flower
petal tube & ovary cut open
× 5

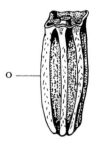

Flowering Shoot
× ½

Flower
× 7

Achene
× 10

Aster Family (Asteraceae)

Asteraceae represents the largest dicot family, with about 25,000 species distributed worldwide. It includes herbs, shrubs, trees, epiphytes, vines, and succulents. Leaves show an extreme range of diversification, with an alternate, opposite, or whorled arrangement. The blades are simple to compound, lobed, needle-like, or scale-like.

If a flower looks like a daisy, it is probably in the aster family. The common characteristic in this family is the head of many modified flowers (a **composite** of flowers). What appears to be a flower is actually a multiple false flower. There is an almost complete loss of sepals (calyx). Another characteristic includes having fused anthers forming a cylinder around the style (**syngenesious condition**). As the stigma and style grows up through the cylinder, it pushes the anthers out so as to facilitate pollen dispersal. This prevents self-fertilization.

The flower's single pistil has an inferior ovary composed of 2 carpels fused to form one chamber with one ovule that matures into an achene fruit without **endosperm**. Some of the flower modifications are shown with the two illustrated plants.

Of interest . . . **food:** *Artemisia dracunculus* (tarragon), *Carthamnus tinctorius* (safflower), *Chichorium endivia* (endive), *C. intybus* (chicory), *Cynara scolymus* (artichoke), *Guizotia abyssinica* (niger seed), *Helianthus tuberosus* (Jerusalem artichoke), *H. annuus* (sunflower seeds, oil), *Lactuca sativa* (lettuce); **medicinal:** *Artemisia annua* (sweet wormwood, source of artemisinin, an anti-malaria drug); **ornamentals:** *Ageratum, Aster, Calendula, Dahlia, Dendranthema* (chrysanthemum), *Echinops* (globe thistle), *Gaillardia, Gazania, Gerbera, Helianthus* (sunflower), *Helichrysum* (strawflower), *Leontopodium* (edelweiss), *Leucanthemum* (daisy), *Liatris* (blazing star), *Solidago* (goldenrod, an attractive fall-flowering addition to the garden, height varies with the cultivar and flowers attract many beneficial insects, commonly assumed that this insect-pollinated plant is the culprit that causes hayfever when the blame goes to wind-pollinated plants such as *Ambrosia*, ragweed), *Stokesia, Tagetes* (marigold), *Zinnia;* **insecticide:** *Tanacetum coccineum* (pyrethrum, painted daisy); **weeds:** *Ambrosia* (ragweed), *Cirsium* (thistle), *Sonchus* (sow-thistle), *Taraxacum* (dandelion), *Xanthium spinosum* (cocklebur), a seedhead with tiny hooklike spines on the surface clings to animal fur to facilitate seed dispersal. George de Mestral invented a fastener combining the hooks like those of the cocklebur seed head and loops like the fabric of his burred pants. He called it "**Velcro,**" combining the words **velour** and **crochet; poisonous pasture plants:** *Eupatorium rugosum* (white snakeroot), *Helenium* (sneezeweed), *Senecio* (ragwort).

Rudbeckia hirta   Black-Eyed Susan, Coneflower

This plant is a biennial or short-lived perennial herb. The habit drawing shows its leaves (a) of variable shape and with some having petioles (b). Long peduncles (c) support flower heads composed of yellow, sterile, **ray flowers** (d) which surround a cone of brown-petaled, fertile **disc flowers** (e). Hairy bracts (f) below the flower head can be seen in the vertical section drawing. Many small disc flowers (e) are borne on a conical receptacle (g).

Evolutionary modifications of Asteraceae flowers include sepals that are reduced to bracts (h) and petals (i) that are fused to form a tube of 5 lobes. Stamen modifications include filaments (j), of the 5 stamens, attached inside the petal tube and lance-shaped anthers (k) joined to form a cylinder. After pollen is shed, the stigma (l) rises through the cylinder. This prevents self-fertilization. Below the double stigmas are the pistil's style (m) and inferior ovary (n) with 1 ovule (o).

Tagetes patula   Marigold

A marigold flower head is supported by a peduncle (p) and has fused bracts (q) below the head. Outer ray flowers (r) are, mainly, unisexually female, while inner disc flowers (s) are bisexual. Both have bracts (t) and both have a pistil with a double stigma (u) and an ovary (v) below other flower parts. The bisexual disc flower, cut open, shows the pistil's ovule (w) attached to the base of the ovary and the style (x) that rises through the cylinder of the stamens' anthers (y) after pollen is shed.

COLOR CODE

green:	leaves (a, b), peduncle (c, p), bracts (f, q)
yellow:	ray petals (d), anthers (k, y)
brown:	disc petals (e, i), bract (h)
white:	receptacle (g), filaments (j), ovule (o,w)
light green:	stigmas (l, u), style (m, x), ovary (n, v),
orange:	ray petals (r), disc petals (s)
tan:	bracts (t)

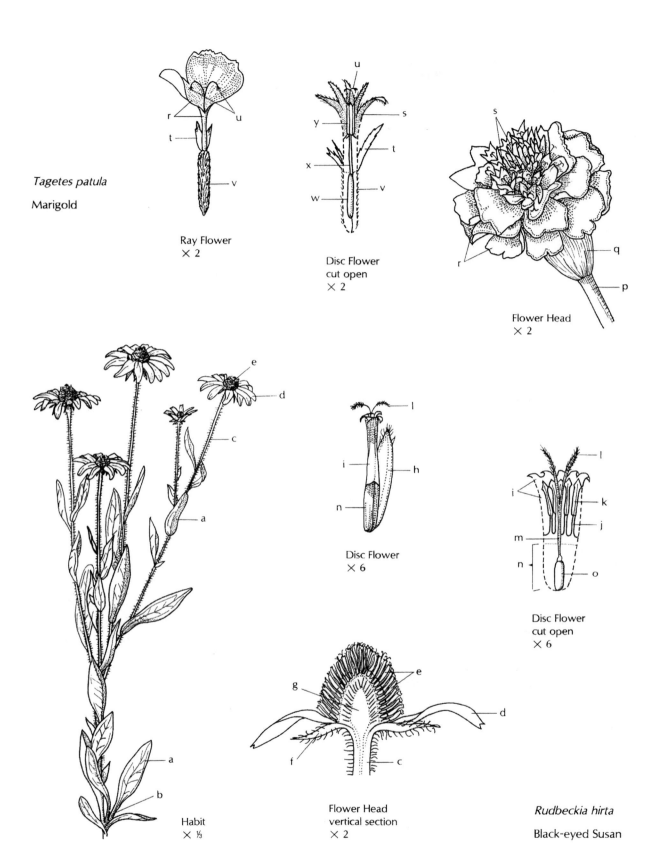

Tagetes patula

Marigold

Ray Flower
× 2

Disc Flower
cut open
× 2

Flower Head
× 2

Disc Flower
× 6

Disc Flower
cut open
× 6

Habit
× ⅓

Flower Head
vertical section
× 2

Rudbeckia hirta

Black-eyed Susan

Water-plantain Family (Alismataceae)

This primitive **monocot** family is composed of aquatic plants that are annual or perennial herbs. Parts of the bisexual or unisexual flower are arranged in whorls of 3 sepals, 3 petals, and 3 to numerous stamens and pistils. A pistil's ovary contains usually one ovule in **basal placentation**. The fruit is an achene. There are 14 genera and about 100 species.

The basally sheathing leaves are either floating or have long petioles that rise above water level. Leaf blades are variable, linear to round, or the bases are arrowhead-shaped (**saggitate**), or arrowhead-shaped with narrow, basal lobes nearly at right angles (**hastate condition**).

Of interest . . . depending on the point of view, these plants are waterweed **pests** and can be very common weeds in rice *(Oryza sativa)* paddies or useful for **ornamental pool gardens:** *Alisma plantago-aquatica* (water plantain), *Echinodorus* (bur-head), *Sagittaria* (arrowhead); **wildlife food:** most genera.

Saggittaria latifolia   Wapato, Arrowhead

Fibrous roots (a) anchor this perennial herb in the mud substrate of aquatic habitats. Older plants have swollen tubers at the ends of **rhizomes** (underground stems). The common name, wapato, was used by Native Americans for the edible tubers. The leaf's sheathing petioles (b) vary in length with the water depth. Arrowhead-shaped leaf blades (c) are narrow, as shown, to wide. The peduncle (d) of the **raceme** bears whorls of unisexual, male flowers (e) at the top and female flowers (f) below. Having male and female flowers on the same individual plant is termed a **monoecious** condition.

An enlargement of the male flower shows more clearly the bract (g) at the pedicel (h) base. Both male and female flowers each have 3 persistent, green sepals (i) and 3 white petals (j), which are shed early.

The male flower has 24 to 40 stamens (k) in whorls. Pollen is shed through slits on the anther sacs (l).

Seen in a vertical section, the female flower has numerous, densely aggregated pistils (n) borne on the receptacle (o). Each pistil has a simple stigma (p), a short style (q), and one basal ovule (r) in the ovary (s). The fruits produced are numerous, flat achenes.

COLOR CODE

white:	roots (a), petals (j), filament (m), receptacle (o), ovule (r)
red-green:	petioles (b)
green:	blades (c), peduncle (d), bract (g), pedicels (h), sepals (i), pistils (n), stigma (p), style (q), ovary (s)
yellow:	stamens (k), anther (l)

Saggitaria latifolia

Arrowhead

Stamen
× 10

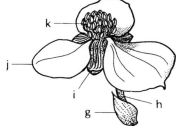

Male Flower
× 2

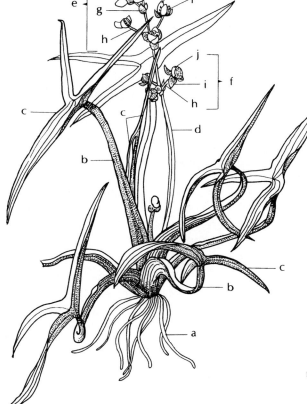

Habit
× ½

Pistil
× 10

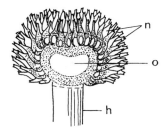

Female Flower
perianth removed
vertical section
× 5

Spiderwort Family (Commelinaceae)

These popular houseplants are annual or perennial herbs derived from tropical and subtropical regions. Closed sheathing leaf petioles and flower parts in 3's are common characteristics. There are 38 genera and about 600 species.

The alternate succulent leaves have entire margins and parallel venation. Leaf-like bracts sometimes partially enclose the flower clusters. The bisexual flower can also occur singly. Parts of the flower include 3 sepals, 3 petals, 6 stamens (or 3 stamens and 3 **staminodes** or one functional stamen), and a single pistil. The stamens' filaments often have hairs. The pistil has a superior ovary of 3 fused carpels, one style, and one stigma. There are no nectar glands. A **loculicidal capsule** is usually the fruit type.

Of interest... **ornamentals:** *Callisia* (striped inch plant), *Commelina* (dayflower), *Cyanotis, Dichorisandra, Gibasis* (Tahitian bridal veil), *Rhoeo discolor* (Moses-in-the-bulrushes), *Tinantia, Trandescantia* spp. (spiderwort, wandering Jew, inch plant), *Zebrina pendula* (wandering Jew); **biology:** The staminal hairs of *Tradescantia* (spiderwort) are commonly used in introductory plant biology labs to observe cytoplasmic streaming (**cyclosis**) under the microscope.

Gibasis geniculata   Tahitian Bridal Veil

The thin, trailing stems (a), small leaves (v, c, d), and tiny flower clusters (e) provide a delicate appearance to this plant. It is often displayed in hanging containers. Because of the sheathing leaf petioles (b), the stem has a jointed appearance. The wavy-margined, alternate leaves are green above (c) and purple (d) below (due to the presence of anthocyanin reddish-purple pigment).

Branching from the peduncle (f), pedicels (g) support the flowers. The flower has three sepals (h), 3 petals (e), 6 stamens (i), and a single pistil (j). The stamen has tufts of hairs (k) on the filament (l) below the anther (m). The pistil consists of stigma (n), style (o), and a 3-lobed, superior ovary (p). Fleshy sepals (h) remain and enclose the capsule fruit.

Zebrina pendula   Wandering Jew

This low, spreading succulent (fleshy) herb has purple (q) and silvery-green (r) striped leaf blades with hairy petioles (s) that sheath the stem (t). *Zebrina* is from the Latin for zebra, and refers to the striped leaves.

COLOR CODE

green:	stem (a, t), upper blade surface (c), peduncle (f), stripes (r), petioles (s)
purple:	petiole (b), lower blade surface (d), pedicel (g), margin and center stripes (q), bract (u, v)
white:	petals (e), hairs (k), filament (l)
light green:	sepals (h), pistil (j), ovary (p)
yellow:	stamens (i), anther (m)
dark pink-lavender:	stigma (n), style (o), petals (w)

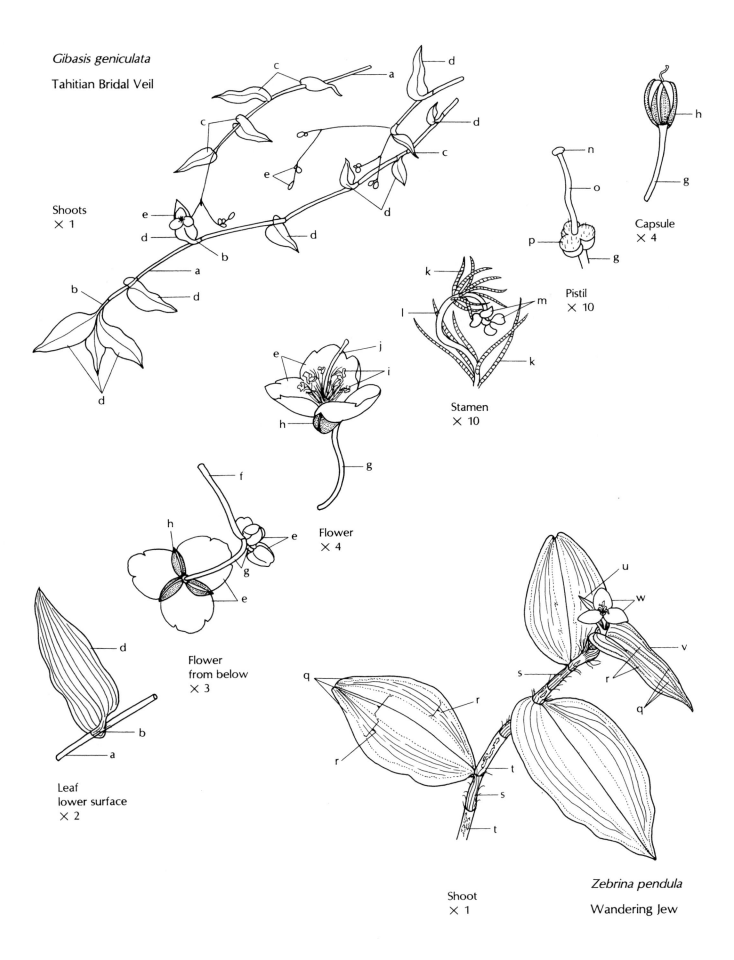

Gibasis geniculata

Tahitian Bridal Veil

Shoots
× 1

Flower
× 4

Flower
from below
× 3

Leaf
lower surface
× 2

Stamen
× 10

Pistil
× 10

Capsule
× 4

Shoot
× 1

Zebrina pendula

Wandering Jew

Sedge Family (Cyperaceae)

Distinguished from grasses, which have hollow, round stems, sedges have solid-pithed, often 3-sided stems. Grasses have leaves arranged in two ranks, whereas sedge leaves usually emerge from a stem node in three directions (three-ranked). As in grasses, each sedge leaf is organized into a blade (lamina) and an enclosing sheath. One bract or scale subtends each flower. There are 100 genera and about 4,000 species.

Sedges are usually bog or marsh plants and grow in clumps or extend from creeping, underground, rhizome-like stems. Minute flowers are arranged in **spikelets**. Sepals and petals are reduced to bristles, hairs, or scales, or are absent. Usually there are 3 stamens and a single pistil with a superior ovary consisting of 2 to 3 carpels fused to form one chamber with one ovule. The pistil's style sometimes forms a beak on the achene-type fruit.

Of interest... **ancient usage:** *Cyperus papyrus* (pith used to make paper), *Cyperus* spp. (mat grass, hay grass, roof thatching), *Eleocharis* spp. (basket making), *Scirpus* spp. (bulrushes for basketwork, mats, chair seats); **food:** *Eleocharis tuberosa* (Chinese water chestnut tubers); **water garden plants:** *Cyperus alternifolius* (umbrella plant), *Carex* spp., *Leiophylum* spp., *Scirpus* spp.; **cattle fodder:** *Carex spp.* (sedges); **wildlife food:** most genera.

Cyperus strigosus   Umbrella Sedge

This perennial sedge has a smooth, three-sided stem (a), leaves (b) at the base, and clusters of flower spike (c) subtended by leaf-like bracts (d). A ray (e) of spikes has loosely arranged, linear spikelets (f). Older plants have a short, knotty rhizome-like underground stem.

A spikelet consists of a double row of overlapping, bisexual flowers, each enclosed by a flattened scale (g, h). In the flower enlargement (shown), 3 stamens have been shed. Remaining are a green-keeled (g), flattened scale with golden, translucent sides (h) and the pistil. The pistil consists of a 3-parted style (i) and an ovary (j). The fruit that develops is a 3-angled, beakless **achene** (k).

Carex hystericina   Porcupine Sedge

Unisexual flowers are formed on separate spikelets of this sedge. Both male (l) and female (m) spikelets have leaf-like bracts (n) below. The female pistil (o) is composed of a 3-branched style (p) and an ovary surrounded by a flask-shaped sac, the perigynium (q), with a 2-toothed beak (r). A small scale (s) subtends the pistil. As in other sedges, an achene (t) fruit is formed.

COLOR CODE

green:	stem (a), leaves (b), bracts (d, n), rays (e), scale keel (g)
gold-green:	spikelets (f)
gold:	scale side (h) covering ovary (j)
tan:	style (i), achene (k, t), male spikelet (l)
light green:	female spikelet of pistils (m, o), perigynium (q, r), scale (s)
brown:	styles (p)

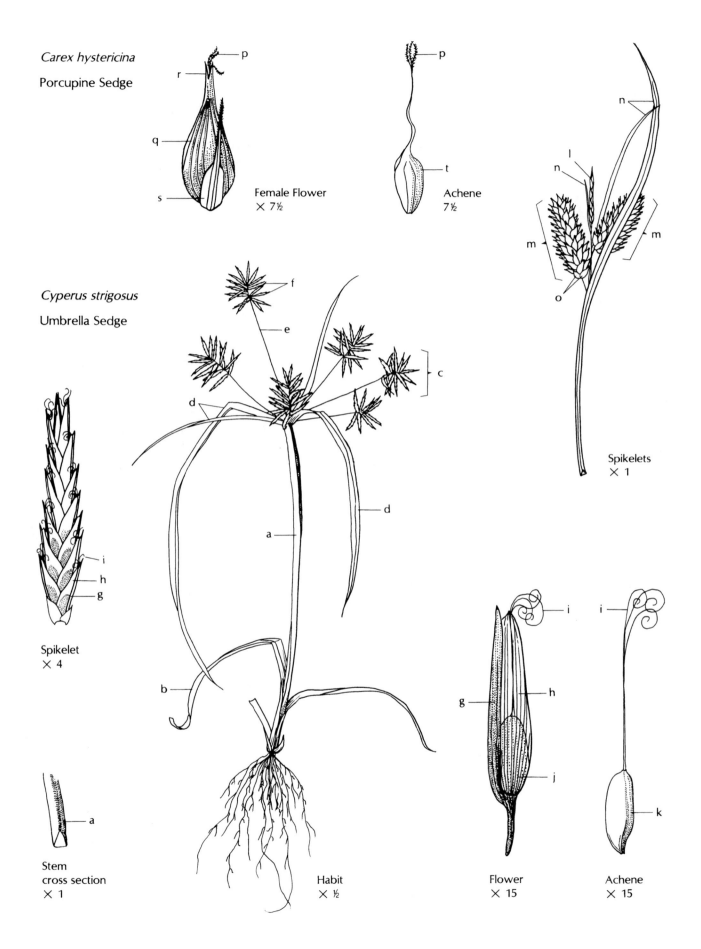

Carex hystericina

Porcupine Sedge

Female Flower
× 7½

Achene
7½

Spikelets
× 1

Cyperus strigosus

Umbrella Sedge

Spikelet
× 4

Stem
cross section
× 1

Habit
× ½

Flower
× 15

Achene
× 15

Grass Family (Poaceae)

The grass flower is unlike any other plant family's flowers. Within a spike (a), the flower unit is called a **spikelet**. An exaggerated separation of the structures is shown in the spikelet diagram. For reference, the diagram and the spikelet drawing of *Secale cereale* (rye) have the same letter for a described structure.

The primary axis of the spikelet is called the **rachilla** (b). At the base of the spikelet are two bracts called first **glume** (c) and second glume (d). The next bracts are called **lemmas** (e) and may have a bristlelike appendage (awn, f). A soft inner bract is called the **palea** (g). The remaining parts comprise a **floret** (h). Each spikelet may have one or more florets, which are unisexual or bisexual.

There are no sepals or petals. Instead, there are 2 or 3 **lodicles** (i). Lodicles enlarge with turgor pressure and cause the lemmas and palea to expand, exposing the stamens and/or stigmas at pollination time.

Grasses are wind-pollinated. There are usually 3 stamens with freely moving (**versatile**) anthers (j) and a single pistil with 2, usually feathery, stigmas (k). The pistil's ovary (l) is superior and contains one ovule. The fruit is usually a grain (**caryopsis**, m) or a berry in some bamboos.

Grasses are annual or perennial herbs except for bamboo, which is woody at maturity. Roots (n) are fibrous and **adventitious** roots arise from stem nodes. The stems, called **culms** (o) are usually hollow and round with one leaf (p) to a node (q). Leaves are alternate in two rows (ranks = a so-called **distichous** arrangement) up the culm. The leaf consists of a blade (**lamina**, r) and a **sheath** (s), which encircles the culm (stem), and has parallel venation (t). Where the leaf blade diverges from the culm is often an appendage, the **ligule** (u), and sometimes, as in rice (*Oryza sativa*) two sickle-shaped **auricles** that clasp the sheath.

Economically, Poaceae is the most important plant family. Of interest . . . **food crops:** *Avena sativa* (oats), *Hordeum* (barley), *Oryza sativa* (rice), *Saccharum officinarum* (sugarcane), *Secale cereale* (rye), *Setaria italica* (millet), *Sorghum bicolor* (sorghum), *Triticum* (wheat), *Zea mays* (corn); **shelter:** *Bambusa* (bamboo); **fodder:** *Agrostis* (bentgrass), *Dactylis glomerata* (orchard grass), *Phleum* (timothy), *Setaria italica* (foxtail grass), *Sorghum, Zea;* **lawn turf:** *Agrostis* (bent

grass), *Cynodon dactylon* (Bermuda grass), *Festuca* spp. (fescue grass), *Poa* (bluegrass), *Stenotaphrum* (St. Augustine grass), *Zoysia* (zoysia grass); **industry:** various grasses (insulation materials, newsprint, ethyl alcohol); **decontamination:** Molasses is made from refined sugar derived from sugarcane (*Saccharum officinarum*). Besides its culinary use, molasses spurs the growth of one type of bacteria that creates favorable growth conditions for other bacteria that digest contaminants in polluted groundwater.

Parts of a Grain

The *Triticum* (wheat) grain diagram shows the source of vitamins, nutrients, and metabolites that we utilize (see also 40, corn grain). Wheat **bran**, composed of **pericarp** (v), seed coat (w) and protein–rich **aleurone layer** (x), contains vitamin B complex (thiamine, riboflavin, niacin, pyridoxine, pantothenic acid), cellulose, phosphorus, calcium, and iron. **Endosperm** (y) is processed into flour that contains mostly starch and some protein. Wheat **germ** refers to the embryo (z) and is a source of vitamin B complex, vitamin E, vitamin A, protein, iron, fat, sodium, copper, zinc, magnesium, and phosphorus.

Wild wheat was domesticated by humans about 10,000 years ago. There are now about 22,000 cultivars. There are two main categories: **hard wheat** to make bread and **soft wheat** to make pasta, pastry, crackers, cakes, cookies and cereal.

Rice (*Oryza sativa*) is sold as either white or brown rice. White rice is "polished" with consequent removal of the protein-rich aleurone layer outside the endosperm. Brown rice has the aleurone layer left intact and is nutritionally far better than starch-rich white rice.

COLOR CODE

yellow:	spikes (a), glumes (c, d), anthers (j), pericarp (v)
green:	lemmas (e), leaves (p), leaf blade (r), leaf sheath (s), ligule (u)
light green:	palea (g), lodicle (i), stigmas (k), ovary (l), culm (o)
tan:	grain (m), roots (n), seed coat (w)
orange:	aleurone layer (x)
white:	endosperm (y), embryo (z)

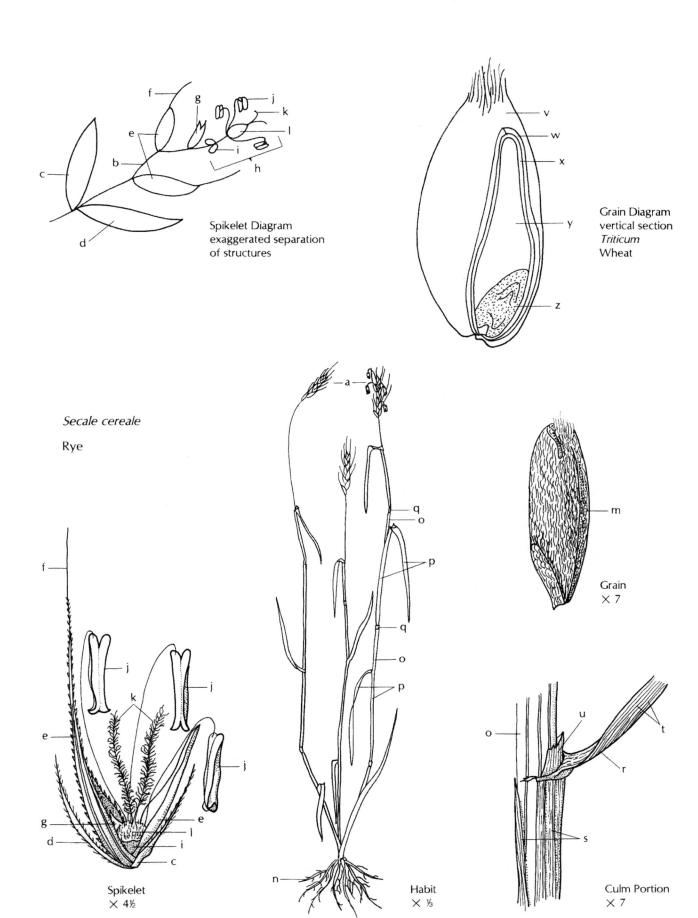

Spikelet Diagram
exaggerated separation
of structures

Grain Diagram
vertical section
Triticum
Wheat

Secale cereale

Rye

Grain
× 7

Spikelet
× 4½

Habit
× ⅓

Culm Portion
× 7

Arrowroot Family (Marantaceae)

Arrowroot plants are tropical, perennial herbs with underground stems (**rhizomes**). Because of the flower structure, this family is considered to be the most advanced in its order, the Zingiberales. Other families included in the order are the bird-of-paradise family (Strelitziaceae), the banana family (Musaceae), the ginger family (Zingiberaceae), and the canna family (Cannaceae).

In the arrowroot family, there are 30 genera and 350 species. The flower is bisexual but the fertile male part has been reduced to one stamen, which consists of one anther sac. One to 5 sterile stamens (**staminodes**) are also present.

The single pistil has an inferior ovary with 3 chambers, although by evolutionary reduction, only one ovule may be produced. The fruit is a **loculicial capsule** or berry-like, and the seeds often have a fleshy covering (**aril**).

The arrow root leaf is composed of a petiole with a sheath covering and a joint where the blade is attached. The joint is a pad-like swelling (**pulvinus**) that controls "sleep movements." Within the pulvinus, the vascular tissue is grouped in the center, surrounded by **sensor** and **flexor parenchyma** cells, which can expand and contract. Leaves are alternate and 2-ranked, emerging in one plane on opposite sides of the stem. Venation is **pinnate** from the midrib and parallel.

Of Interest... **food:** *Maranta arundinacea* (West Indian arrowroot) cultivated for the starchy rhizome: **ornamentals:** *Calathea mackoyana* (peacock plant), *Calanthea zebrina* (zebra plant), *Calathea ornata, Maranta leuconeura kerchoveana* (prayer plant, red-spotted arrowroot, rabbit tracks), *Thalia dealbata* (water canna).

Maranta leuconeura massangeana   Prayer Plant

The common name, prayer plant, is derived from the orientation of the leaves at dusk. During daylight the leaves are in a horizontal position. As light fades at sunset, the leaves take a vertical position that appear like hands in prayer.

This small herbaceous plant is cultivated primarily as a foliage-type houseplant for its brightly colored leaves. The stem (a) has a sheath (b) at nodes and the leaf petiole has a winged sheath (c). Where the petiole joins the blade is a short, jointed segment, a pulvinus (d), where blade movement occurs in response to light and dark (nyctinastic response).

The blade coloring is unusual, consisting of red parallel lines (e) that follow the pinnate venation pattern, a band (f) of yellow-green on both sides of the midrib area, with the remaining blade area (g) a dark green. The lower blade surface (h) is purple-red. In bud (i), the leaves are rolled.

Tiny flower bud (j) clusters (**panicles**) have **pedicels** (k) enclosed by bracts (l) at the base. Each flower has 3 sepals (n), a fused petal tube consisting of 3 lobes (o) and 3 petal-like **staminodes** (p). A fourth staminode (r), fused to the petal tube, extends from the colored staminodes. It has one fertile anther sac (s) attached to the side of a hood (r) that encloses the pistil's stigma (t).

With physical disturbance of the hooded staminode covering, the stigma (t) springs free as the style (u) coils down to its fusion point (v) on the staminode. At the same time, the few large pollen grains are propelled from the anther sac. The pistil's inferior ovary (s) is below the sepals.

COLOR CODE

green:	petiole sheath (c), petiole pulvinus (d), peduncle (m)
purple-green:	stem sheath (b), bracts (l)
red:	vein lines (e)
yellow-green:	midrib band (f)
dark green:	stem (a), remaining blade area (g)
purple:	lower blade surface (h), leaf bud (i), shaded area of staminodes (q)
lavender:	flower buds (j), staminodes (p)
light green:	pedicel (k), sepals (n), ovary (w)
white:	petals (o), staminode hood (r), anther sac (s), stigma (t), style (u)

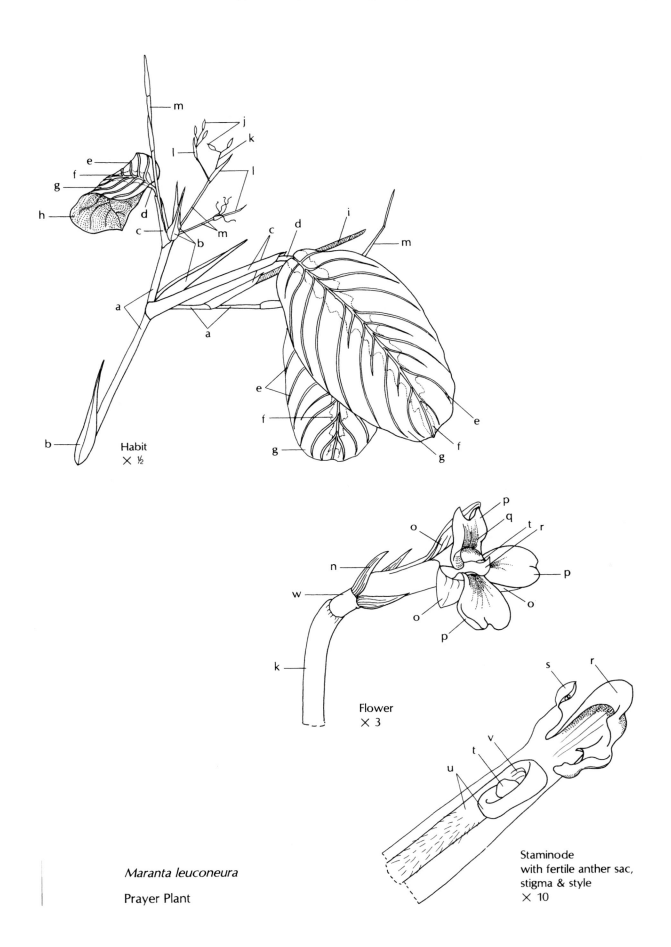

Habit
× ½

Flower
× 3

Staminode
with fertile anther sac,
stigma & style
× 10

Maranta leuconeura

Prayer Plant

Palm Family (Arecaceae)

Palms are second to the grasses (see 123) in economic importance. In the tropics, they may be a source of food, clothing, shelter, and fuel. The palms, numbering about 3,500 species, mostly consist of trees. Shrubs and vines are also represented.

Lodoicea maldivica has the world's largest seed, a "double coconut," made up of a two-lobed **drupe**. *Raphia fainifera* has the largest flowering plant leaf, up to twenty meters long.

Palms have a single apical bud, called "heart of palm." When it dies or is removed, the plant dies. Tree-like forms have an unbranched trunk with a terminal crown of leaves, commonly called "**fronds**," which emerge one at a time from the apical bud. Lignin, deposited in cell walls of stem (trunk) tissues provides sturdiness. As monocotyledons, palms have no vascular tissue in the stem; so there is no secondary growth (see monocot stem, 15).

The leafy frond is made up a blade, a petiole, and a sheathing base. Blade types are: fan-shaped with feather-like (**pinnate**) veins, as in *Lodoicea* (a); fan-shaped with veins arising from one point (**palmate**), as in *Sabal* (b); feather-shaped (**pinnately compound**), as in *Chamaedorea* (c); or feather-shaped, twice-divided (**bipinnately compound**) as in *Caryota* (see 126).

Regardless of shape, the young leaf (e) looks like a rod with a length-wise strip that peels down like a zipper, to free the one to many leaflets that unfold like a fan. The leaf petiole may be smooth or toothed on its margin.

Small flowers are usually formed in loose clusters, called **panicles** (f), which have one or more bracts at the base. Commonly, the plants have separate male and female flowers (unisexual) on the same plant (**monoecious**), or on separate plants (**dioecious**), while others have flowers with both male and female parts within one flower (bisexual). Flower parts are usually in 3's, being separate or fused. Flowers are wind-, insect- or self-pollinated.

Usually the fruit has one seed and is a berry or drupe type. The outside wall of the fruit can be fleshy, fibrous or leathery. Storage tissue (**endospem**) within the seed is oily or fatty rather than starchy. In coconuts *(Cocos nucifer),* it is liquid encased within the solid coconut "meat" that makes up the solid part of the endosperm.

Of interest . . . **economic plants:** *Areca catechu* (betel nut palm), *Calamus* and *Caemonorops* (rattan cane), *Cocos nucifera* (coconut palm), *Copernicia* (carnauba wax), *Elaeis guineesis* (oil palm), *Metroxylon* (sago palm), *Phoenix dactylifera* (date palm), *Paphia pendunculata* (raffia, used as twine to tie tall plants to supports); **ornamentals:** *Arecastrum* (queen palm), *Arenga pinnata* (sugar palm), *Caryota mitis* (fishtail palm), *Chamaedorea elegans* (parlor palm), *Chamaerops* (European fan palm), *Chrysalidocarpus* (Madagascar feather palm), *Cocothrinax argentea* (silver palm), *Erythea* spp. (Mexican fan palms), *Howeia* spp. (curly palm, sentry palm, flat palm), *Jubaea spectabilis* (coquitos palm), *Livistona* spp. (fan palms), *Metroxylon* (sago palm), *Rhapidophyllum* (needle palm), *Rhapis* (lady palm), *Roystonea regia* (royal palm), *Sabal palmetto* (cabbage palmetto), *Serenoa* (saw-palmetto), *Trachycarpus fortunei* (Chinese windmill palm), *Thrinax* (peaberry palm), *Washingtonia filifera* (sentinel palm); **food:** heart-of-palm; **building materials:** large palm leaves are used as "thatch" on roofs of houses/huts in the tropics.

COLOR CODE

yellow-green:	leaf (a)
green:	leaf (b)
dark green:	leaf (c), stem (d), new leaf (e)
orange:	panicle (f)

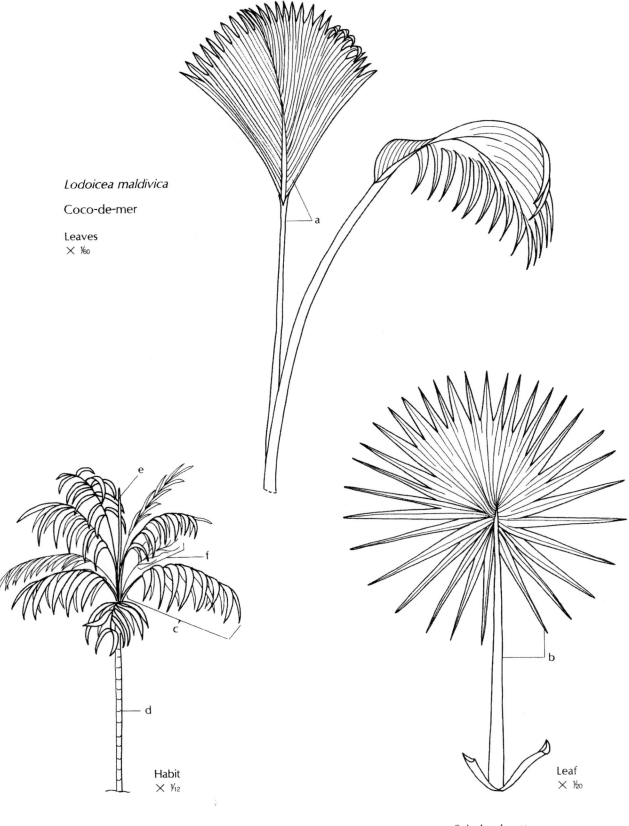

Lodoicea maldivica

Coco-de-mer

Leaves
× 1/60

a

Chamaedorea elegans

Parlor Palm

e

f

c

d

Habit
× 1/12

Sabal palmetto

Cabbage Palmetto

b

Leaf
× 1/20

Palm Family (continued)

Sabal palmetto   Cabbage Palmetto

This palm is native to the coastal regions of southeast United States, the West Indies, and Venezuela. It may grow up to 21 meters high. The leaf (see 125) is fan-shaped with **palmate venation**. Long stems of many **panicles** arise with the leaves. A single panicle (a) with small, white bisexual flowers (b) is shown. The flower is without a stem (**sessile**) and consists of 3 bracts (c) at the base, 3 fused sepals (d), 3 petals (e), 6 stamens (f), and a single pistil (g) with a superior ovary having 3 fused carpels.

Chamaedora elegans   Parlor Palm, Collina, **Neathe Bella**

The parlor palm, native to South America, is often used as a potted houseplant. The **fronds** (leaves, see 125) are feather-shaped. Arising with the leaves at the crown are branched racemes (**panicles**, h) of flowers which develop into black **drupes** (i). Bracts (j) enclose the base of the panicle.

Chamaedora ernesti variety *augusti*

Instead of a single trunk, like the parlor palm (see 125) this Mexican species has many leafy stems arising from a common base. The panicle (k) has many bracts (l) enclosing the base and only a few stalks of flowers (m).

Caryota mitis   Fishtail Palm

Twice divided (**bipinnately compound**) leaves distinguish this palm. The sub-leaflets (n) are fishtail-like. Leaves emerge along the stem instead of in a crown at the top. Panicles of flowers also arise along the length of the trunk. Flowers mature into fruits (o, p) starting from the end of the stalk (q). The drupes drop off as they mature.

COLOR CODE

white:	flowers (b), petals (e), pistil (g)
green:	panicle stalk (a), sub-leaflets (n), immature drupe (o)
tan:	bracts (c, j, l), stalk end, shaded (q)
light green:	sepals (d), stalk, upper portion (r)
orange:	panicle stalks (h)
black:	drupe (i), mature drupe (p)
dark green:	panicle stalk (k)
yellow:	stamens (f), flowers (m)

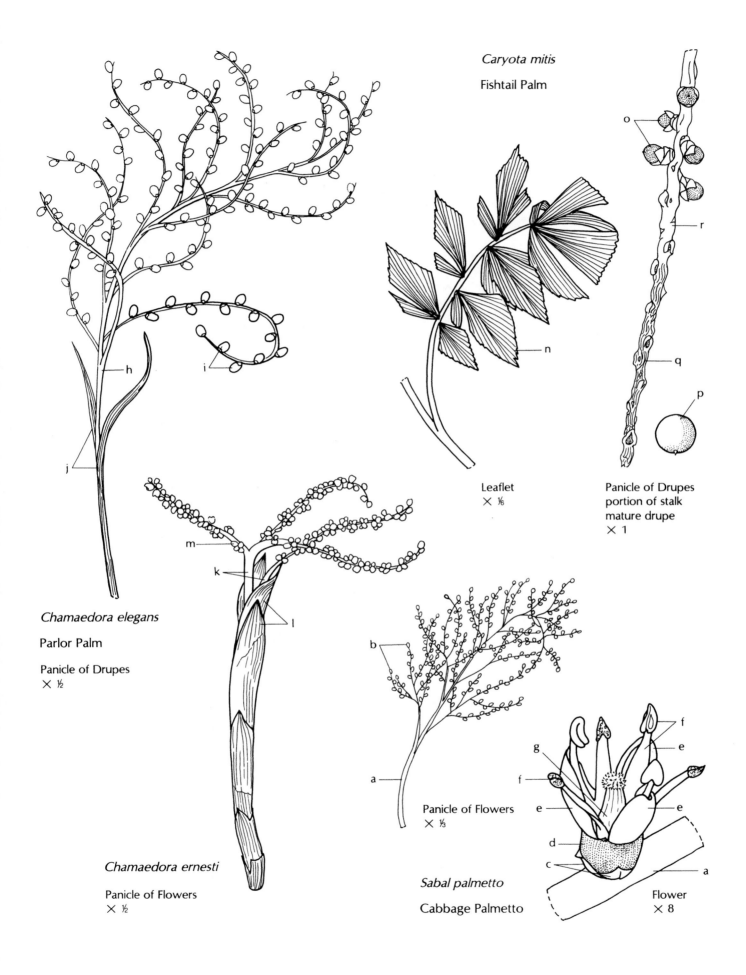

Caryota mitis

Fishtail Palm

o

r

n

q

p

Leaflet
× ⅙

Panicle of Drupes
portion of stalk
mature drupe
× 1

h

i

j

Chamaedora elegans

Parlor Palm

Panicle of Drupes
× ½

m

k

l

b

a

Panicle of Flowers
× ⅓

f

g

e

f

e

e

d

c

a

Sabal palmetto

Cabbage Palmetto

Flower
× 8

Chamaedora ernesti

Panicle of Flowers
× ½

Arum Family (Araceae)

A unique characteristic of this family is the elaborate reproductive structure consisting of a fleshy spike of flowers, the **spadix,** subtended by a leaf-like bract, the **spathe.** These mostly tropical and subtropical plants are usually terrestrial herbs growing on the forest floor. Because of their low-light tolerance, they are popular as houseplants in cool climates. Poisonous calcium oxalate crystals are present in plant tissues and there may be a milky sap (latex).

The leaves are simple or compound, occur at the base or are alternate on the stem, and have petiole bases that sheath the stem. The small flowers on the **spadix** are bisexual or unisexual. The flowers have an odor of rotten flesh, which attracts fly pollinators (see 34). There are 1 to 6 stamens and a single pistil with one to many carpels. Sepals and petals that look alike (**tepals**) may be present. The fruit type is a **berry.** There are about 110 genera and about 2,000 species.

Of interest... **ornamentals:** *Aglaonema* (Chinese evergreen), *Amorphophallus* (voodoo lily), *Anthurium* (flamingo flower), *Caladium* (elephant's-ear), *Dieffenbachia* (dumb-cane), *Helicodicerus, Monstera deliciosa* (Swiss-cheese plant, cut-leaf philodendron), *Philodendron, Pistia* (water lettuce), *Scindapsis* (pothos, devil's-ivy), *Spathiphyllum* (flame plant), *Zantedeschia aethiopica* (calla lily); **food:** *Alocasia* (taro), *Colocasia* (taro), *Xanthosoma* (tannia); **native wild plants:** *Arisaema triphyllum* (Jack-in-the-pulpit), *Calla palustris* (wild calla), *Orontium* (golden club), *Peltandra* (arrow arum), *Symplocarpus foetidus* (skunk cabbage, see 34).

As skunk cabbage shoots develop in early spring in northern temperate regions, they melt the snow around each shoot. This heat is generated through high respiration rates in the developing spadix, in sufficient amounts to melt the snow and ice.

Spathiphyllum   Flame Plant

This perennial herb has simple leaf blades (a) with sheathing petioles (b). The **spadix** (c) of bisexual flowers is subtended by a white petal-like **spathe** (d). Each flower (e) has 6 tepals (f) enclosing 6 stamens. The anther (g) of the stamen is shed form a fleshy filament (h) after pollen is released. The pistil (i) has a flattened 3-part stigma (j) with ovules (k) in **basal placentation** (see 28).

Arisaema triphyllum   Jack-in-the-pulpit

This North American wild flower is a perennial herb that arises from a solid bulb-like structure (**corm,** see 12). The compound leaf blade (l) is divided into 3 leaflets (hence, *triphyllum*). The petiole (m) is enclosed by a sheath (n) at the base. Plants with one compound leaf have male flowers on the spadix, while plants with two compound leaves have female flowers.

This unusual plant has unisexual flowers on the same plant at different times. Young plants with one compound leaf have only male flowers at the base of the spadix. In subsequent years, two leaves develop from the corm, and in a developmental sex change, the spadix produces only female flowers. Energy is expended in the production of fruit; so food produced in the leaves and translocated to the corm must be sufficient to assure that shoot and reproductive (flowering and fruiting) development occurs the following year.

In autumn, on plants with female flowers, a cluster of red berries (s) is revealed as the spathe and leaves (t) wither.

COLOR CODE

green:	blades (a), petioles (b, m), leaflets (l), sheath (n)
white:	spadix (c), spathe (d), tepals (f), pistil (i), stigma (j), ovules (k), filaments (h)
yellow:	anthers (g)
light green:	outer spathe (o), spadix (q), peduncle (r)
purple:	inner spathe (p)
red:	berries (s)
tan:	peduncle (u), leaves (t)

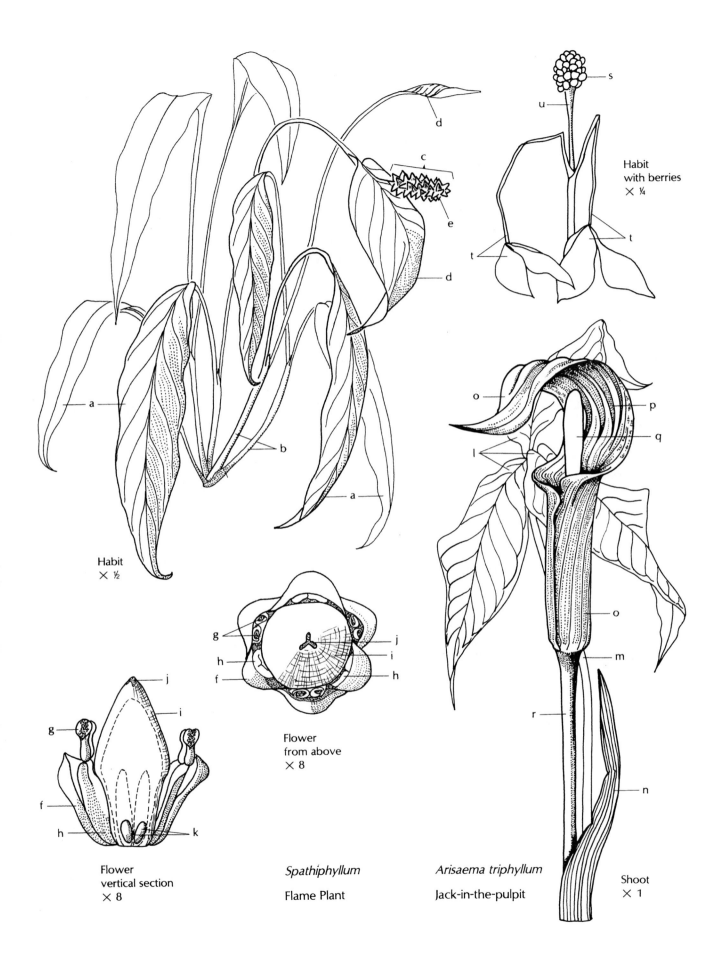

Habit
× ½

Habit
with berries
× ¼

Flower
from above
× 8

Flower
vertical section
× 8

Spathiphyllum

Flame Plant

Arisaema triphyllum

Jack-in-the-pulpit

Shoot
× 1

Lily Family (Liliaceae)

Lily flowers have 3 sepals and 3 petals that look alike (**tepals**), 6 stamens, and a single pistil with a superior ovary composed of 3 fused carpels. Fruit types are **septicidal** or **loculicidal capsules** or **berries**.

This large family of about 250 genera and about 4,600 species consists mostly of perennial herbs with narrow, parallel-veined leaves and underground storage organs such as rhizomes, bulbs, corms, or tubers. Some plants are evergreen succulents, as in *Aloe* and *Haworthia* or vines as in *Smilax* (see 16).

Leaf characteristics vary greatly from basal and linear with parallel veins to **cauline** (growing on a stem), as shown in the illustration. Some have net (**reticulate**) venation *(Trillium)*. Others have leaf-tip tendrils *(Gloriosa)*. *Asparagus* leaves are scale-like.

Of interest... **food:** *Allium* spp. (onion, shallot, leek, garlic, chives), *Asparagus officinalis* (asparagus); **medicine:** *Aloe vera* (sap from leaves is used to treat burns and relieve the pain of sunburn), *Colchicum* (colchicine), *Urginea* (red squill); **ornamentals:** *Allium, Aloe, Asphodeline* (Jacob's rod), *Agapanthus* (lily-of-the-Nile), *Calochortus, Colchicum autumnale* (autumn crocus), *Convallaria majalis* (lily-of-the-valley), *Eremurus* (foxtail lily), *Fritillaria, Gloriosa* (gloriosa lily), *Haworthia, Hemerocallis* (daylily), *Hosta* (hosta), *Hyacinthus* (hyacinth), *Kniphofia* (torch lily, red-hot-poker), *Lilium* spp. (lilies), *Muscari* (grape-hyacinth), *Sansevieria (snake plant, bowstring-hemp, leopard lily), Scilla* (squill), *Smilax* (greenbriar, see 16), *Tulipa* (tulip); **native wild plants:** *Erythronium* spp. (dogtooth violet, trout lily), *Maianthemum* (false lily-of the valley), *Lilium, Medeola virginiana* (Indian cucumber-root), *Polygonatum* (Solomon's seal), *Smilacina* spp. (false Solomon's seal), *Trillium* spp., *Uvularia* (bellwort),

Veltheimia viridifolia); **poisonous:** *Amianthemum* (stagger-grass), *Convallaria majalis* (lily-of-the-valley), *Ornithogalum umbellatum* (star-of-Bethlehem), *Veratrum* (false hellebore), *Zigadenus Nuttallii* (death camas, poison camas, merryhearts).

Lily is a confusing common name. Examples of a few plants that are called lilies include lilies (*Lilium*), and daylilies *(Hemerocallis)* in the lily family (Liliaceae). Then, there are waterlilies *(Nymphaea)* in the waterlily family (Nymphaeae, 79), peace lily *(Spathiphyllum)* in the arum Family (Araceae, 127), blackberry lily *(Belamcanda chinensis)* in the iris family (Iridaceae, 129), and spider lilies (*Lycoris* spp.) in the amaryllis family (Amaryllidaceae).

Lilium michiganense   Michigan Lily

This wild lily is a perennial herb that grows to two meters in height. It is native to the midwestern prairie region of the United States. Whorls of leaves (a) occur along the stem (b). At the top are flowers with 6 orange or, rarely, orange-red, recurved tepals (c) that arise on **peduncles** (d).

The flower also has 6 stamens (e) with each filament (f) attached in the center of the anther (g), providing for free movement. The single pistil (h) has a superior ovary composed of 3 carpels (i) with numerous ovules (j) in **axile placentation**, a long style (k), and a 3-lobed stigma (l). Sweat bees, butterflies and hummingbirds are the pollinators.

A **loculicidal capsule** (m) is formed from the ovary. One of the 3 locules shows a stack of flat, winged seeds (n).

COLOR CODE

green:	leaves (a), stem (b), peduncle (d), pistil (h)
orange:	tepals (c), anther (g)
brown:	stamens (e), capsule (m)
light green:	filament (f), style (k), stigma (l)
white:	carpels (i), ovules (j)
tan:	seed (n)

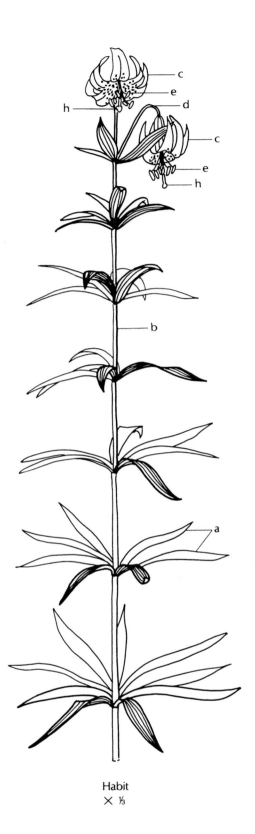

Habit
× ⅓

Lilium michiganense

Michigan Lily

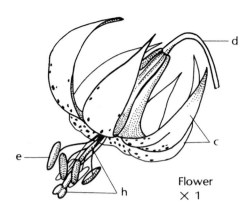

Flower
× 1

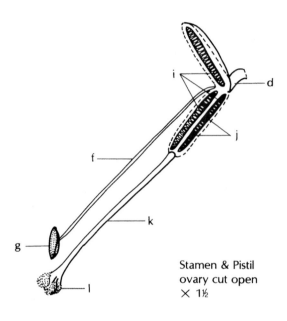

Stamen & Pistil
ovary cut open
× 1½

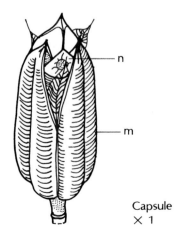

Capsule
× 1

Iris Family (Iridaceae)

Plants are perennial herbs or subshrubs with underground storage organs of rhizomes, bulbs, or corms. The leaves sheath at the base, overlapping each other transversely to form a flat fan-like arrangement (**equitant**) unique to this family. Venation is parallel.

The 3 sepals of flowers in this family are colored and petal-like. They may differ in size, shape, and color from the 3 petals. The bisexual flower has 3 stamens and a single pistil with an inferior ovary composed of 3 fused carpels. The fruit type is a **loculicidal capsule**. There are about 70 genera and about 1800 species.

Of interest... **ornamentals:** *Antholyza, Belamcanda chinensis* (blackberry or Chinese lily), *Crocus, Eustylis, Freesia, Gladiolus, Iris, Ixia, Moraea, Nemastylis, Neomarica, Sisyrinchium, Tigridia;* **commercial:** *Crocus sativus* (saffron flavoring, saffron dye), *Iris* rhizomes (orris root).

Iris germanica   Flag, Bearded Iris, German Iris

Iris plants are perennial herbs with a rhizome (a) storage organ and **adventitious roots** (b). Sword-shape leaves (c) overlap at the base in a fan-like arrangement.

Flower color is wide-ranging in cultivated hybrids. It varies from white, yellow, orange, blue, purple, red, pink, to tan and various bicolors. A pair of green bracts (d) occurs below the elaborate flower. Bracts cover a flower bud (e). The 3 petal-like sepals (f), "**falls**," are bearded with hairs (g). The 3 upright petals (h), "**banners**," curve inward. Above each sepal is a stamen (j).

In the flower structure drawing, an upright stamen made up of filament (k) and anther (l) can be discerned clearly. Nectar glands (m) occur where the filament emerges from the sepal tube (f).

Where is the pistil? The bulge below the sepal tube is an obviously inferior ovary (n) with indentations denoting the three fused carpels. In an ovary cross-section, each carpel (o) has a chamber (**locule**, p) with ovules (q) in **axile placentation.**

In the flower structure drawing, only one portion remains unidentified. What looks like a petal with crested wings above is one of 3 styles (r) with a flap of tissue, the stigma (s).

Pollination involves insects. When a bee (t, u) alights on the sepal, it is guided by the beard hairs and color markings to the nectar location. As it works its way through a tunnel between style (v, r) and sepal (f) to the nectar glands, pollen on its body from another flower is rubbed off onto the stigma flap (w, s). Then, as the bee backs out of the flower, its hairs are dusted with pollen that is shed outwardly from anther sacs (j, l). At maturity, the sepals raise upward against the styles, closing off the tunnels.

The iris fruit is a **loculicidal capsule** (x) with seeds.

COLOR CODE

tan:	rhizome (a), roots (b), capsule (x)
green:	leaves (c), bracts (d), bud (e), peduncle (i)
yellow:	beard (g), bee's thorax (t) and abdomen (u)
optional:	1 color from text: sepals (f), petals (h), styles (r, v)
white:	stamen (j), filament (k), anther (l), nectar glands (m), ovary (n), carpel (o), ovules (q), stigma (s, w)

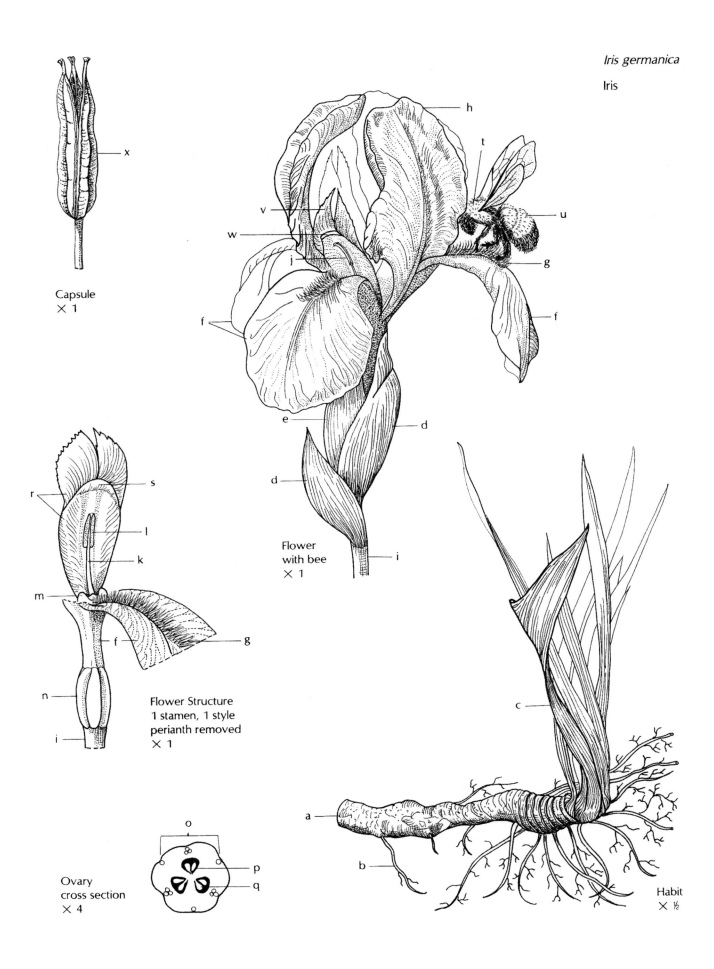

Iris germanica

Iris

Capsule
× 1

Flower
with bee
× 1

Flower Structure
1 stamen, 1 style
perianth removed
× 1

Ovary
cross section
× 4

Habit
× ½

Orchid Family (Orchidaceae)

The orchid flower is so specialized that the illustrated species is needed for a description of its structures. In general, the flower is usually bisexual, has 3 sepals that may resemble the 3 petals in color and form, a column of fused stamens and stigmas, and an inferior ovary composed of 3 carpels. Flower colors and color patterns vary widely. Nectar, odor, and form of the flower attract pollinators (specific in many cases). The fruit is a capsule that contains very tiny seeds. Since the endosperm aborts as seeds mature, a mutualistic fungal relationship has evolved that provides a source of metabolites for seed germination.

The plants are generally leafy, sometimes leafless. Leaves are alternate, very rarely opposite, whorled or reduced to scales. They are simple, thickened, usually linear, strap-shaped or round, and basally sheathing the stem. These perennial herbs are distributed worldwide and comprise the largest plant family, with an estimated 30,000 species. They are terrestrial and **saprophytic**, deriving nutrients from soil and dead organic matter. Or they are **epiphytic**, attached to the surface of another plant, where they obtain nutrients from the atmosphere and debris accumulations among the roots. Orchids have fungus-root (**mycorrhizal**) association.

Nutrient storage organs may be swollen stems (**pseudobulbs**, see 16) or swollen root-stem tubers, which the Greeks called "orchis" for their testiculate appearance; hence, the orchid name. Roots of most epiphytes are covered with a layer of white-colored dead corky cells called **velamen**. Habitats vary enormously from dry sand to acidic bogs and wet meadows, from temperate forest and mangrove swamps to tropical cloud forests. There is even an underground orchid, *Thizanthella gardneri.*

Of interest... **flavoring:** *Vanilla* (vanilla extract from capsules, commonly called beans); **wild:** most wild orchids are considered to be endangered species and are legally protected. In the United States, there are about 62 native genera; **ornamentals:** *Anagrecum, Brassavola nodosa* (lady-of-the-night, named for release of scent at night when moth pollinators will be attracted), *Brassia spp.* (spider orchids), *Cattleya, Cymbidium, Dendrobium, Epidendrum, Habenaria* spp. (fringed orchids), *Laelia, Miltonia* spp. (pansy orchids), *Odontoglossum, Oncidium, Paphiopedilum* spp. (lady's slipper orchids), *Phalaenopsis* spp. (moth orchids), *Stanhopea,* and *Vanda* are a few examples. In the quest for more exotic specimens, growers combine genera through plant breeding to form multigeneric types. For example *Sophrolaeliocattleya* is a hybrid composed of *Sophronitis, Laelia,* and *Cattleya,* abbreviated as Slc. Further tinkering has produced Slc. Jewel Box 'Scheherazade,' with *Sophronitis coccinea,* used to produce red hybrids; *Laelia* is used to produce free flowering and a fleshy substance, and *Cattleya* is used to produce size and modify form of foliage and flower.

Cattleya bowringiana

This plant is a tropical perennial epiphyte. Pseudobulbs (a) provide nutrient storage. Reduced leaves (b) cover the stem. Expanded leaves (c) are alternate and strap-shaped. A sheath (d) encloses the base of the peduncle (e). In this genus, the flower's sepals (f) are strap-shaped, whereas, the 3 petals (g) are ruffled.

In orchids, the central petal, called a **lip**, is usually larger and highly modified, sometimes occurring in quite bizarre forms. The lip encloses the **column**, resulting from a fusion of male and female parts. At the tip of the column is an anther cap (h) with 4 masses of pollen, called **pollinia** (i), tucked into 2 pocket-like structures. A pollinium has a sticky pollen sac (j) and a hooked **caudicle** (k). The remaining end of the column is formed by 3 fused fertile stigmas (l) with the end of the terminal stigma forming a sterile, sticky flap, the **rostellum** (m).

An insect follows the nectar guides (n) in order to locate the nectar tube (o), dislodges the anther cap, and carries pollinia to another flower, where they stick to the rostellum. The 3-carpelled ovary (p), surrounding the nectar tube, contains minute ovules. The remaining part of the column (q) is stamen-style tissue, and is therefore, bisexual. A capsule (r) fruit is formed.

COLOR CODE

tan:	pseudobulb (a), reduced leaves (b), withered petals (s)
green:	expanded leaves (c), sheath (d), peduncle (e), capsule (r)
pink-purple:	sepals (f), petals (g)
yellow:	pollinia (i), pollen sac (j)
white:	anther cap (h), caudicle (k), stigmas (l), rostellum (m), nectar tube (o), ovary (p), column (q)
purple:	nectar guides (n)

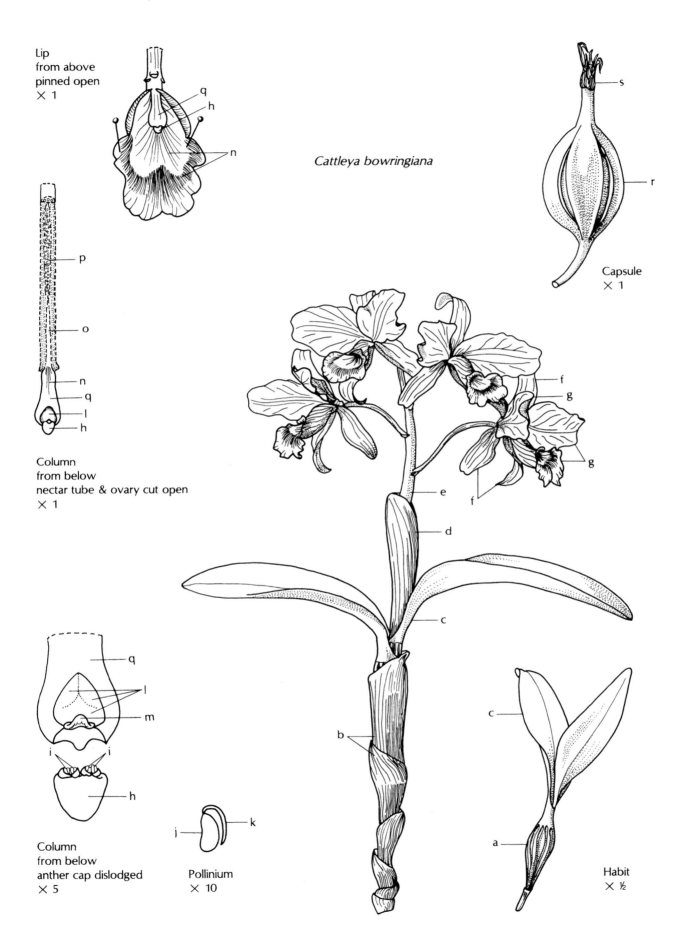

Lip
from above
pinned open
× 1

Cattleya bowringiana

Capsule
× 1

Column
from below
nectar tube & ovary cut open
× 1

Column
from below
anther cap dislodged
× 5

Pollinium
× 10

Habit
× ½

Bibliography

Adams, Preston, Jeffrey J. W. Baker, and Garland E. Allen. *The Study of Botany.* Reading, ME: Addison-Wesley, 1970.

Adams, II, Richard M. *Native Lilies,* American Horticulturist, August 1984, p. 28–31.

Alexopoulos, Constantine John. *Introductory Mycology,* 2nd Ed. New York: John Wiley & Sons, 1962.

Bailey, Liberty Hyde and Ethel Zoe Bailey, *Hortus Third, A Concise Dictionary of Plants Cultivated in the United States and Canada,* revised and expanded by the Staff of the Liberty Hyde Bailey Hortorium, A unit of the New York State College of Agriculture and Life Sciences, a Statutory College of the State University of Cornell University, Macmillan Publishing Company, New York, 1976.

Barnes, Burton V. and Warren H. Wagner, Jr. *Michigan Trees. A Guide to the Trees of Michigan and the Great Lakes Region.* Ann Arbor: The University of Michigan Press, 1981.

Bold, Harold C. and Michael J. Wynne. *Introduction to the Algae. Structure and Reproduction.* Englewood Cliffs, N.J.: Prentice-Hall, 1978.

Britton, N. and H. A. Brown. *An Illustrated Flora of the Northern United States and Canada,* 3 Vols. New York: Dover, 1979.

Camp, Wendell H., Victor R. Boswell, and John R. Magness. *The World in Your Garden.* National Geographic Society, Washington, 1957.

Craig, James. *Production for the Graphic Designer.* New York: Watson-Guptill, 1979.

Cronquist, Arthur. *The Evolution and Classification of Flowering Plants.* U.S.A.: Allen Press, 1978.

Davidson, Treat. "Moths that Behave like Hummingbirds," Washington: *National Geographic* 127(6):770–775, June 1965.

de Beer, Sir Gavin. *Atlas of Evolution.* London: Thomas Nelson & Sons, 1964.

Esau, Katherine. *Plant Anatomy,* 2nd Ed. New York: John Wiley & Sons, 1967.

Galston, Arthur W., Peter J. Davies, and Ruth L. Satter. *The Life of the Green Plant,* 3rd Ed. Englewood Cliffs, NJ: Prentice-Hall, 1980.

Gleason, Henry and Arthur Cronquist. *Manual of Vascular Plants of Northeastern United States and Adjacent Canada.* New York: Van Nostrand Reinhold, 1963.

Glimn-Lacy, Janice. Class notes, The University of Michigan. Instructors: Barnes, Burton V., *Woody Plants,* 1972; Hesseltine, Douglas, *Book Design,* 1980; Kesling, Robert, *Organic Evolution,* 1971; Overmyer, Duane, *Lettering,* 1980; Shaffer, Robert, *Mycology,* 1973; Sparrow, Frederick K., *Non-vascular Plants,* 1972; Wagner, Jr., Warren H., *Systematic Botany, Field Botany, Woody Plants, Pteridology, The Natural History of Butterflies,* 1971–76.

Good Housekeeping Illustrated Encyclopedia of Gardening, 16 Vols. New York: Hearst Magazines, Book Division, 1972.

Guthrie, H. A. *Introductory Nutrition,* 3rd Ed. St. Louis: C. V. Mosby, 1975.

Hale, Mason E. *The Biology of Lichens.* London: Edward Arnold, 1967.

Harlow, William M. and Ellwood S. Harrar. *Textbook of Dendrology,* 5th Ed. New York: McGraw-Hill, 1969.

Harrison, S. G., G. B. Masefield, Michael Wallis, and B. E. Nicholson. *The Oxford Book of Food Plants.* London: Oxford University Press, 1975.

Heywood, V. H. *Flowering Plants of the World.* New York: Mayflower Books, 1978.

Hurlburt, Allen. *The Grid.* New York: Van Nostrand Reinhold, 1978.

Kaufman, Peter B., John Labavitch, Anne Anderson-Prouty, and Najati Ghosheh. *Laboratory Experiments in Plant Physiology.* New York: Macmillan, 1975.

Kaufman, Peter B., editor, Thomas Carlson , P. Dayanandan, Michael Evans, Jack Fisher, Clifford Parks, and James Wells. *Plants: Their Biology and Importance.* New York: Harper & Row, 1989.

Kaufman, Peter B., T. Lawrence Mellichamp, Janice Glimn-Lacy, and Donald La Croix. *Practical Botany.* Reston Publishing Company, Inc., A Prentice-Hall Company, Reston VA, 1983.

Keshishian, Kevork K. *Romantic Cyprus, 11th edition revised,* Nicosia, Cyprus, 1963.

Lacey, Elizabeth P. "Seed Dispersal in Wild Carrot (*Daucus carota*),"*Michigan Botanist* 20(1):15–20, January 1981.

Lawrence, George H. M. *Taxonomy of Vascular Plants.* New York: Macmillan, 1951.

Lehman, Stan. *Damage to Amazon rain forest set record in '95.* Associated Press, Indianapolis Star, Indianapolis, Indiana, January 27, 1998.

Lellinger, David B. *A Field Manual of the Ferns & Fern-Allies of the United States and Canada,* Smithsonian Institutional Press, Washington, D. C., 1985.

Margulis, Lynn and Karlene V. Schwartz. *Five Kingdoms, An Illustrated Guide to the Phyla of Life on Earth.* Third Edition. New York: W. H. Freeman and Company, 1998.

Mauseth, James D., *Botany, an introduction to plant biology,* Philadelphia, Saunders College Publishing, 1991.

Mitchell, Katherine and Margaret Bernard. *Food in Health and Disease.* Philadelphia: F. A. Davis, 1954.

Muenscher, W. C. *Poisonous Plants of the United States.* New York: Macmillan, 1975.

National Geographic Magazine Map, *A World Transformed.* September, 2002.

Norstog, Knut and Robert W. Long. *Plant Biology.* Philadelphia: W. B. Saunders, 1976.

Odum, Eugene P. *Fundamentals of Ecology,* 3rd Ed. Philadelphia: W. B. Saunders, 1971.

Pocket Pal. International Paper Co., 1979.

Pohl, R. W. *How to Know the Grasses.* Dubuque: Wm. C. Brown, 1968.

Porter, C. L. *Taxonomy of Flowering Plants.* San Francisco: W. H. Freeman, 1967.

Pratt, Nancy C. and Alan M. Peter. *Pollination: The Art and Science of Floral Sexuality,* Zoogoer, Smithsonian National Zoological Park Publication, July/August, 1995.

Preston, Richard. *Climbing the Redwoods, A scientist explores a lost world over Northern California,* A Reporter At Large, The New Yorker, February 14 & 21, 2005.

Proctor, Michael and Peter Yeo. *The Pollination of Flowers.* New York: Taplinger, 1972.

Ray, Peter M. *The Living Plant,* 2nd Ed. New York: Holt, Rinehart & Winston, 1972.

Ray, Thomas S. "Slow-motion World of Plant 'Behavior' Visible in Rain Forest," Washington: *Smithsonian,* 121–130, March 1979.

Rhoades, Robert E. *The Incredible Potato,* National Geographic **161**(5), p. 668–694, May 1982.

Rosen, Ben. *Type and Typography, The Designer's Type Book.* New York: Van Nostrand Reinhold, 1976.

Salisbury, Frank B. and Cleon Ross. *Plant Physiology.* Belmont, CA: Wadsworth, 1969.

Scagel, Robert F., Robert J. Bandoni, Glenn Rouse, W. B. Schoefield, Janet R. Stein, and T. M. C. Taylor. *An Evolutionary Survey of the Plant Kingdom.* Belmont, CA: Wadsworth, 1967.

Smith, Alexander H. *The Mushroom Hunter's Field Guide Revised and Enlarged.* Ann Arbor, Mich.: The University of Michigan Press, 1973.

Smith, Helen V. *Michigan Wildflowers.* Bloomfield Hills, MI: Cranbrook Institute of Science, 1966.

Specter, Michael, *Miracle in a Bottle,* The New Yorker, February 2, 2004, p. 64–75.

Stebbins, G. Ledyard. "Plant Evolution," Encyclopedia of Science and Technology, 5th Ed., Vol 10: 398–400. New York: McGraw-Hill, 1982.

Steffey, Jane, *The Olive Family, American Horticulturist,* December, 1982, p. 10–13.

Taylor, N. *Taylor's Encyclopedia of Gardening.* Boston: Houghton Mifflin, 1948.

Time-Life Encyclopedia of Gardening, 30 Vols. New York: Time-Life Books, 1977.

Vietmeyer, Noel, *Future Harvests,* Horticulture Magazine, pages 24–29, April, 1983.

Vignelli, Massimo. *Grids: Their Meaning and Use for Federal Designers.* Federal Design Library, National Endowment for the Arts, Second Studio Seminar.

Voss, E. G. *Michigan Flora. Part I Gymnosperms and Monocots.* Bloomfield Hills, MI: Cranbrook Institute of Science, 1972.

Wagner, Jr., Warren H, "Fern," *Encyclopaedia Britannica,* 237–248, 1974.

Wagner, Jr., Warren H. "Systematic Implication of the Psilotaceae," *Brittonia* **29**(*1*):54–63, January–March 1977.

Weatherbee, Ellen Elliott and James Garnett Bruce. *Edible Wild Plants of the Great Lakes Region.* P.O. Box 8253, Ann Arbor, MI: 1979.

Whittaker, R.H. and Lynn Margulis. "Protist Classification and The Kingdoms of Organisms," *Bio Systems,* Vol. 10:3–18, Elsevier/North Holland: Scientific Publishers Ltd., 1978.

Wigglesworth, V. B. *The Life of Insects.* New York: Mentor Book, New American Library (Times Mirror), 1964.

Wilson, Adrian. *The Design of Books.* Salt Lake City: Peregrine Smith, 1974.

Wood, Phyllis. *Scientific Illustration.* New York: Van Nostrand Reinhold, 1979.

Zimmer, Carl, *The Web Below,* Discover, p. 44, November, 1997.

Glossary of Word Roots

a	not, without	(acephala)	gamet	spouse, sex cell	(gamete)
ab	away from	(abscise)	gamo	union, marriage	(oogamy)
aleuron	flour	(aleurone)	gen	origin, birth	(gene)
andro	male	(androecium)	genus	race	(genus)
angio	box, vessel	(angiosperm sporangium)	geo	earth	(epigeous)
			glab	bald	(glabrous)
anth	flower	(perianth)	gland	secretory spot	(glandular)
antheros	flowery	(antheridium)	glut	glue	(glutinous)
anti	against	(antipodal)	gymn	naked	(gymnosperm)
apertur	opening, hole	(triaperturate)	gyn	female	(gynoecium)
api	tip, apex	(apical)	habitus	condition, physique	(habit)
arche	beginning	(archegonium)	hal	salt	(halophyte)
asco	sac	(ascocarp)	hemi	half	(hemiparasite)
ate	provided with, formed into	(perfoliate)	hetero	other, different	(heterospory)
			hex	six	(hexagonal)
auto	self	(autotrophic)	holo	entire	(holoparasitic)
basidio	small base	(basidiocarp)	homo	same, alike	(homospory)
bi	two, double	(biloba)	hydro	water	(hydroponics)
cap	head	(capitate)	hyph	web	(hyphae)
capill	hair	(capillitium)	hypo	below	(hypogeous)
carp	fruit	(schizocarp)	idium	diminutive ending	(antheridium)
caul	stem	(cauline)	il	diminutive ending	(lentil)
cephal	head	(acephala)	infra	below	(infrared)
chloro	green	(chlorophyll)	inter	between	(intercalary)
chrom	color	(chromosome)	intra	within	(intracellular)
cid	cut, kill	(septicidal)	iso	equal, homogenous	(isogamy)
circ	circle	(circinate)	karyron	nut, nucleus	(eukaryote)
cleisto	closed	(cleistothecium, cleistogamous)	lab	lip	(bilabiate)
			lamel	plate, layer	(lamella)
coll	glue	(collenchyma)	lamin	blade	(*Laminaria*)
coma	hair	(comose)	lance	lance, blade	(lanceolate)
cyan	dark blue	(anthocyanin)	leuco	white	(leucocyte)
cyst	bladder, bag	(pneumatocyst)	lig	strap, ribbon	(ligulate)
cyt	cell	(leucocyte)	lith	stone	(*Lithops*)
dendr	tree	(dendrite)	loc	a small place, cell	(locule)
dent	tooth	(dentate)	ligne	wood	(lignin)
derm	skin	(epidermis)	logos	discourse, study	(ecology)
di	two, separate	(dicot)	macro	large, long	(macroscopic)
dictyo	net	(*Dictydium*, dictyosome)	mega	large	(megaspore)
			meiosis	reduction	(meiosis)
dichotomos	cut into two	(dichotomous)	meri	part, segment, component	(mericarp, meristem)
el	diminutive ending	(pedicel)			
endo	inside	(endocarp)	meso	middle	(mesophyll)
epi	upon	(epiphyte)	micro	small, tiny	(microphyll)
equinus	horse	(*Equisetum*, equitant)	mitos	thread	(mitosis)
			mono	one, single	(monocot)
erythr	red	(phycoerythrin)	morph	form, structure	(dimorphic)
eu	true, good	(eukaryote)	myco	fungus, mushroom	(Basidiomycota)
ex	from, beyond	(excise)	myx	slime	(Myxomycota)
exo	outside	(exocarp)	nat	born, borne	(circinate)
fer	bearer	(conifer)	nom	name	(binomial)
fil	thread-like	(filiform)	ode	like	(phyllode)
fissio	splitting	(fission)	oec	household	(dioecious, ecology)
flagellum	whip	(flagellum)	oid	like	(rhizoid)
flav	yellow	(riboflavin)	ol	little	(petiole)
flor	flower	(inflorescence)	orth	straight	(orthotropism)
fol	leaf	(perfoliate)	ostiol	little door	(ostiole)
fungus	mushroom	(fungus)	ov	egg	(ovule, ovary)
funiculus	a small cord	(funiculus)	paleo	ancient	(paleobotany)

palm	hand	(palmate)
palustr	swamp	(palustris)
para	beside	(parasite)
pect	comb	(pectinate)
ped, pod	foot, stalk	(stylopodium, peduncle)
pelt	shield	(peltate)
per	through, by means of	(perfoliate)
peri	around	(perianth)
phellos	cork	(phelloderm)
phil	love	(*Philodendron*)
phloos	bark	(phloem)
phore	bearer	(antheridophore)
phyle	tribe	(phylogenic)
phyll	leaf	(sporophyll)
phyco	seaweed	(phycoerythrin)
phyte	plant	(epiphyte, phytoplankton)
pinn	feather	(pinnate)
plast	thing molded	(protoplast)
ploid	fold	(polyploid)
plume	feather	(plumose, plumule)
pneum	lung, air, gas	(pneumatocyst)
poly	many, much	(*Polyporus*)
por	small opening	(pore, poricidal)
prim	first	(primordial)
pro	before	(prophase)
protiston	the very first	(*Protista*)
proto	first, original	(protoplast)
pseud	false	(pseudobulb)
psilos	bare, mere	(*Psilotum*)
pterid	fern	(*Pteridophyta*)
pyle	gate	(micropyle)
rad	root	(radicle)
reticulum	a small net	(reticulate)
rhap	needle	(*Rhapidophyllum*)
rhiz	root	(mycorrhizae)
rhodon	rose color	(*Rhodophyta*, *Rhododendron*)
ros	rose	(rosette)
sagitt	arrow	(sagittate)
sapro	rotten	(saprophyte)
schiz	deeply divided	(schizocarp)
scis	cut	(abscise)
sect	cut	(dissect)
semi	half	(*Semibegoniella*)

septum	partition, wall, enclosure	(septate)
sessilis	fit for sitting	(sessile)
soma	body	(somatic)
soros	heap	(sorus)
species	kind	(species)
sperm	seed	(angiosperm)
sphen	wedge	(sphenophyll)
sporo	seed, spore	(sporophyte)
stell	star	(stellate)
strobilos	twisted object, top, pinecone	(strobilus)
stoma	mouth	(stomate)
stroma	cushion, mattress	(stroma)
stylo	pillar	(stylopodium)
sub	under, less than	(subtropical)
super, supar	above, over	(superior)
syn, sym	together, with, at the same time	(synergid)
taxis	arrangement, order	(taxonomy)
terra	earth	(terrestrial)
tetra	four	(tetraspore)
thall	shoot	(thallus)
tri	three	(triaperturate)
trichom	hair	(trichome)
trop	bend	(tropism)
troph	nourish, food	(autotrophic)
tuber	knob, swelling	(tuberous)
ul	little	(pinnule)
uni	one	(unisexual)
vor	eat	(insectivorous)
xanthos	yellow	(xanthophyll)
xer	dry	(xerophyte)
xylon	wood	(xylem)
zygo	yoke, pair	(zygote, zygospore)

Singular (Plural) Endings

alga (algae)
mitochondrion (mitochondria)
nucleus (nuclei)
phytum (phyta)

Metric Equivalents

nanometer (nm) = 10^{-9} meter
1 micrometer (micron, μm) = 10^{-6} meter
1 millimeter (mm) = 0.001 meter, 0.04 inch
1 centimeter (cm), 10 mm = 0.01 meter, 0.39 inch
2.54 centimeters = 1 inch
1 decimeter, 10 cm = 3.9 inches
0.91 meter = 1 yard
1 meter (m), 100 cm = 3.28 feet